AF602139

A Biologic Approach to Environmental Assessment and Epidemiology

A Biologic Approach to Environmental Assessment and Epidemiology

Thomas J. Smith and David Kriebel

2010

UNIVERSITY PRESS

Oxford University Press, Inc., publishes works that further
Oxford University's objective of excellence
in research, scholarship, and education.

Oxford New York
Auckland Cape Town Dar es Salaam Hong Kong Karachi
Kuala Lumpur Madrid Melbourne Mexico City Nairobi
New Delhi Shanghai Taipei Toronto

With offices in
Argentina Austria Brazil Chile Czech Republic France Greece
Guatemala Hungary Italy Japan Poland Portugal Singapore
South Korea Switzerland Thailand Turkey Ukraine Vietnam

Published by Oxford University Press, Inc.
198 Madison Avenue, New York, New York 10016
www.oup.com

Oxford is a registered trademark of Oxford University Press

Library of Congress Cataloging-in-Publication Data
CIP data on file
ISBN 978-1-10-514156-6

To Mary and Margaret

PREFACE

Environmental chemical hazards remain a highly contentious topic in modern life. Rightly or wrongly, they occupy high priority in many people's perceptions of the quality of their lives. Nearly every nation on earth has its own environmental crises and also shares perspectives on the possibility of global catastrophes. There are many things that make the environmental issue different from the other big worries that occupy our minds, fill our newspapers, and busy our governments. One of these unique aspects is the scientific nature of the information on which opinions about environmental hazards are formed. Regardless of whether one does or does not believe that we have brought global warming upon ourselves, or that electromagnetic fields cause cancer, or that synthetic pesticides threaten the ecosystem, nearly everyone agrees that the debates revolve around scientific information. Among the key sciences contributing to these debates are the closely linked fields of environmental epidemiology and exposure assessment. These two fields are still given separate names because of their separate historical roots and scientific traditions, but they are seen increasingly as inseparable aspects of the same investigation into the health risks of environmental exposures to synthetic chemicals.

This is a book about improving the way environmental chemical risks are quantified and hence about providing more accurate scientific information to the environmental debates. Our central thesis is that important advances in the quantification of environmental risks can come only through a synthesis of epidemiology and exposure assessment. Further, this synthesis requires a sound theoretical base—it is not simply a problem of poor communication, of different languages or traditions that often divide scientific disciplines. We have found it useful to explicitly elaborate a dynamic biologic representation of the temporal processes leading from exposure to an agent through uptake and metabolism to disease development, what we are calling a *disease process model*. When exposure assessor and epidemiologist agree from the start about this model, then the conceptual framework for the study they design and the analyses they carry out are much more likely to yield useful exposure-risk information. Thus an explicit biologic model of the apparent processes linking exposure to disease should form the basis for any study seeking to quantify risk from environmental chemicals. The processes are apparent from the time course of the disease or of the adverse effects—the way the disease unfolds over time in response to its mechanism. By observing these developments we can understand some of the rate-limiting processes that underlie the disease. Of course, we are often called on to study hazards whose mechanisms are poorly understood, and so the desired biologic model is not known with any certainty. But even in these cases, we argue that it is often possible to make reasonable assumptions about certain key aspects of the process based on available information—for example, whether damage is largely reversible or irreversible, whether the agent accumulates in the body, what the target tissue may be, and whether damage seems to accumulate gradually, such as in noise-induced hearing loss, or instead seems to come on rapidly, such as in lung cancer. Failing to explicitly specify an assumed underlying biologic model does not "protect" the investigator from the errors of having the wrong model; those errors are simply harder to detect.

We hope this book will be useful to environmental health researchers, as well as risk assessors, who use epidemiologic studies as their "raw material."[1] Our approach has been influenced by the work of some risk assessors who are developing the use of explicit biologic models of disease processes (toxicodynamic processes; Clewell and Andersen 2004; Andersen, Dennison et al. 2005). We have been equally influenced by the epidemiologic perspectives of important theoreticians of our generation (Thomas 1987; Greenland and Robins 1988; Robins and Greenland 1989; Rothman and Greenland 1998; Checkoway, Pearce et al. 2004). We owe a debt also to the seminal work of S. A. Roach, who, in the 1960s and 1970s, showed that simple biologic models could clarify understanding of how to measure and control airborne contaminants in the workplace (Roach 1966; Roach 1977). This cannot be a textbook covering all relevant methods in epidemiology and exposure assessment; that would be too much for one volume. Rather, we present a synthesis of existing methods, already described by ourselves and others.[2] No such synthesis currently exists, and we believe that it can improve the application of environmental epidemiology and exposure assessment to current and future environmental chemical risks.

GOALS OF THE BOOK

In this book, we:

1. Present our approach to the integration of exposure assessment and epidemiology

2. Contrast this approach to current practices
3. Explore the implications of the approach for the design and analysis of field studies
4. Illustrate the approach with a wide range of examples
5. Discuss the implications of the approach for future research directions, for risk assessment, and for disease prevention.

In developing the conceptual framework for the approach, we first develop the notion of a basic biologic model of the exposure-disease process in individual human beings. Later we discuss how to apply these models in epidemiologic studies of populations.

HISTORICAL ROOTS OF ENVIRONMENTAL EPIDEMIOLOGY

Epidemiology has a long and proud history as the principal quantitative science of public health (Rosen 1993; Morabia 2004). Prior to the last half of the twentieth century, its primary contributions were in the study of mortality and in the transmission of infectious diseases. Perhaps the most important understanding that the early epidemiologists gained was the realization that diseases could profitably be studied in populations, rather than simply in the afflicted individuals. When viewed from the distance that the calculation of population rates permitted, diseases and mortality were seen to behave quite rationally, despite the terrible randomness of an epidemic of plague or cholera or the tragic risks of childbirth.

Early epidemiology (essentially prior to the Second World War) was a largely descriptive and synthetic discipline, practiced almost exclusively by physicians (Morabia 2004). They used a wide variety of different sources of evidence to formulate and then test hypotheses about the distribution and determinants of diseases. They calculated rates but also carefully studied individual patients. They observed air, water, and food, climatologic data, and any other relevant information. They did not, in general, concern themselves with quantitative exposure-risk relationships.

Modern epidemiology arose out of these historical roots as a more quantitative and formally constructed set of methods for detecting and quantifying exposure-disease relationships. This history has been well reviewed by others (Morabia 2004). It is sufficient for this brief background to environmental epidemiology to note a few key historical developments and trends. First, epidemiology was strongly influenced by the success of early clinical trials, perhaps most notably the tremendous inferential power of the great polio vaccine trials of the 1950s (MacMahon and Pugh 1970). It was seen that what rendered the randomized trial so much more powerful than the observational study of populations of, say, smokers or aniline dye workers was the control of potential confounding factors that randomization provided. The clear explication of what confounding was and how it functioned was a critical step in the development of modern epidemiology (Rothman and Greenland 1998). This insight led quickly to the realization that epidemiologic studies in which the exposure was not to the polio vaccine but instead to tobacco smoking or other "exposures" could be seen as providing reliable information on which to make judgments about health risks.

The formal comparison of an exposed and an unexposed population, again inspired by the treatment and placebo groups of the clinical trial, led also to the standardization of the concept of the measure of association. Also called a measure of effect and most frequently represented by a "relative risk," or RR, this was seen as the key statistic capturing the essential information about the magnitude of the risk to be attributed to a particular exposure.

Statistical control of confounding and quantification of measures of association are largely accomplished through the fitting of statistical models to large masses of health and exposure data. This process was greatly facilitated in the 1970s by the widespread introduction of computers. Without substantial, inexpensive computing power, epidemiology would likely have remained a much less quantitative and more descriptive discipline. Thus the theoretical advances in epidemiology occurred nearly simultaneously with the technological revolution that made implementation of the new statistical methods and modeling techniques possible.

It is a truism that epidemiologists study disease in populations, but it is not often made clear exactly why this is so. Disease, like other complex biologic processes, is a stochastic process. This means that in any single individual, whether and when a disease will develop can rarely be specified with accuracy. It therefore follows that the causes of a single case of disease are very difficult to know with certainty. We seek to prevent disease, and so the transition from health to disease is the object of study. The analogous object of study for a population is the incidence rate, or the risk.[3] Risk is a population phenomenon; populations experience increased or decreased risk because of some causal factors. The individuals in the populations either get the disease or they do not in some time period. We may assign a population estimate of a risk to an individual, saying, for example, that a lifelong smoker has a risk of developing lung cancer of about 16%. But this assignment requires that we accept certain assumptions that are never knowable—such as that this particular smoker is sufficiently like the other smokers from whom the 16% figure has been calculated. Sometimes it is said that epidemiologists must study populations because one needs large numbers of people in order to see small risks. But this is not quite correct; risk can be studied only in populations.[4]

In an ideal world, an epidemiologic study would be conducted in the following way: A population would be exposed to some agent and followed for the development of disease. Then the hands of the clock would be turned back, and the same population would be followed over the same interval of time with everything occurring in exactly the same way, except that the exposure would be lacking. Because this type of study is impossible, the evidence that we seek from epidemiologic studies is called *counterfactual* (Rothman and Greenland 1998). The alternative means at our disposal for approximating this ideal evidence is to compare the rates or risks of disease in an exposed and an unexposed population. The unexposed population is studied to approximate the risk that the exposed population would have experienced had it not been exposed. The trouble is we can never be sure that the unexposed population experienced the disease risk that the exposed population would have had it not been exposed.

What concerns us here is the point that, fundamentally, epidemiology compares populations; environmental epidemiology is about comparing populations that differ with respect to some environmental agent of health concern.

HISTORICAL ROOTS OF EXPOSURE ASSESSMENT

To identify the origins of exposure assessment, it is perhaps not necessary to look as far back in history as it is for epidemiology. Exposure assessment specialties have their roots in the work of engineers, chemists, physicists, and toxicologists. The earliest work was done in the 1800s by K. B. Lehmann and J. B. A. Chevallier, who both studied the toxic effects of gases on animals (Patty 1958). These studies provided the basis for explaining or anticipating toxic effects in humans. The evaluation of environmental conditions had to await the development of analytical chemistry in the 1900s. The terrifying use of poison gases in the First World War was an important catalyst for the development of analytical tools to measure toxic airborne exposures. The measurement of air and water pollutants in the general environment can be traced to the sanitary engineering movement, reborn as environmental engineering after the Second World War. Industrial hygiene arose out of the occupational health movement and followed a roughly parallel, though quite separate, track. For the most part, these fields do not concern themselves with the quantification of chemical exposures for the purpose of informing epidemiologic studies. Instead, primary prevention through exposure control is their first priority. In the United States, the national environmental laws of the early 1970s created a demand for exposure measurement in order to monitor compliance with pollution exposure regulations. This in turn stimulated the development of instrumentation and environmental chemistry techniques to inexpensively measure small concentrations of pollutants in various environmental matrices.

Unlike epidemiology, the measurement of exposures is not fundamentally a study of populations. The goal is to characterize the hazard presented by an exposure, usually by comparing the measured level against an allowable exposure standard. Thus it is often desirable to measure individual exposures to a hazardous chemical. But initially environmental exposure assessment did not primarily concern itself with group- or population-level estimates, even if these were ignored primarily for practical rather than theoretical reasons. The early monitoring devices were large, heavy, and noisy, and the chemical measurement techniques were relatively insensitive, so large amounts of sample were required. For example, the trash-can-sized high-volume filter sampler was (and with some modifications still is) used for ambient air pollution particulate monitoring on the roofs of buildings in the center of a city. This type of sampler necessarily had to be placed in a central location and its data assigned to all those living nearby.

Prior to about 1960, industrial hygienists also had to measure exposures for jobs or areas in a manufacturing plant using bulky fixed monitors or cumbersome handheld devices. They would then assign those exposures to all the workers in the job or work area analogously to the air pollution assessments. The development of the personal sampler by Sherwood in the 1950s made it possible, at least in principle, to measure exposure individually for each worker. Exposure assessment for ambient air pollution studies, too, has become more individualized through the use of personal monitors.

Sampling for air and water contamination has also become more and more automated, and direct-reading instruments can provide real-time data so that large numbers of data points can be gathered. Sometimes the technology has run ahead of the science that could use these data. It has not always been clear how to use highly detailed data for health studies; the data are often simply averaged, and much of the detail is lost.

STUDYING POPULATIONS VERSUS STUDYING INDIVIDUALS: AN UNEASY TENSION

Beginning in the 1980s, epidemiologists and environmental exposure assessors began to work closely together to produce epidemiologic studies with quantitative exposure data. The promise of this combination was tremendous; epidemiology could be greatly improved by comparing groups whose exposure levels could be measured and compared quite accurately. Thus the crude exposed-not exposed framework of traditional epidemiology was replaced by empirical quantitative exposure-response models—at its simplest, an *x-y* regression model. This allowed two improvements. First, quantitative exposure-response data permitted scaled risk determinations—the estimation of how much risk each additional unit of exposure carried. Second, this marriage of epidemiology and exposure assessment made the whole exercise a more powerful risk detector, identifying hazards that had heretofore been undetectable because of the inevitable misclassification of yes-no exposure definitions.

Although quantification is a sine qua non of all physical sciences, it is worth reviewing briefly what we mean by the quantification of risk and what quantification of environmental hazards can provide to policy makers and concerned citizens. In a sense, epidemiology has always been quantitative, but until recently only risk was quantified—the exposure half of the story was often qualitative or descriptive. Linking exposure assessment to epidemiology introduces the possibility of combining quantitative exposure data and quantitative risk data. This is the basis of risk assessment: the ability to estimate how much risk a given level of exposure will carry. But there is more; quantitative exposure-risk relationships have at least five specific strengths:

1. Studies quantifying both exposure and risk in populations may be able to detect smaller risks than studies with simple, categorical definitions of exposure. A study seeking to compare a "very low exposed" to an "unexposed" population will need very large numbers to detect this difference. But, if we are willing to make some kind of assumption about the shape of the underlying exposure-response curve (at a minimum, assuming monotonicity—consistently increasing risk with increasing exposure), then information on risk at higher levels of exposure lends statistical power to the detection of low-level risks.
2. Studies in which exposure is quantified may contain less misclassification of exposure than studies with categorical definitions of exposure, thus providing more precise estimates of the risk.
3. Studies that provide evidence that increases in exposure carry increases in risk are believed by epidemiologists to provide stronger evidence of causality than studies showing simply that, as a class, the exposed are at higher risk than the unexposed (this is one of the well-known "Bradford Hill criteria";(Bradford Hill 1965)).
4. Studies in which exposure is quantified on a reproducible scale of measurement are more generalizable because the exposure measurement scale allows one to estimate what the risk might be for another group, so long as their exposure levels can be measured on the same scale.
5. Finally, exposure quantification also improves the ability of epidemiologic data to provide information on the shapes of underlying exposure-response relationships. Categorical exposure estimates do not have the resolution to determine whether a relationship with an effect is linear or of some other form—quadratic, for example—which can be very important for projecting risks at low exposures.

There are now numerous examples of hazards that have been effectively studied by the marriage of epidemiology and exposure assessment that we advocate. The link between fine particulates in urban air and mortality, the association between radon and lung cancer, and the evidence for an exposure-response relationship between benzene and leukemia are all fruit of studies that have linked quantitative exposure assessments to epidemiology.

But this seemingly perfect match has problems. Many times researchers in the two fields find themselves working at cross-purposes. For example, epidemiologists too often choose their statistical models based on software availability and familiarity of use rather than on an explicit assessment of the proper form for the exposure-risk model. But a logistic regression model (to cite one example widely used by epidemiologists studying many different disease processes) contains critical implicit assumptions about the nature of the disease process and effects of modifying factors that may or may not be consistent with the underlying biology of the disease process being studied. In the same way, an exposure assessor often chooses a measurement strategy based on technical feasibility, existence of standardized analytic methods, cost, and so on. We argue that there are numerous ways that a mismatch between research methods and underlying biology can introduce errors into the estimation of exposure-risk relationships. As researchers increasingly study smaller and smaller risks, these errors become all the more important to avoid.

Here are just a few of the challenges that face those who would conduct environmental epidemiology studies with quantitative exposure assessments:

- How should individual exposure measures over time be summarized for use in a statistical epidemiologic model? What averaging time interval should be used for exposure sampling and risk calculation?
- How should errors in exposure estimation be assessed? How should these errors be incorporated into the epidemiologic model?
- Epidemiologists, essentially studying population average effects, are often comfortable with considerable error at the individual level, assuming that it will have little effect on population risk estimates; whereas exposure assessors judge their own accuracy in terms of how well an individual's exposure is measured.
- Quantitative, individual exposure assessments for an epidemiologic study can often cost more than all of the rest of the study.
- Epidemiologists are often uncomfortable with complex exposure metrics and may prefer to use simple measures that the exposure assessor "knows" to be inaccurate.
- The epidemiologist will often choose to do the study in a population in which bias and confounding can be minimized, whereas the exposure assessor will seek the population in which exposure estimation is more feasible or in which exposures are highest.

We have found that by building a common biologic model of exposure, physiologic response, and disease, many of these questions and issues can be resolved. The biologic approach also can make it easier for epidemiologists and exposure assessors to share common perspectives on their research strategies, improving their effectiveness and reducing many of the difficulties that we and our colleagues have experienced in conducting studies.

ORGANIZATION OF THE BOOK

In the first chapter, we introduce the basic argument for an underlying biologic model for both exposure assessment and epidemiology and provide a brief overview of how this model can be used to link practice in the two fields. The remainder of the book is divided into three sections.

Section A, consisting of Chapters 2 through 6, provides information on the assessment of both exposure and disease in individuals. Chapters 2, 3, and 4 form a sequence on environmental exposures—their fundamental characteristics (Chapter 2), methods of exposure measurement (Chapter 3), and methods of studying the entry of exposures into biologic tissues using toxicokinetic models (Chapter 4). These chapters constitute an overview of biologically based exposure assessment. Chapter 5 is an overview of the use of biomarkers in environmental assessment. Finally, Chapter 6 presents the four principal disease process models which lie at the heart of our approach.

Section B switches gears to present the epidemiologic approach to studying exposure and disease in populations rather than in individuals. Chapter 7 is a review of the basic epidemiologic method and key concepts needed to understand how and why disease risk is the fundamental object of epidemiologic investigation. Chapter 8 is an explanation of the many sources of uncertainty in epidemiologic studies and how this uncertainty can be managed. Chapter 9 shows how dosimetry can be used in epidemiology to link exposure assessment to epidemiology through the use of biologically based models.

Section C is a series of six chapters covering practical applications of disease process models. The first three chapters cover what are called proportional disease processes, and the final three cover discrete disease processes. In these chapters, the details of building and applying a disease process model for an epidemiologic investigation are presented and illustrated through examples. A final chapter looks to the future, considering opportunities and challenges to applications of the approach.

NOTES

1 Risk assessment is a formal discipline of decision science, which is used to make decisions about hazards in the presence of missing or insufficient data. Consult the bibliography at the end of the book for recommended readings.
2 See the bibliography for recommended readings.
3 Incidence rate and risk are actually slightly different concepts, but for now they may be thought of interchangeably.
4 There are some kinds of epidemiologic studies that measure risk in individuals. This is accomplished by studying the same individual repeatedly, gathering, for example, information on asthma attacks and air pollution levels on each of a series of days. In this type of study, which might be called *intrademiologic*, the unit of analysis is not the individual person but rather the person-day. These designs are useful for studying changes in function (lung function, for example) and for diseases that are completely or largely reversible (asthma attacks, colds, bouts of back pain, etc.).

Ashland, Massachusetts T.J.S.
Gloucester, Massachusetts D.K.

ACKNOWLEDGMENTS

There were many people who helped in the development of our ideas and the production of this book. We especially wish to thank a few whose help was particularly valuable. Students and colleagues at Harvard School of Public Health and the Department of Work Environment, University of Massachusetts Lowell who read early drafts and asked particularly good questions include: Dhimiter Bello, Jaime Hart, Mindy Powers, Rebecca Gore, Ariana Zeka and David Skinner. The students of EH 269 *Exposure Assessment for Epidemiology* in 2000 read some of the early chapters and provided critical feedback. We thank Margaret Quinn for thinking of the concept of biologically-based exposure assessment. Noah Seixas provided critical comments and inspiration at several key points. We thank Tom's long-term commuter train conductor, Eddy, who protected his writing time coming home on the train, at the price of constant teasing about the slow pace of the writing. Finally, we thank our editors at Oxford University Press, for their patience and encouragement during the book's lengthy gestation.

CONTENTS

1 Introduction: Relating Disease to Exposure 3

Section A. Exposure and Disease in Individuals

2 Characteristics of Exposure and its Measurement 27

3 Applications of Exposure Assessment in Epidemiology 65

4 Personal Exposure-Tissue Concentration Relationships 101

5 Biomarkers as Indicators of Exposure 137

6 Disease Process Models 158

Section B. Exposure and Disease in Populations

7 Epidemiologic Evaluation of Environmental Hazards 179

8 Uncertainty in Measuring Risk 198

9 Dosimetry in Epidemiology 223

Section C. Practical Applications of Disease Process Models

10 Modeling Proportional Disease Processes 251

11 Reversible Proportional Disease Processes: Effects of Ammonia and Ozone on Respiratory Symptoms 278

12 Irreversible Proportional Diseases Processes: Neurobehavioral Effects of Mercury, Popcorn Workers' Lung 304

13 Modeling Discrete Disease Processes 320

14 Reversible Discrete Disease Processes: Asthma and Indoor Air, Dermatitis and Metalworking Fluids 347

15 Irreversible Discrete Disease Processes: Silica and Lung Cancer 366

16 Where Do We Go From Here? 383

Annotated Reference Books 391

References 393

Index 415

A Biologic Approach to Environmental Assessment and Epidemiology

1 Introduction: Relating Disease to Exposure

1.1. SURPASSING JOHN SNOW

John Snow's identification of the link between polluted water and cholera is probably the most famous success story of environmental epidemiology (Morabia 2004). His work is still cited 150 years later to illustrate the power of epidemiology to detect risks with astute powers of observation, very simple math, and no known causal mechanism (or, putting it more plainly, an incorrect causal mechanism). Today, with greatly advanced understanding of disease mechanisms and sophisticated tools for measuring pollutants, epidemiologists should be able to do an even better job than John Snow of identifying environmental health hazards. And although we believe there are many examples of this advance, there is still much to be learned about how best to design studies and analyze environmental and health data in order to identify hazards sooner and at lower levels of exposure. We believe that in many of the relevant sciences, improvements in our measurement tools have outpaced our understanding of the theoretical framework which underlies efforts to detect small risks to population health. This book is a contribution toward building that framework. We believe that exposure-risk relations can best be detected and quantified when the entire enterprise of data gathering and analysis is based on a clearly stated hypothesis about the biological processes that link exposure to disease.

John Snow's ability to detect a link between polluted water and cholera illustrates both a strength and a limitation of epidemiology. The strength is more often discussed—the idea that it is not necessary to have a hypothesis about the mechanism underlying an exposure-disease association in order to prevent a disease. But this strong empiricism is also a limitation because weak associations may be missed. If, for example, we do not measure the actual causal agent (the vibrio cholerae bacterium in Snow's work) but, out of ignorance, study a factor that is more or less well correlated with it (the private company providing drinking water to the different wells of London), then we run the risk of failing to detect the association through dilution of the true effect (epidemiologists call this *misclassification*). From this perspective, we would argue that Snow succeeded because the link between vibrio cholerae and the water company was very strong: water from the Southwark and Vauxhaul Company was highly polluted, whereas that from the competing Lambeth Company was not. Snow would not have been so successful had the difference in cholera contamination been less striking. The modern epidemiologist is often faced with shades of difference in risk—nothing like the stark pattern observed by Snow. We believe that the "black box" epidemiology inspired by John Snow is reaching its limits, and this book proposes a systematic approach to pushing back those limits by incorporating biologic or toxicologic evidence into the design and conduct of exposure assessments and epidemiologic investigations.

1.2. INSIDE THE BLACK BOX: EXPOSURE → [BLACK BOX] → RESPONSE

When we look inside the black box, we are examining the complex and dynamic biological processes that lead to disease. The observable quantitative relationship between exposure and disease is a consequence of a series of temporal processes—a chain or web of linked steps occurring through time, such as cell damage and repair. Figure 1.1 is a cartoon representing a simple temporal relationship between an external exposure and a pathologic response—often a disease. In this example, exposure to an environmental toxic agent leads to its uptake through inhalation or skin or gastrointestinal absorption. Within the body it is transported to the target tissues, perhaps after metabolic activation. The concentration in the target tissue causes damage, such as cell death or inflammation. If the exposure causes sufficient damage or response, then preclinical effects may be detectable by some tests. When sufficient damage has occurred, a clinical disease diagnosis may occur. The toxic damage may also trigger secondary processes, such as apoptosis, or repair of the damage. Once exposure stops, some or all of the adverse effects may be repaired, whereas others, once initiated, may progress without further exposure or damage. Throughout this book we often move back and forth between description of pathologic processes within an organism and the population distribution of some response or disease outcome. At the individual level, the terms *response*, *outcome*, and *effect* are often used interchangeably. At the population level, the "risk" or probability of some particular effect or outcome is most often the object of study.

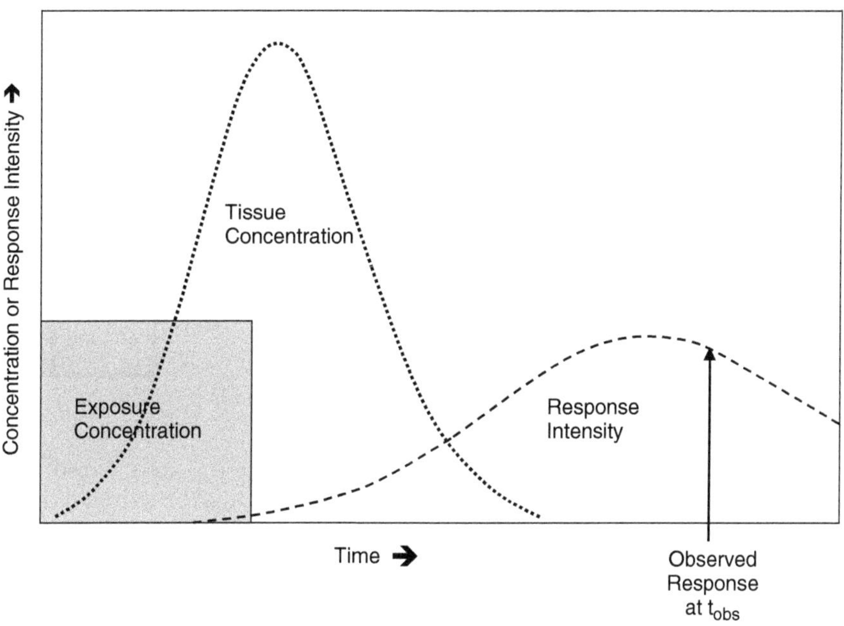

Figure 1.1 Simplified schematic showing how an exposure across time (shaded box = dose) produces a tissue concentration (dotted line), which in turn produces a tissue response (dashed line). Because these are temporal processes, a different exposure-response relationship will be observed depending on when we measure the response.

Exposures and effects are linked by dynamic processes occurring across time. These processes can often be usefully decomposed into two distinct biologic relationships, each with several components:

1. *The exposure-dose relationship* ((Gargas, Medinsky et al. 1995; Sexton, Callahan et al. 1995), consisting of
   a. *uptake* of a toxic material from the environment and into the body via one or more of three common routes: inhalation, ingestion, and skin absorption.
   b. *transport* of the agent to the target tissue and possibly also metabolic activation into a toxic form.
2. *The dose-effect relationship (Becking 1995; Menzel 1995)* consisting of
   a. the temporal pattern or time course of the concentration of the active agent in the target tissue, which is the direct cause of the effect.
   b. *early or intermediate pathophysiologic responses,* which are the results of damage processes, that may be repaired or lead to secondary effects.
   c. a *clinical response* detected as signs and symptoms, loss of function, or the onset of disease, once a sufficient dose has been reached.

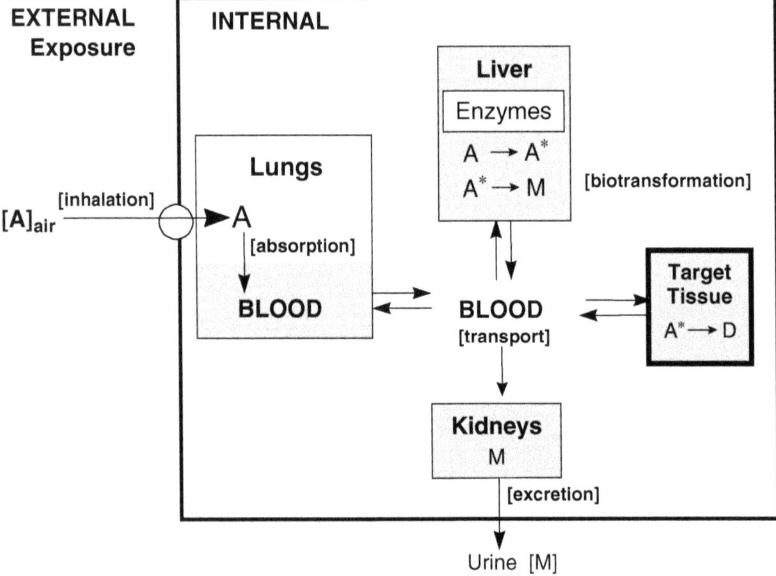

Figure 1.2 Typical example of toxicokinetic processes of uptake, transport throughout the body, biotransformation, and excretion. Boxes are tissues, and arrows are transport or transformation processes. Toxin A enters through inhalation, is transformed into active agent A* in the liver, which leads to tissue damage D in the target tissue. A* is also broken down into a metabolite M, which is eliminated through the kidneys.

These two components are sometimes represented by two different models: a toxicokinetic model (Figure 1.2) and a disease process model (Figure 1.3). A disease process model is a description of the pathophysiologic mechanisms that lead to the observable adverse effects over time. This biological model will usually have several components, such as damage and death of target cells and tissue repair that replaces dead cells; the "damage-repair model" is one common generic type of disease process model. Often the various steps in the disease process do not occur at the same rate; some of these processes are "fast," such as cell killing, whereas others are "slow," such as damage repair. It is often the case that a few slow steps in a process become limiting to the overall rate, setting the temporal pattern for the entire exposure-response relationship. We may be able to define a relatively simple but useful disease process model (Figure 1.3) just by defining the slow-rate limiting steps, because they define the overall pace of the process.

This book goes into considerable detail about toxicokinetic and disease processes linking exposure to disease and argues that it is better to be explicit about them than to hide them in epidemiology's black box. We believe that formal disease process models represent the best opportunity to improve the sensitivity of epidemiology for detecting new and emerging environmental risks where there is

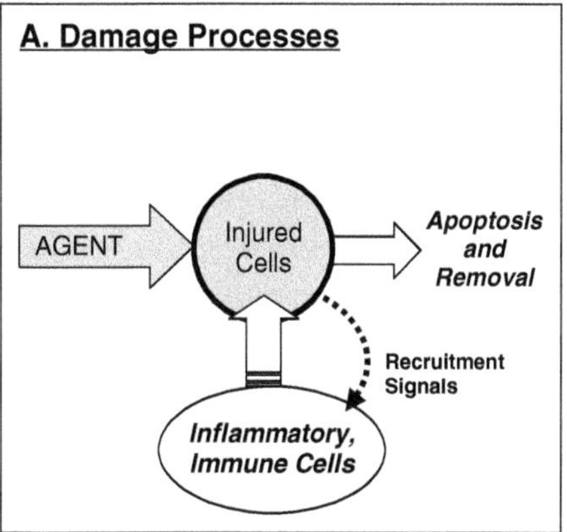

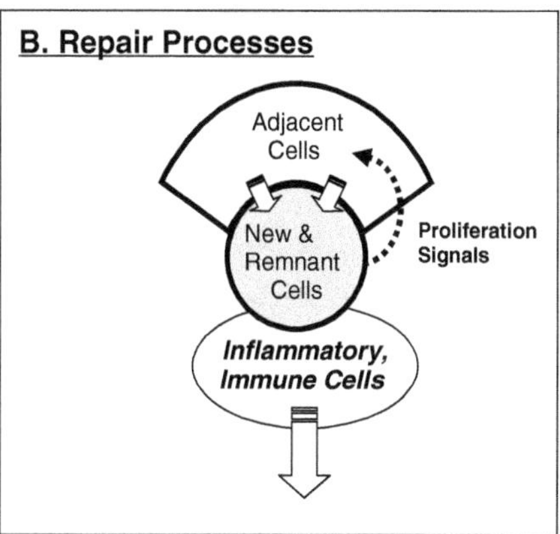

Figure 1.3 A simplified damage-repair model. All cellular processes are controlled by cell signaling, which initiates and stimulates them. Feedback signals (not shown) can limit or prevent some damage processes. Part A: damage to molecular targets, such as DNA or cell wall proteins, lead to injury or death of cells. Cell injury/death produces signals that recruit inflammatory cells of the immune system to the area. A strong inflammatory response can directly injure cells and lead to a feedback loop, which can prolong the damage response. Part B: represents the principal repair processes involving cell proliferation and rebuilding of tissue architecture. Inflammatory cells die or leave the area, and repair signaling decreases to normal levels. Misrepair and exaggerated stimulation of cell proliferation are important causes of long-term irreversible effects.

limited mechanistic information. But making this potential a reality means explicitly defining exposure-disease hypotheses in simple models. In our approach, these models are often used to create exposure or dose metrics (summary measures of the time profile of exposure), which are in turn used in epidemiologic models to estimate dose-disease associations.

The dose metrics can be thought of as weighted summaries of the exposure data for each study subject. The dose metric is hopefully more closely associated with the disease outcome than a simple average of the original exposure data. There are many ways to develop these weights, and the approach we advocate bases them on an explicit biologic hypothesis. Sometimes the weights that go into the dose metrics can be determined from animal experiments or previous epidemiologic studies, and sometimes they can be estimated empirically from the study data themselves.

Suppose, for example, that we want to study an acute illness that occurs in response to chemical exposures within the preceding 24 hours. Knowing this (or correctly hypothesizing it), one would not choose to study the lifetime average exposure, because this will probably show little or no relationship to the risk of the acute effect. Instead we would construct a dose metric that summarized the recent exposures – perhaps the average of the past 24 hours. This is a simple example, but oftentimes researchers are not systematic in their choice of metrics, allowing their biases or implicit assumptions to guide their choices rather than formal statements of hypothesized mechanisms.

One of the key concepts in this book is that it is not necessary to know the full mechanism of effects to guide selection of an exposure-response model or exposure metric. Because of the strong influence of the rate-limiting steps, often it is necessary only to have observations on the approximate time course of effects, whether the effects appear to be reversible or irreversible, and whether damage progresses proportionately with each unit of exposure (actually dose, but more on this later) or instead occurs suddenly and seemingly without regard to the amount of exposure, such as an asthma attack. These fairly simple qualitative characteristics of exposure-response relationships can provide useful guidance to the entire investigation of the health effects of an environmental hazard. More specifically, we have observed that, to begin modeling, there are two critical elements about the disease process that must be identified: (1) whether the *time course* of the disease or damage process shows it is reversible or irreversible and (2) whether the *type of response* is proportional to the tissue concentration of the toxin or, alternatively, that the adverse effect develops in a discrete step function, from absent to present, with the strength of response not proportional to the concentration of the causal agent in the tissue. Combining these two characteristics, there are four distinct models which we propose can be used as the starting point for studying the majority of environmental disease processes (Table 1.1). More complex disease processes can be modeled by combining two or more of these four in series. These time-course and response types define profoundly different pathologic processes and exposure-response relationships that are the subject of the remainder of the book.

Table 1.1 The four principal disease process models can be classified based on their time course (reversible or irreversible) and their response type (proportional or discrete)

Time Course	Response type	
	Proportional (Pathophysiologic responses are proportional to dose)	**Discrete** (Pathophysiologic responses are quantal, and not proportional to dose)
Reversible (damage/repair processes in which there is recovery after exposure ceases)	Airway inflammation, gastritis, mechanical injury to musculoskeletal system and skin	Asthma attacks Dermatitis
Irreversible (process with limited repair that lead to the accumulation of effects)	Pneumoconioses, hearing loss, macular degeneration	Carcinogenesis, immune sensitization

1.3. TIMING IS EVERYTHING

The response or effect we observe depends on a specific agent's pathophysiologic mechanism, which is the combination of molecular, cellular, tissue, and systemic processes operating over a period of time to ultimately cause the effect. The tissue concentration of the active agent and its variability over time are the causes of those effects. Effects are rarely instantaneous, although time scales for some effects may be measured in seconds, whereas others are measured in years. The observed response at a point in time (marked on Figure 1.1) is usually the net result of an array of both damage and repair processes, because these generally operate concurrently (Mehendale 2005). We develop an approach for describing the quantitative relationship between the observed response at t_{obs} and the intensity and duration of exposure. The repair process is critical for the shape of the response-time course. For example, if there was no repair, then the response line in Figure 1.1 would plateau after the tissue concentration curve came down to zero again. Thus the behavior of the response or disease over time clearly indicates whether there is repair or not.

Research on toxicologic mechanisms shows these processes to be quite complex (for an example that illustrates some of the complexity that links silica exposure and silicosis, see Figure 10.4 in Chapter 10). The complicated branching cascades of signaling pathways and feedback loops invariably observed by molecular biologists when they look closely at toxicologic mechanisms are an almost perverse denial of Occam's razor.[1] One wonders what a modern Occam would think, picking up a textbook of molecular biology! However, when we say that we should start our investigation from a hypothesis about the disease process, we do not mean all of the hundreds of steps that occur at the cellular level. We mean only a simple model determined by the apparent steps of the disease process that ultimately define the observable time course of the outcome. It is also often true that very different mechanisms often produce similar temporal patterns for their effects. For example, the simple product of duration and intensity of noise exposure is approximately proportional to the amount of hearing loss, and so too is the product of duration and intensity of cigarette smoking related to the loss of lung function. This parallelism occurs despite very different disease processes.

Another important and useful principle is the independence of local repair processes. For example, consider a small paper cut on your finger; the damage is done in a fraction of a second, whereas the repair takes a week or longer. If you were unfortunate enough to get a second similar cut on your finger while the first was still being repaired, the repair of the second one would take about the same time and would be unaffected by the presence of the first cut, which illustrates the principle of local independence that occurs for many nonsystemic processes. If, on the other hand (sorry for the pun!), you received a more severe cut, it would probably require more time to repair, and if you get hit by a bus . . . well, the damage may overwhelm the repair process. In summary, each instance of damage is repaired independently (when the effects are not overwhelming), and the amount of repair per unit time is roughly proportional to the amount of damage, an approximately fixed percentage.

Discrete effects, such as asthma attacks or other immunologically mediated events, are another distinct type of adverse effect. They are quite different in their temporal dynamics from the basic damage-repair model we have been discussing. The important characteristic of this type of effect is that, once triggered, the strength and duration of response are not determined by the magnitude of the triggering exposure or dose. An asthma attack involves triggering a complex cascade of molecular and neural signals that leads to the response. The clinical signs and symptoms are finally triggered by some ultimate discrete event in the cascade. More generally, the severity of discrete (yes/no) responses such as an asthma attack are not proportional to the initial triggering event or exposure. There *is* a proportionality that is observable at the population level: the *probability* of a response will be proportional to some function of the triggering exposure and its duration. The principles of different disease types and exposure-response relationships are discussed in more detail throughout the book, and especially in Chapter 6.

1.4. DYNAMIC VERSUS STATIC MODELS: A KEY DISTINCTION

This book develops the idea of using biologically based models of disease processes that occur dynamically through time. Like the standard statistical models widely used in epidemiology, our models also are meant to describe relationships between exposure and disease in populations. But there are important differences in our approach. In general, statistical epidemiologic models are not intended to represent disease mechanisms but to be empirical representations of the "black box" relationships. In particular, these models implicitly assume that there is a single proportionality constant describing the relationship between exposure and risk or response. This proportionality is assumed to be invariant over time—it is static—and the same among all the members of the study population. A relative risk in a case control study or a "slope" parameter in a linear regression model estimating rate of loss of pulmonary function per unit of exposure are examples of such proportionality constants. To distinguish these static statistical models from the biological process or dynamic models discussed in this book, we call models with invariant proportionality constants linking exposure and response *static*. The word *linear* is also sometimes used, but *static* may be more appropriate because *linear* has many different interpretations and because it is not so much the "straightness" of the exposure-response relationship that distinguishes these models as the time invariant nature of the proportionality between exposure and response.

The idea that each pack-year of smoking causes a fixed amount of loss in pulmonary function implies a static exposure-response model. Models of noise-induced hearing loss are static if they are based only on the assumption that hearing loss is proportional to the number of hours of exposure to noise above 90 dB(A), for example, without considering that one's history of noise exposure affects a biological process leading to hearing loss. Repeated short periods of intense noise may cause much more hearing loss than a lower steady exposure. Most models that have been used in cancer epidemiology are static, such as that pack-years of smoking cause a fixed amount of risk, with the exception of certain multiple stage cancer models (see detailed discussion in Chapter 15), in which the risk conveyed by a unit of exposure may vary with the stage of the disease process and with the pattern of exposure in earlier time periods.

Biologic Models

Physiologic and toxicologic processes, in contrast to epidemiologic models, are not static. The degree of response of an individual to a given unit of exposure to a toxin can vary over time. Of particular interest for those trying to elucidate exposure-response relationships is the possibility that the proportionality parameter that describes the risk of disease per unit of exposure may itself be determined by the individual's history of exposure (and by the history of response to that exposure). Models that exhibit this behavior are dynamic. Epidemiologists are quite familiar with exposure-response relationships that change with age or have secular trends. These might be considered dynamic in one sense, but we reserve the term for models in which the parameter(s) defining the effects of exposure are themselves temporal functions of environmental conditions and/or of the individual's response to that exposure. Two basic classes can be distinguished: (1) those in which the most recent unit of exposure carries the same risk to everyone, no matter what the individual's past history of exposure has been, and (2) those in which the entire history of exposure/response can determine the effect of the next unit of an individual's exposure.

The former is the standard, static assumption of most epidemiologic models. Suppose that an epidemiologist has chosen to summarize each participant's exposure history as "cumulative exposure" (intensity multiplied by duration; described in more detail later) and is regressing risk on cumulative exposure with an appropriate multivariate model (a Cox model, for example). Implicit in these choices is the assumption that every individual in the study with the same cumulative exposure faces the same risk (from exposure)—it matters not whether the pattern of exposure was, for example, many years at low intensity or a few years at very high intensity. The same holds true with other standard summary measures of exposure (average exposure, duration, peak exposure) and other standard exposure-response models.

Now imagine that we could look inside the black box that this epidemiologist is studying and that what we saw was that previous exposures inhibit normal clearance or repair processes in the target tissue. The result would be to increase the magnitude of the effect of later units of exposure compared with earlier ones. This would violate the assumption underlying the static model. It would also be violated when the toxic agent is a metabolite produced by an enzymatic process that is saturable, so that exposures which do not exceed the saturation level will have a larger effect per unit than exposures that saturate the metabolic system. Asthma is another example for which a dynamic model would be more appropriate: the asthmatic exhibits an extreme sensitivity to an antigenic agent because of an immunologic response mechanism not present in that individual before the development of asthma. If the asthma was caused by a toxic exposure, then the incremental response to a unit of exposure after asthma onset will be much greater than in the same individual before onset. The airway responses to ozone represent another example for which a static model may be inappropriate. (Ozone's effects are described in some detail in Chapter 11.) Ozone causes a "simple" inflammatory effect in the airways as a result of the oxidation of epithelial cell membranes leading to the leakage of inflammatory mediators and the stimulation of irritant receptors in the airway walls. No immunologic mechanisms are believed to be involved, but it is clear that the degree of response to an increment of ozone exposure depends on the recent history of ozone exposure and the degree of inflammation of the airways (Kriebel and Smith 1990). Volunteers exposed a few hours each day to a moderately irritating level of ozone (300 to 600 ppb) show the largest lung function effect on one of the first days of exposure, and normally by the fourth or fifth day no drop in lung function is observed between pre- and postexposure measurements. A change in responsiveness to ozone also occurs: bronchoconstriction seen in pulmonary function tests is often greater on the second than the first day of a week-long protocol of brief (2-hour) exposures. Thus any model which seeks to represent the full range of ozone's effects on the airways should be capable of showing both enhanced and diminished responses, depending on the precise timing and intensity of repeated exposures.

As a general principle, we can say that any disease in which repair mechanisms are stimulated by injury would be expected to exhibit dynamic behavior. Cancer, heart disease, low back pain, neurological or respiratory symptoms, infectious diseases, and immunologic and inflammatory diseases are examples of diseases that fall into this category.

Many of the dynamic physiologic processes described in the preceding involve complex "interactions" between exposure and pathophysiologic responses over time. Traditional epidemiologic statistical models are structured as if individuals experience just one sudden transition—from health to disease, or from alive to dead. The whole history of the individual leading up to the moment of incidence of the disease is summarized in a series of independent variables, some of which may be measures of temporal factors such as time since first exposure or cumulative exposure. This seems an overly narrow perspective when we realize how fundamental time is to the biologic processes that we seek to describe with our analytic models. Time is the dimension in which the toxin and the organism interact to produce the disease state. To put it another way, it is the temporal dynamics of exposure and response that must be studied to fully understand a disease process.

Time Lags in Response

Time lags are common in biologic processes, and their existence is well known to epidemiologists and toxicologists. However, time lags resulting from biologic processes may lead to far more complex exposure-response patterns than can be accommodated by standard static methods such as ignoring the 10 or 20 years of exposure immediately prior to cancer diagnosis, or associating response at t_1 with exposure at $t_{1\text{-lag}}$. Suppose, for example, that exposure to a respiratory irritant causes an immediate bronchoconstrictive response, as well as a delayed recruitment of inflammatory cells into the airway walls from the bloodstream. This latter response also causes bronchoconstriction, but more slowly than the direct irritant effect. Suppose further that the amount of delay between exposure and in-migration of cells is a function of the degree of initial inflammation (an assumption well supported by experimental data) (Kriebel and Smith 1990). Under these conditions, assessment of the bronchoconstrictive response to brief pulses of exposures might show little or no consistent trend over varying exposure intensities, and a simple lagged exposure variable would not perform well either because the effect is being driven by the slow inflammatory response.

A set of graphs may help to contrast alternative ways that time can be handled in environmental epidemiology (Figures 1.4 to 1.6). In the standard approach, a summary measure of exposure, cumulative exposure (CE) for example, is calculated for each subject, as shown in Equation 1.1:

$$CE = \sum_{i=1}^{N} C_i \times t_i$$

Equation 1.1 Calculation of cumulative exposure

At the end of the study period, the moment at which the outcome is evaluated, or the moment of disease onset, subjects are compared to evaluate differences in CE, the summary measure of exposure, between diseased and nondiseased groups. If the summary measures of exposure do not differ between diseased and nondiseased, then it is concluded that exposure is not associated with disease. Figure 1.4 illustrates the standard approach. In this example, the data consist of annual average exposures, so all t_i values are 1 year. The two subjects in Figure 1.4 have identical values of CE, determined at the end of the 10-year time interval shown. Notice, though, that subject 1 has a different temporal pattern of exposure than subject 2; the former's early years were rather heavily exposed, whereas the latter's history is characterized by fairly constant exposures. Do these differences in the pattern of

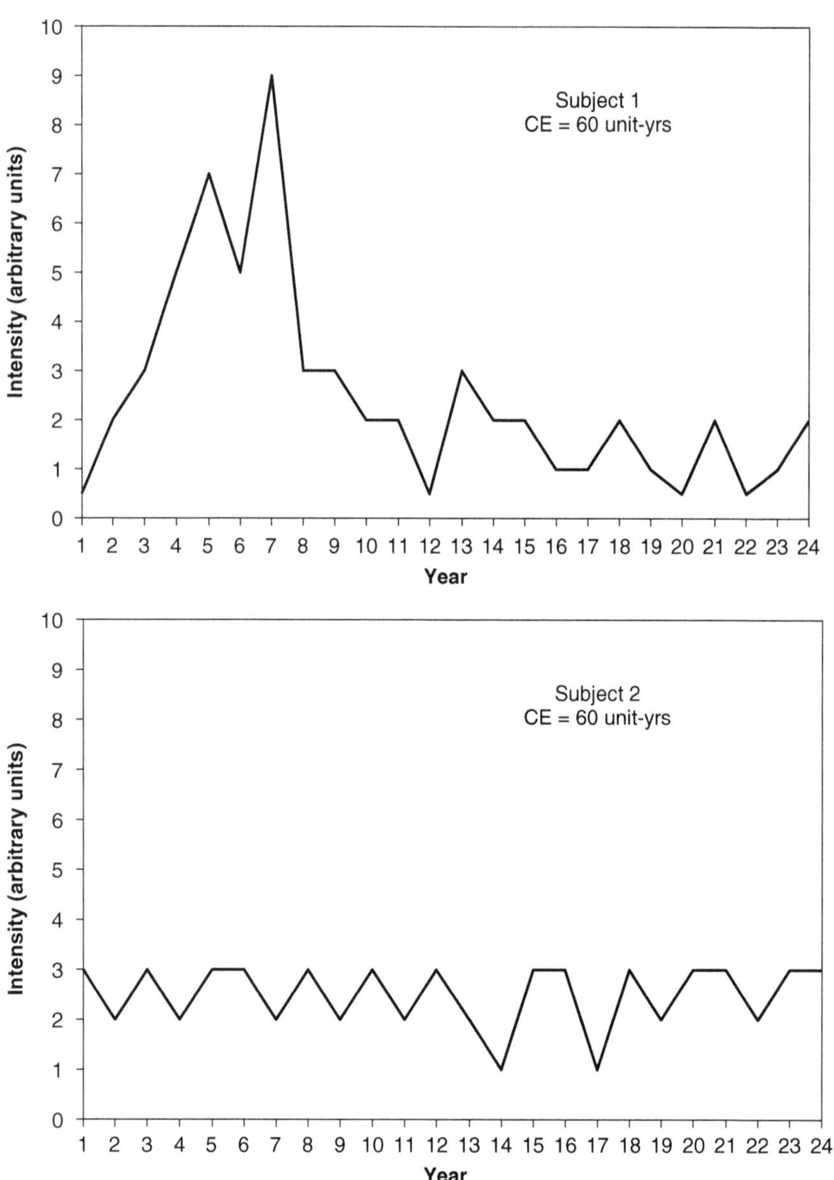

Figure 1.4 Illustration of alternative treatments of the temporal dynamics of exposure and response in epidemiologic models. The exposure profiles (arbitrary units) of two subjects with the same cumulative exposures are shown.

exposure affect the risks faced by these two individuals? Perhaps, and perhaps not—the answer clearly depends on the disease process. But if the investigator chooses to summarize exposure using CE, then implicit in this choice is the assumption that the patterns *do not matter*—that only the final value of CE affects risk. Thus the choice of summary measure of exposure carries with it implicit assumptions about the underlying disease process.

Time Windows of Susceptibility

This limitation of cumulative exposure is well recognized (Smith 1992; Checkoway, Pearce et al. 2004), and "time windows of exposure" have been advocated as a solution (Rothman and Greenland 1998; Checkoway, Pearce et al. 2004). The exposure history of each subject can be partitioned into time windows (usually constructed *backward* from the time of disease onset), and then the exposure in each window is separately summarized with a summary measure such as CE. This approach is illustrated in Figure 1.5. For the two subjects in our example, this method has the advantage that in the earliest time window subject 1 has a higher CE than subject 2. Now instead of one summary measure of exposure there are three, and when these are fit to the outcome data, the early exposure differences between these two subjects may be found to affect their risks.

The time-windows approach relaxes somewhat the implicit assumptions of a single summary measure such as CE, but there are significant limitations as well. How many windows should be used? How should the cut points be determined? Which summary measure of exposure should be calculated within each window? The use of time windows artificially divides a constantly varying lifetime exposure history into "independent" periods, which carry their own implicit assumption that exposure arriving just prior to a cutoff date for one window may have a very different risk than exposure occurring just a few months later, after the start of the next window.

Exposures in different time windows prior to an outcome are recognized as important for some adverse effects, such as pregnancy outcomes. Windows can be defined empirically by dividing a time period into some number of equal periods, or a priori biological data can be used to define the windows of increased or decreased susceptibility. Empirical time windows should be distinguished from the less common practice of separately studying exposures in windows of inherently different susceptibility; as, for example, in different periods of pregnancy or in life stages such as pre- or postmenarche (Agalliu, Eisen et al. 2005; Thompson, Kriebel et al. 2005).

Time windows defined empirically have another limitation: they do not address the possibility that the risk or effects from a particular unit of exposure may be altered by the pattern of exposure that has preceded it. Suppose that the exposure received by subject 1 in Figure 1.5 in the second time window carries

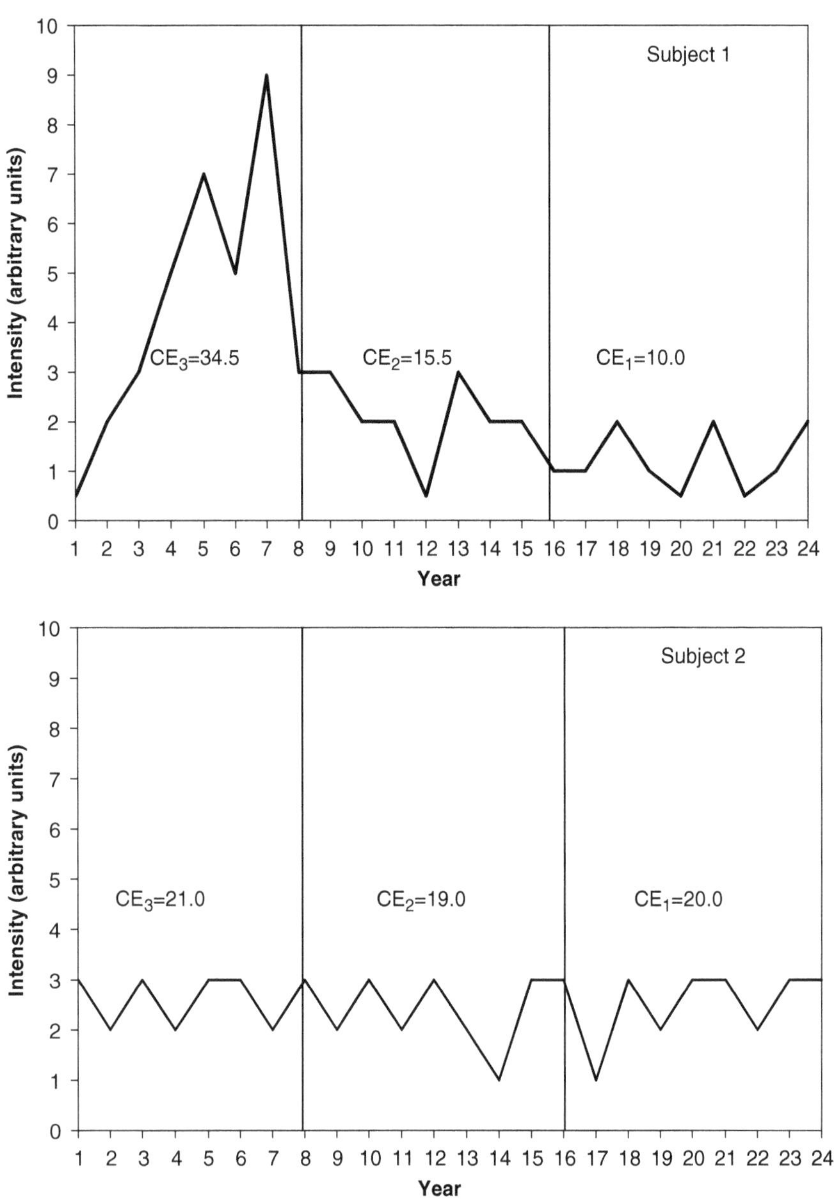

Figure 1.5 Illustration of alternative treatments of the temporal dynamics of exposure and response in epidemiologic models. The same exposure profiles as in Figure 1.4 are partitioned into three time windows. The first subject has a higher exposure in the early window, whereas the second subject has fairly constant exposures throughout.

much higher risk because he experienced a period of high exposure in the first time window. The closest we could come to capturing this behavior using the time-windows approach would be to hypothesize that there was a *statistical interaction* between exposures occurring in different windows. This interaction would probably be very difficult to evaluate in regression models because of the problem of collinearity of the main effects of the time windows with their product term. But even if we could do it, it would likely be an unsatisfactory solution. The standard method of constructing interaction terms captures only one rather simplistic hypothesis for a two-variable interaction (Lewontin 1972) (Checkoway, Pearce et al. 2004) (Vineis and Kriebel 2006). This pattern may not capture the multiplicity of possible biological interactions and may miss an important relationship.

Dynamic Biological Process Models

An alternative to the time-window approach is a dynamic model in which one or more biological processes operate through time to produce the observed pathology. In the most general case, a disease process operates in response to exposures, and the body responds to the damage with repair processes. As a result, the prior history of exposure affects the response, as does the repair process, to define what we see when we measure the damage at a point in time (Figure 1.1). The particular pattern of changing response over time will depend on the exposure pattern and on the details of the disease process model. These are topics of Chapters 3, 4 and 6. The temporal pattern of response generated by a disease process model driven by the same exposure time profile of Figures 1.4 and 1.5 is shown in Figure 1.6. The details of this model are postponed until Chapter 6, but essentially it represents two independent proportional opposing processes, a damage process and a repair process, operating over short intervals, Δt. The model can be represented as Equation 1.2:

$$Damage(at\ t+\Delta t) = Damage(at\ t) - Repair(during\ \Delta t) + Damage(during\ \Delta t)$$

Where

$$Repair(during\ \Delta t) = k_{rep} Damage(at\ t)$$

$$Damage(during\ \Delta t) = k_{dam} Exposure(during\ \Delta t)\ and$$

$$k_{dam}\ and\ k_{rep}\ are\ proportionality\ constants$$

Equation 1.2 A dynamic biological model expressed as a difference equation

Representing the continuous processes as discrete changes over short periods of time Δt makes the mathematics simpler and allows us to use difference equations rather than differential equations. Consider what happens within one time interval: First a portion of the damage at the beginning of the interval is repaired. The amount of damage repaired during the interval is a percentage of the total at

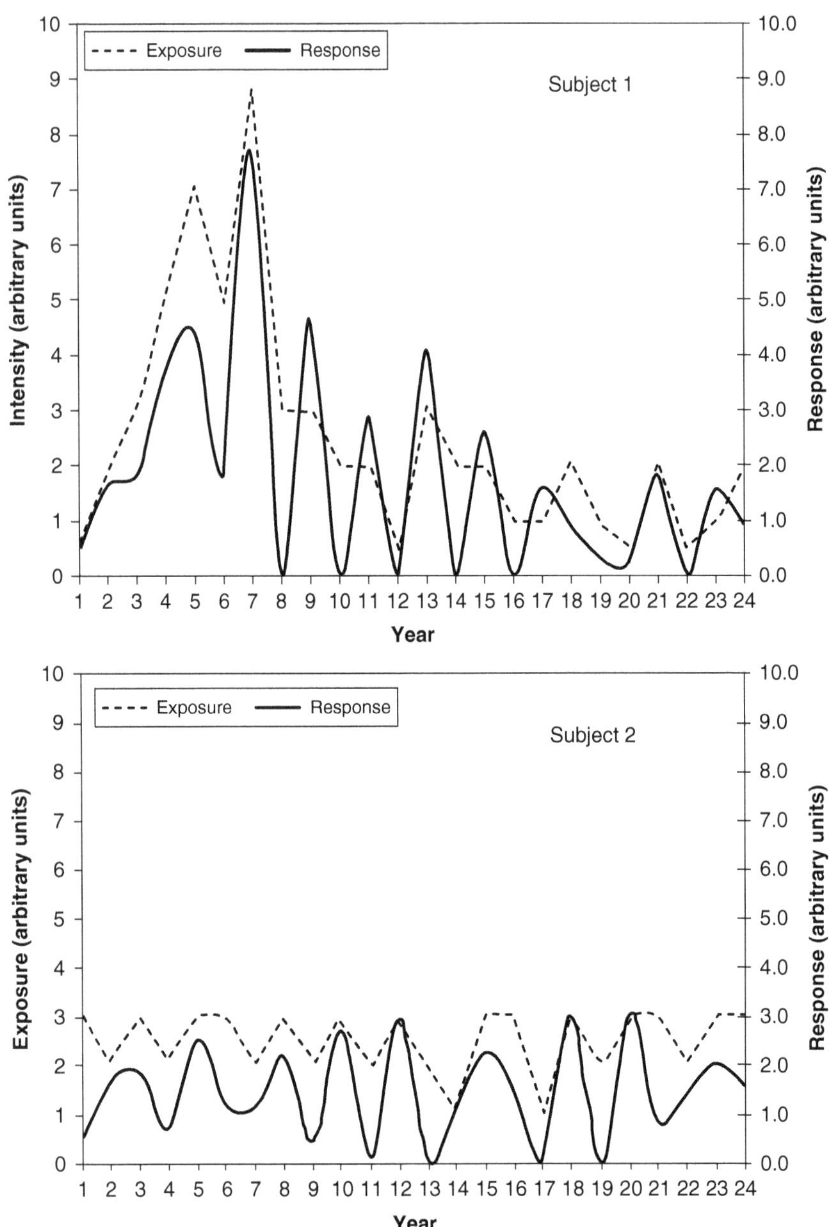

Figure 1.6 Illustration of alternative treatments of the temporal dynamics of exposure and response in epidemiologic models. The predicted response from a hypothetical disease process model (shown in Figure 1.7) is shown for the subjects in Figures 1.4 and 1.5. The model predicts different risks for these two, because it is iteratively recalculated over each short interval of time, continually changing the weights on each annual exposure as a function of the previous exposures.

the start. A new amount of damage is defined by the average exposure intensity during Δt times a weighting or conversion factor (the response per unit of exposure) during the interval. At the start of the next interval, the process is repeated. Like an inchworm, the process works its way along the time line of exposure, making a calculation for each interval (Figure 1.7).

In Figure 1.6, the solid line represents the value of a response or effect variable calculated from the damage-repair model for the same exposure history illustrated in Figures 1.4 and 1.5 (and shown as dotted lines in Figure 1.6). According to the particular disease process acting in this hypothetical example, the high exposures in the early years for subject 1 did not have the effect of increasing his response at the end of the period because those early effects were all repaired. At the end of the time interval, the value of this response (on an arbitrary scale) is nearly twice as large for subject 2 as for subject 1.

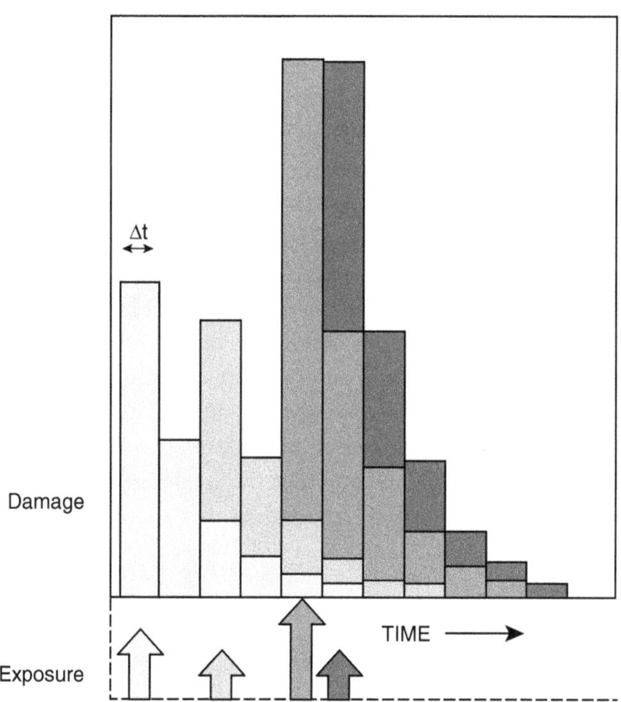

Figure 1.7 Temporal sequence of responses produced by a simple damage-repair model following four short duration exposures (different shading)—the height of the added damage (same shading as exposure) is proportional to the intensity of exposure. The duration of the time increment, Δt, was chosen to equal the half-time of repair. At each time interval the residual amounts of damage are added to each other. The total damage we would observe is the sum of the new and all residual damage from the prior exposures.

The challenge for those who would like to try this method is the design of the disease process model. Our basic approach is to gather some data about the nature of the exposure agent, observe the time course for the development of the effects or disease, and then develop a disease process model that captures the key temporal patterns of the exposure-response relationship. Although biologic processes are extremely complex, certain basic patterns are common, such as the damage-repair process model, and we can use them to provide the basic structure for a disease process model. We propose that four basic disease process models can represent most of the common diseases and adverse effects caused by environmental exposures.

1.5. FOUR DISEASE PROCESS MODELS

As a result of examining a number of different adverse effects and diseases, we have observed that there are two critical elements of the disease process that must be identified: (1) whether the *time course* of the disease or damage process shows that the process is reversible (shows evidence of repair) or irreversible; and (2) whether the *type of response* is proportional to the tissue concentration or, alternatively, whether the adverse effect develops in a discrete step function, from absent to present. Combining these two characteristics, there are four distinct models that we propose can be used as the starting point for modeling the majority of environmental disease processes (Table 1.1). These time course and response types define profoundly different pathologic processes and exposure-response relationships. Fortunately, these features can often be identified using descriptive and readily observable information about the time course of development and recovery from the adverse effects. As more information is gathered, more complex mechanisms and models can be developed as needed to describe the observed effects. Multistep process models can often be developed by linking together separate basic models, for example to represent the transition from a reversible effect to one in which damage begins to accumulate. Let's look in detail at these two issues.

Reversible versus Irreversible Time Course

When a smoker first picks up the habit, smoking produces reversible proportional respiratory airway symptoms. If smoking ceases within a few months or years, all or almost all of the damage may be repaired with no long-term effects. However, if exposure continues, irreversible cell damage accumulates in the small airways, often leading to chronic obstructive pulmonary disease, an example of an irreversible proportional response but associated with a much slower process. Long-term smoking also increases the risk of lung cancer through a series of different types of genetic damage, some of which may be reversible. However, once the final transformation occurs and makes a malignant cell, further cigarette smoke exposure is irrelevant, and the individual has made the discrete transition to a case of lung cancer—an example of an irreversible discrete response. This illustrates that a single type of exposure (tobacco smoking) may cause several different effects/diseases and that each should be studied using an appropriate disease process model, defined by the temporal behavior of that specific outcome.

It is a matter of considerable importance to know whether the short-term, mostly reversible effects of an agent contribute to long-term irreversible damage or whether these are independent phenomena. For example, it would be very useful to know whether someone who responds acutely to air pollutants that cause short-term airway inflammation is at higher risk of developing chronic lung disease. Despite considerable research, this question remains largely unanswered (Becklake 1995). Often the option of detecting a reversible or irreversible time course is made implicit by the length of the study period: if one is studying the effects of air pollution over a period of hours or days, then necessarily one is able to detect only effects that will probably be largely reversible. On the contrary, a study following people for many years usually can detect irreversible effects, although it may also detect short-term reversible effects, confusing the data analysis and its interpretation if care is not taken to separate the two patterns. By explicitly stating in advance what time course we think we are studying, we decrease the chances for confusion and misinterpretation of the results.

Proportional versus Discrete Response

Whether a response is proportional or discrete may in some cases depend on the time scale of observation. In many of the examples we cite, a very close view of the evolving disease process on a very short time scale (minutes or even seconds) might reveal a "gradual" development of a response. But when viewed from a time scale of clinical relevance—hours or days for an allergy or years for a cancer—the response will appear "instantaneous": one moment the subject was healthy, and the next she was sick.

In the remainder of the book, we demonstrate the use of biological models to describe dynamic (temporal) disease processes in much more detail. Using examples, we show how such models may be formulated, how different biological model types behave, how they may be used to design studies, and how they can be used to evaluate competing alternative hypotheses about mechanisms of adverse effects. Our goal is to formulate strong tests of our exposure-disease hypotheses in which a hypothesis is developed in as much biological detail as it can be, expressed in a suitable dynamic (temporal) model, and tested by its fit with a rich data set so that its flaws and misperceptions of reality are fully displayed. Rejecting such a fully developed biologic hypothesis is more informative than either rejecting or failing to reject a generic or vaguely defined hypothesis. For example, the hypothesis "truck drivers have more risk of lung cancer than non-drivers" (Platt 1964) is of limited usefulness for prevention because truck drivers have many environmental exposures; "lifestyle" attributes such as smoking, diet, and exercise; and socioeconomic characteristics, as well. Hypothesizing that a particular chemical agent is associated with lung cancer—whether the hypothesis is refuted or supported by data—is more likely to lead to successful prevention activities.

1.6. ON THE LIMITS OF EPIDEMIOLOGIC DATA AND BIOLOGIC MODELS

Before embarking on such an ambitious journey, we should acknowledge some of the challenges that we will face. The observational nature of epidemiology and the complexity of pathophysiologic processes place limits on the degree to which epidemiologic data can directly and confidently be used to quantify environmental health risks. When epidemiologists speak of models, they generally mean empirical statistical models of the population risk, not biologic models. Their goal is to find a mathematical relationship between the risk and some set of covariates, such as age, gender, genetics, exposure, and so forth, that can account for a major part of the variation in risk among the population members. These mathematical models are necessary "restrictions on the possible states of nature" (Robins and Greenland 1986)—restrictions that are needed because epidemiologic data can only be expected to distinguish among a very limited set of alternative states (Robins and Greenland 1986), (Vandenbroucke 1987). Although we agree with this rather sobering view of the limits of epidemiology, we believe that the choice of models against which to compare the data should, so far as possible, be guided by explicit hypotheses about the underlying biologic processes. In other words, you can get as much as possible from epidemiology by starting from well-thought-out hypotheses that are formalized as mathematical models into which the data will be placed. The disease process models can serve this purpose.

In such an investigation, there will always be many important sources of error, and in Chapter 8 we help the reader understand how these errors can be evaluated. Epidemiologists are familiar with the importance of avoiding biases and confounding when conducting observational studies. They are also well aware of the imprecision that arises when studies are too small, because then chance can be an important determinant of an outcome. To evaluate this potential role of chance, epidemiologists are taught to calculate confidence intervals, or *p*-values (from which the overused and misunderstood concept of "statistical significance" is assessed). But because of the large number of unverifiable assumptions, implicit and explicit choices in study design made by the investigators, and the likelihood of unknown biases and confounders, these quantitative measures of precision should not generally be interpreted as formal tests of hypotheses, much as we might wish for the satisfaction of the crisp conclusion: "the result was/was not statistically significant." A similar view was stated succinctly by Heij (Heij 1989):

> The aim [of modeling] is to determine a simple model which is supported by the data. In general no simple relationships are satisfied exactly by the data. This discrepancy between observed data and simple relationships is often modeled by introducing stochastics [what epidemiology texts call "the role of chance"]. However, instead of stochastic uncertainty it is in our opinion primarily the complexity of reality which often prevents existence of simple exact models. In this case model errors do not reflect chance, but arise because a simple model can only give an approximate representation of complex systems.

If, as we believe, observational studies are seriously limited by unknown biases and confounders, how should environmental scientists best contribute to the understanding of disease processes? Two solutions have been proposed: abandon modeling altogether (Vandenbroucke 1987) or use biologically based mathematical models with several stated caveats: (1) the modeler's prior beliefs are clearly stated, (2) a range of plausible models are tested, (3) the sensitivity of the results to changes in model parameters are investigated, and (4) the modeler presents as complete a description of the uncertainty in the results as possible (Robins and Greenland 1986); (Thomas 1983); (Vandenbroucke 1987). This book provides guidance for those who would choose the second solution.

1.7. SUMMARY

The goal of environmental health is the prevention of disease through identification and elimination of environmental hazards or the reduction of exposures. Exposure assessment and epidemiology can contribute to continued progress in achieving this goal, but we believe that this will not occur without integrated, biologically based methods of study design and analysis. With the rapid proliferation of molecular and cell biology, the mechanisms underlying pathologic processes leading to adverse effects are increasingly well understood. This new knowledge should be used more systematically as the theoretical basis of exposure-response modeling. We show how disease process models can provide a simple but useful starting point and a framework for added complexity as new information is discovered.

The purpose of this book is to promote biologically based investigations of exposure-dose-response relationships. This means both the relationship between exposure and internal dose to the target tissue, the "exposure-dose relationship," and the relationship between the tissue dose and the adverse effect, the "dose-response relationship." We begin with a discussion of the personal nature of exposure and the individual's responses to it, and then move to the epidemiological realm of populations and their responses. Mathematical models derived from one of four simplified disease processes are developed and illustrated to encourage readers to explore new methods for improving the study of environmental health hazards.

NOTE

1. William of Occam was a 14th-century philosopher and theologian who was concerned with avoiding unnecessary complexity in philosophical explanations, preferring instead the simpler or parsimonious explanation.

SECTION A

Exposure and Disease in Individuals

Section A provides information on the assessment of both exposure and disease in **individuals** and consists of Chapters 2 through 6. Chapters 2, 3, and 4 are a sequence on environmental exposures—their fundamental characteristics (Chapter 2), methods of exposure measurement and estimation suitable for epidemiologic applications (Chapter 3), and methods of studying the entry of materials from exposures into biologic tissues using toxicokinetic models (Chapter 4). Chapter 5 is an overview of the use of biomarkers in environmental assessment. These chapters constitute an overview of biologically based exposure assessment. Finally, Chapter 6 presents the four principle disease process models that lie at the heart of our approach.

2 Characteristics of Exposure and its Measurement

"What is the nature of exposure?" asked the Philosopher.
"Smoke gets in your eyes," laughed the Jester.

2.1. WHAT DO WE MEAN BY "SOMEONE IS EXPOSED TO A HAZARD"?

Exposures to chemicals and other hazards have an intuitive meaning based on common experiences of environmental conditions: exposure is the composition (defining the agent), concentration (or intensity), and duration at the point of entry. In this chapter we review and emphasize chemical hazards, but the concepts and approaches also apply equally well to other types of hazards, such as biological agents, radiation, noise, or mechanical energy (ergonomic and biomechanical hazards). Each has some unique features, but they have much in common: they all have a fundamental agent (bacteria, radiation and noise frequency, mechanical energy) that can interact with tissues, all have a quantitative measure that defines intensity, and duration is needed to define the hazard. We use chemical exposures to illustrate a comprehensive approach for the epidemiologically relevant aspects. We leave it to the reader to extend these ideas to other types of hazards.

Consider the common hazard of combustion emissions. Nearly everyone has had the experience of being exposed to smoke from a campfire outdoors. There is the wonderful, characteristic smell of burning wood, which stimulates memories of cool evenings gathered around the radiant warmth of the cheery fire, perhaps while roasting marshmallows or hot dogs on a stick. That is, until the wind shifts and brings the concentrated smoke into your face for more than a moment. Then you have the burning eyes, scorched throat, and coughing that are also characteristic of exposure to smoke from burning wood, especially green wood. Fire studies have shown that wood smoke contains a number of potent irritants, such as formaldehyde (Hedberg, Kristensson et al. 2002). Thus, if you sit near a campfire, you will be exposed to wood smoke. Right? Well, maybe not. It depends on how big the fire is, what is burning, and which way the smoke is blowing.

If an investigator classifies epidemiologic subjects into "exposed," based on sitting near a campfire, versus "unexposed," sitting at a picnic table 100 yards from the fire, we may discover some misclassification. A person sitting upwind of the fire and not too close can certainly sit by a campfire without ever having the smoke blow into his or her face. If so, is he or she still "exposed"? An exposure assessor would say he or she had potential for exposure, but none occurred. Ideally we would like to define our exposed group such that we do not include any unexposed, but that may be difficult to ensure. An epidemiologic investigator must be careful to distinguish potential from actual exposure because there may be a substantial difference in the number potentially versus truly exposed and in the doses they receive. Later, in Section 2.6, we discuss potential exposure in more detail.

Exposure may be momentary as you walk past a campfire, or it may consist of hours and days spent sitting around campfires during a long camping trip. Your risk of health effects from the campfire smoke is not the same for a momentary exposure as for one lasting hours, because your internal dose is very different. Duration is a critical feature that is generally considered. Eye irritation may result from a momentary exposure, but deep lung irritation may require hours of exposure.

Campfire Study Example

Consider a simple study of people sitting around a campfire. We designate them as the "exposed" group. We discover their symptoms and other effects by asking them questions, such as, Have you ever had eye or throat irritation, or a cough, or a headache while sitting by a campfire? Then, in a classic epidemiologic design, we compare their responses to those of "unexposed" people sitting at picnic tables in a nearby area. We hypothesize that campfire sitters will have more of all types of symptoms but that not all would be affected. After asking individuals in both groups the same questions, we find that 8 out of the 10 exposed reported symptoms of eye and throat irritation, but most symptoms were mild. Only 2 of 10 picnic table sitters reported symptoms, and all were very mild. We are concerned that fire smoke is an irritant, but why do some of the "exposed" have no symptoms? If nothing was known about the conditions during campfire sitting, then it might be concluded that some fire smoke may not be irritating or that some individuals were very resistant to irritation from smoke. The symptoms of the picnic table sitters would probably be attributed to background levels of symptoms.

Now consider the nature of the exposure to campfire smoke. Because the sitters were largely surrounding the campfire, it is likely that some campfire sitters were not exposed because smoke never blew toward them. Conversely, if the picnic tables were located where some of the smoke from the fires may have been carried by the wind, then some picnic table sitters were exposed without being near a fire. If we had really bad luck during our study, a steady light wind might have been blowing toward the picnic tables so that most campfire sitters, who sat on the upwind side of the fire, were unexposed and most people at the picnic tables were exposed. The large misclassification of exposed and unexposed subjects would make the results meaningless relative to the hypothesis. The point of this exercise is to show that the exposure situation must be considered in detail to define exposure in a way relevant to the effects if we are to obtain useful findings.

The goal of this chapter is to provide some background on the nature of exposure and how it may be characterized and measured. More detail on many of the topics covered can be found in the reference books listed at the end of the book.

Dimensions of Chemical Exposure

Exposure is defined as the concentration of contaminant at the point of entry into the exposed individual's body. Exposure has four dimensions: composition, physical form (solid, particulate, liquid, or gaseous), environmental concentration (intensity), and their variation across time. Each of these has important characteristics.

Composition and chemical characteristics of environmental contaminants define the potential hazardous agents, which are specific for the adverse effects (see (Klaassen 2001) for more detail). Changing the composition changes the hazard. The chemical behavior of the agent is important for point-of-entry penetration and absorption and tissue interactions. Every chemical has intrinsic properties that affect its toxicity: acid-base properties, polarity (water solubility), lipid solubility, and tendencies to react with biological materials, such as amino, hydroxyl, carbonyl, or sulfhydryl groups in tissue constituents.

Physical form of a substance is determined by such characteristics as the melting and boiling point temperatures and vapor pressure at the ambient temperature. Airborne materials may be gaseous (gases or vapors) or aerosols (solid particles or liquid droplets suspended in air). Aerosols are usually complex mixtures of particles with a distribution of sizes (described by geometric mean and geometric standard deviation diameter) and shapes, and they often contain gaseous materials, including vapors of solids or liquids. Fibers, such as asbestos, are an important subgroup of airborne particles, having much larger length relative to diameter. Bulk liquids may be mixtures containing dissolved materials, sometimes also containing suspended solids, and may have separate phases, such as gasoline floating on top of water. Liquids may also be retained within solids, such as organic solvents in soil. Solids may have varying degrees of aggregation ranging from massive solid blocks to very finely divided dusts and powders and nanoparticles whose size can overlap with that of large viruses (Lippmann, Cohen et al. 2003).

Environmental concentration at the point of entry is sometimes thought of as the single fundamental definition of exposure, whereas we emphasize that it is just one of the four fundamental dimensions. Three routes are important common entry points: inhalation of air at the mouth and nose, absorption of material on the skin, and gastrointestinal (GI) absorption of materials in food and drink. Because passage through the respiratory tract leads to deposition and absorption depending on the physical and chemical characteristics of the substance, respiratory contact is defined at the point of entry. Not all particles entering are deposited; some are exhaled. Particles with diameters of about 0.3 μm (for detail on how particle diameter is measured, see Hinds, 1999), are the least effectively retained in the lungs (about 10%), whereas both larger and smaller particles have higher retentions. Airborne particles and droplets with sizes in the range of 4–100 μm are deposited in the upper airways and moved into the GI tract, where soluble components will be absorbed. Similarly, skin in different locations has different particle and liquid penetration and absorption characteristics (Semple and Cherrie 2003).

Time is the fourth dimension of exposure, and variation in the other three dimensions across time is critical for biological interactions and disease processes, such as uptake, distribution, and reactions with cellular materials. Exposure can also vary with physical location as a result of proximity to different exposure sources and heterogeneity of environmental transport and chemical processes, but this variability will lead to temporal variability at the point of entry into the body, and so we do not count it as a primary dimension of exposure.

When assessing an exposure, all of the dimensions of exposure need to be considered. When one or more are important sources of exposure variation, they should be characterized for consideration in dose modeling.

Epidemiologic Importance of Exposure Variation

Variation in exposure is very important for epidemiologic studies because it is the tool we will use to detect the effects of exposure. We introduce the topic here and elaborate on it in Chapter 3. Ultimately we want to compare groups with different exposures to see whether they experience different disease risks. There are two critical types of variation: variations among individuals and variations across time within individuals. Also, we want to distinguish systematic versus random differences in exposure, usually across time. No two individuals will have identical exposures, even though they have the same jobs and employer, live next door to each other and have the same water supply, and have the same diet and buy their food at the same grocery store. But despite this, they may have *similar* exposures, and the differences between them may not carry important differences in disease risk.

Variation among Individuals and Groups

Because epidemiologic studies compare risks among groups of individuals, we wish to form groups within which exposures are similar and between which exposure differences are large. It is very rare that an exposure has only one toxic component whose concentration is stable across time, a situation akin to the laboratory experiments in which animals receive carefully calibrated doses. In observational studies, there is always variability in exposure intensity over time within individuals and among individuals at any point in time. Temporal variation in personal exposures, when measured and linked to response variations on the same time scale, perhaps with a lag, can represent a probe of the biological system of the exposed individual. One of our premises in this book is that averaging exposure across time and groups should be done with care and attention to what potentially useful information may be lost.

2.2. HOW TO THINK ABOUT EXPOSURE: SOURCE-RECEPTOR MODEL

Exposure is a process that happens to individuals, and there is a very simple but useful way to conceptualize an exposure situation: the source-receptor model (in this case, the "receptor" is the exposed person, not to be confused with biochemical receptors). Exposure originates from one or more sources of environmental emissions, indoors or out. Emissions from a source enter the environment as concentrated contaminants, which are then diluted by diffusion and mixing with air or water. Some of the emissions may be removed by control devices at the emission source. In the environment, over the time scale of minutes to days, some substances may be lost by chemical reactions, or new things may be formed, such as through photochemistry. The contaminants eventually reach the individual (called the "receptor"). These may be the agent(s) of effects or their precursors (see Figure 2.1). From an individual's immediate environment, contaminants can enter the body by one or more of three common routes of entry noted earlier. This model is useful because it clearly shows the processes that relate an individual's exposure to the sources of chemicals in the environment. It also suggests how exposure can be analyzed to identify the important features defining an exposure situation.

Source Characteristics

Reconsidering the campfire example, the fire is the source. The type of fire (including its relative heat), its fuel characteristics, and its size will all determine the emission rate and composition of emissions. A small campfire with green wood will burn relatively cool, releasing a relatively small amount of dense irritating smoke at ground level. A large bonfire made with dried logs will burn hot, releasing large

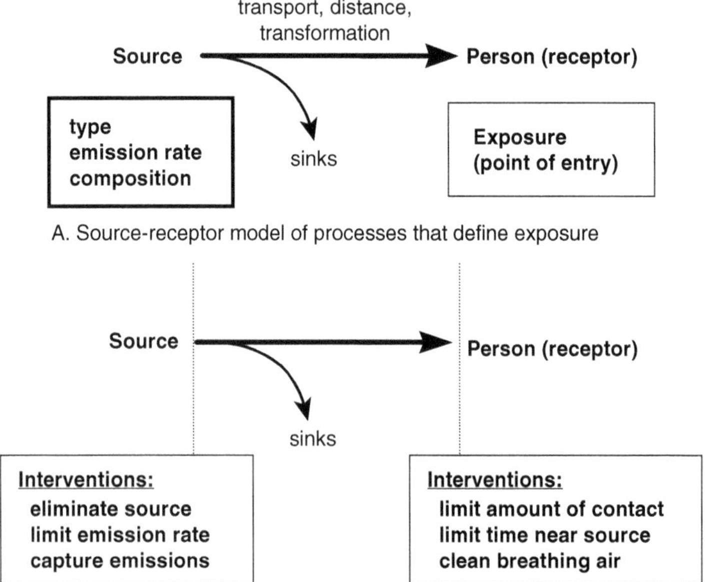

Figure 2.1 Simple source-receptor model of exposure: the basic model (A) and the model with points of intervention noted (B). Note that a receptor is a person and a sink is a removal process.

amounts of less dense smoke at an elevation (because of the buoyancy of hot gases). Fires burn down after a period of time, so the heat output rate and the composition of emissions are not constant over time. The transport of the smoke to the person is affected by several factors. The wind is the transport vehicle, which has speed and direction, so the relative direction of the person to the fire is important. The wind speed and distance from the fire to the person define the amount of time available for dilution, dispersion, and transformations (particle removal, chemical reactions, etc.).

The source-receptor model is a powerful concept for structuring the investigation and description of environmental exposures. A similar model can be used to represent occupational exposures, in which there may be several sources and in which there is usually less transport and less time for chemical transformation and removal because of a worker's close proximity to a source. This simple but powerful model is readily applicable to many situations. Consider water contamination: some source of pollutants releases a contaminant into a stream, river, lake, or groundwater, where it can be transported with mixing and dilution to a point of contact with people. Food contamination is generally more direct, usually without an environmental transport process, but one may occur, such as when radioactivity deposited on vegetation is eaten by cows, leading to contamination of milk. The food distribution system brings the food to people. Often the contamination is very heterogeneous: methyl mercury levels differ by fish species, size, age, diet, and location. Skin contamination may result from direct contact with a contaminated surface. It also may result from environmental processes, liquids splashing onto the skin, dust deposition, or contact with contaminated water. The release of a contaminant into the environment may result in several concurrent routes of contact. For example, contamination of groundwater with a volatile solvent can lead to ingestion contact via drinking water, skin contact during bathing, and vapor inhalation contact during showering.

Contact with the Receptor (Person)

An individual's contact with contaminants may be passive when materials are transported to the individual's location. For example, residence in a certain location may result in contact with contaminated water. Alternately, a person's activities may produce emissions and subsequent exposure. Certain job titles, for example, coke oven worker, are associated with work activities that create carcinogenic materials and that bring the worker into contact with the airborne carcinogens from coke production.

Variation in Contact across Time and Space

As noted earlier, two other general characteristics of exposure are its variation across time and across space or location, which are often the result of variations in emission and transport processes. The activities of people will, of course, also affect the exposure intensity that they receive.

Again, we can use the behavior of smoke from a campfire to nicely illustrate the characteristics of exposure variation by time and space. Sitting near a small campfire on a cool evening, one can watch the smoke coil, loop, and spread as it leaves the fire. In some places the smoke will be very dense, but in other places it will be wispy and nearly invisible, and this difference reflects the variation in the concentration of smoke particles in small parcels of air at a moment in time. This variation is the result of turbulent mixing whereby the dense emissions from the fire are mixed with the surrounding air, diluting and dispersing them as they are transported away from the fire. Gases will also disperse by diffusion, which is not important for most airborne particles. The turbulent mixing process is highly uneven and produces declining, variable concentrations with distance. Diffusion and turbulent mixing are general processes that occur in all air and water systems. They operate on different spatial scales (diffusion for small scales, 1–100 cm, and mixing for larger scales, 1–100 meters, with some overlap at intermediate scales). Because the mixing occurs while air or water is moving, transport clearly leads to spatial variability: the average concentration declines as you move downwind from the fire, which is visually evident in campfire observations. Low-speed light winds produce less mixing than gusty strong winds. Similarly, lightly colored material in water from a sewer outfall going into a river can be seen swirling and dispersing in the different colored water of the river flowing downstream.

If you sit at a fixed location downwind from a fire, you can observe the variation in exposure with time: the smoke will blow toward you from the fire for awhile and then shift away from you, as the wind shifts about its predominant direction. Consequently, the concentration at a fixed location, such as your breathing zone, will vary across time, even though you do not move around. If you sit closer to the fire, the concentrations will be higher because there has been less dilution. This variability across time and space is a basic characteristic of environmental fluid transport and dispersion processes (see Lippmann et al. 2003 for a more detailed discussion).

Exposure Variation and Intake Processes

How does the variability in airborne campfire smoke affect the uptake of exposure components? We inhale a breath of air every 5 to 6 seconds. The amount of smoke in air from your breathing zone will vary across time in a highly variable, random fashion. If you change your location relative to the fire and the wind, you will change the likelihood that smoke will be present, as well as its concentration in your inhaled breaths. By moving, you may reduce or even eliminate your exposure, but you may also increase it, because the wind direction may also change. People who are more sensitive to the smoke will move more than those who are not bothered by it. Thus an individual's contaminant concentration at the point of contact is a complex function of the characteristics of the exposure situation and the subject's response to it.

Exposure to contaminated water also has variation across time and space, and individuals may avoid drinking discolored, cloudy, or unpleasant-smelling water. Spilled pesticide entering a reservoir may affect only a portion of a city and may be uneven because the pesticide is not fully mixed into the reservoir. Lead from old plumbing pipes will be a hazard only for older homes, and flushing the pipes by letting the water run for a minute before you take a drink can substantially reduce exposure, because the lead only slowly leaches from the pipe walls and takes time to accumulate in stationary water. Gasoline contamination of groundwater can move through subsoil structures in unpredictable patterns, so that only nearby wells are contaminated.

Skin exposure varies because the amount of contamination reaching various sites on the skin varies over time, and its residence on the skin may be different in different skin locations because the individual may be more or less likely to clean it off. A homeowner spraying his roses with pesticide may wear gloves that keep the pesticide off of his hands, but some spray may still land on his face and clothing.

He also may contaminate his hands when he removes the gloves. Washing his hands will not remove the pesticide from his face or from the shirt that contacts skin on his arms.

2.3. BIOLOGICALLY BASED EXPOSURE ASSESSMENT

The goals of exposure assessment for epidemiology are to determine whether hazardous agents are present and to determine their concentrations at the point of contact during the time period of interest. This is not simply an analytical chemistry question. Often one of the most difficult analytical questions you can ask is: What chemical hazards are present? This question can be asked with some urgency when the speaker is holding a bottle partially filled with a black, foul-smelling goop that has perhaps come from his back garden or local river. Although there are powerful analytical chemistry tools that can eventually provide an exquisitely detailed list of components in the black goop and their concentrations, doing this might require lots of money and lots of time for just a single sample. In addition, the list may not provide many useful clues to which of the hundreds of specific chemicals may be hazards, either alone or in combination. Environmental contamination problems have often been approached as though they were only analytical chemistry or physical science questions, but they are fundamentally biological questions about toxic risks, requiring a different approach to guide the application of analytical chemistry.

The source-receptor analysis of an exposure situation can help identify the types of materials released into the environment and give clues to their concentrations. The assessment of the hazard from those emissions requires an identification of the possible toxic agents and their concentrations. The key question for exposure assessment is not what is present, but rather: What aspects of the exposure are biologically relevant?

Biological Relevance of Environmental Contaminants

Biological relevance is defined by the ability of a compound to cause or aggravate adverse health effects. Because many environmental exposures are complex and contain many potentially toxic chemicals, we wish to narrow the scope of our assessment. Consider a bicycle repair worker in a small shop who reports numbness and tingling in his hands (peripheral neuropathy) after washing bicycle parts with a commercial degreaser over several weeks. The degreaser is a complex mixture of unspecified volatile hydrocarbons. The application of the degreaser produces substantial skin contact and vapor in a small, poorly ventilated workroom. The worker is also exposed to grease used during assembly activities and to solvents from spray paints used on the bicycles.

How do we decide what is biologically relevant about an exposure? First, we need to determine which exposures present the highest concentration at a route of entry. Because eating is not allowed in the work area, ingestion is unlikely. Degreaser solvents are intended to evaporate rapidly from cleaned surfaces, and the label on the can or bottle gives some information about what is used. Similarly, the solvents in sprayed paints are intended to evaporate from the painted surface. The paint droplets in the air also are a major source of evaporating solvent. Our source-receptor model suggests that evaporation of solvents in a confined, poorly ventilated space will produce a high air concentration. Hand washing of bicycle parts with a solvent will result in both extensive skin contact and evaporation into the breathing zone of the individual doing the hand washing. Thus solvent components are at high concentrations both in the breathing zone and on the skin of the hands and forearms, and they may enter by inhalation and skin absorption. Chemicals in paint (printed on the label or reported by the manufacturer), aside from the solvent, are typically large polymers, pigments, and fillers that are unlikely to be absorbed through the lungs or skin but that might cause local effects with prolonged contact. The bicycle grease is a high molecular weight hydrocarbon with low volatility and low penetration through the skin. So the paint and grease are not first-order concerns but may be reconsidered later if they appear relevant for some other condition.

Next we need to determine whether any of the chemicals at high concentrations are associated with relevant health effects. The labels on the degreaser and paint spray cans both indicate that they contain "petroleum distillates," which is not a very useful description. Chemical analysis of the degreaser indicates that n-hexane is a significant component, along with toluene. Checking a general toxicology reference text (Klaassen 2001), we find that both n-hexane and toluene are readily absorbed by inhalation and through the skin. However, only n-hexane has been strongly linked to peripheral neuropathy. Toluene is reported to have central nervous system effects but limited peripheral nerve effects. Although an interaction cannot be ruled out, a good working hypothesis is that n-hexane is the agent of the health effects and should be the focus of the exposure assessment. Some limited evaluation of toluene also might be made in case it contributes. This example illustrates how suspected agents can be identified to guide the development of an exposure assessment.

Biological Factors Limiting Access to the Body

A number of protective barriers must be breached for a material to enter the body. These barriers vary depending on the route of entry. A simple example will make this clear.

Lead metal is highly toxic when dissolved in body fluids. A half-centimeter ball of lead presents little hazard (unless fired from a gun): it cannot be inhaled, and little if any will be absorbed from the ball rubbing against the skin or even by briefly placing it in the mouth. However, if the lead ball is melted and strongly heated, it will form a very finely divided powder (particles with diameters of less than 0.1 μm) in the air in the breathing zone. This would present a major hazard, because the powder can be easily inhaled, and it will be deposited deep in the lungs, where it can be solubilized and absorbed directly into the bloodstream. Thus physical form is very important for defining a hazard.

All materials cannot gain entrance into the body equally well. We wish to focus our exposure assessment on those components of exposure that are most likely to enter the body and to be retained and absorbed into the tissues and that are likely to cause the adverse effects of interest. As a result, we often do not need a full chemical and physical description of every environmental contaminant, because some are unlikely to enter the body or cannot be absorbed.

The three primary routes of entry each have different barriers to environmental agents.

Inhalation Exposures

Inhaled materials must pass through the nose, mouth, and airways before they reach the gas exchange area of the lungs, which limits access for some substances. The moist surfaces of the upper airways will absorb water-soluble gases and vapors before they can penetrate deep into the lungs. The branching geometry of the airways leads to deposition and removal of most large airborne particles (> 3.5 μm dia.) before they reach the deeper lungs. Alveolar macrophages will engulf and destroy or remove some of those particles that do penetrate. Materials that are absorbed from the alveolar or gas exchange area of the lungs pass into arterial blood and are transported directly to the tissues. This makes them particularly dangerous.

Ingestion Exposures

Ingested materials must be stable in strong acids and must be solubilized before they can be absorbed. Gastrointestinal tract absorption is somewhat selective, and absorbed materials must pass through the liver, where they come into contact with detoxifying enzymes, before they enter the venous blood circulation.

Skin Exposures

The skin is a physical barrier that prevents absorption of aqueous and ionic materials. Small, fat-soluble molecules, such as petroleum-based solvents, can readily penetrate the skin and be absorbed into venous blood. Very small insoluble particles may penetrate as well, particularly if the skin has been damaged. These and other important characteristics of the routes of entry are discussed in more detail in the next chapter.

When examining the health effects associated with specific contaminants in an environment, consideration of these factors also can help to define which contaminants might be hazards and to design an appropriate exposure assessment. The route of entry has obvious implications for the choice of exposure measurement methods. These will be discussed in Chapter 3.

2.4. NATURE OF COMMON PERSONAL EXPOSURES

In this section we focus on chemical exposures to illustrate in more detail the nature of personal exposures using the source-receptor model coupled with route-of-entry considerations. This will be done by types of environmental exposure: air and water contamination, food contamination, skin contamination, and more unusual routes of exposure, such as cigarette contamination. The goal of this discussion is to identify the most important features of various types of common exposures. The details of exposure measurement methodology are left to the many excellent texts that cover this topic. Following that discussion, we investigate the statistical characteristics of temporal and spatial variability and sampling strategies used to characterize them.

Air Contamination Exposures

Air contaminants come in two types, gaseous and particulate, and often both are present. Note that some differences are found between the common-use definitions and technical definitions, such as those for the terms a *fume* is a solid condensation particle from strongly heated metals or plastics and *vapor* comes from materials that are liquids or solids at room temperature. The difference between gases and vapors is that *gases* are gas phase at room temperature, and as a result vapors are much more likely to condense on surfaces than gases are—compare, for example, water vapor with carbon monoxide. *Aerosol* is the general scientific term for all types of particles suspended in air, not just the materials sprayed from cans onto frying pans and under your arms (Hinds 1999). A *mist* consists of liquid droplets suspended in air. When the droplet size is very small (< 20 μm dia.), the aerosol behavior of droplets is identical to that of solid particles, although only the former will evaporate. Fortunately, the meanings of most terms are relatively consistent with general usage. More extensive discussions of the chemistry and physics of air contaminants can be found in textbooks on environmental chemistry and aerosol science.[1]

Emission Sources

The example of the campfire exposure can be used to illustrate the important processes that affect the formation and dispersion of air contaminants. Combustion and hot industrial processes are very common emission sources. When a process is very hot, most of the initial emissions are released in a gaseous phase. However, when hot gases and vapors mix with ambient air, they will cool, and vapors will condense into clouds of very small droplets or solid particles (< 1 μm dia.). This is analogous to the formation of steam (tiny water droplets suspended in the air). Because the concentration is very high and the particles are close together, they bump into each other and stick together, agglomerating into larger particles. As the smoke or emissions leave the area near the source, they continue to be diluted by mixing with air, which reduces and eventually stops the agglomeration. When the smoke becomes very dilute, then the more volatile liquids (those with high vapor pressure) begin to evaporate from the particles, which is one of the processes that cause smoke and fog to dissipate. As a result of these processes, the physical form, size distribution, and concentration of air contaminants change as the emissions move away from the source. The changes that occur with time after emission are called "aerosol aging," and the particular characteristics of an exposure to emissions from a hot source depend on how close a person is to the source.

Another set of atmospheric processes operate for large-scale urban air pollution, which is the result of the accumulation of emissions from many sources and the atmospheric processes that modify them. The same processes of condensation and agglomeration occur for combustion emissions from millions of cars and trucks in urban traffic. However, once the emissions are in the atmosphere, sunlight can cause additional photochemical processes that produce smog, such as in Los Angeles, where these processes were first described (Lippmann, Cohen et al. 2003). As a result of these complex chemical reactions, many additional toxic chemicals are produced, including ozone, nitrogen oxides, peroxides, free radicals, oxygenated hydrocarbons, and organic particles. Many of these are potent oxidizers of biological materials and cause symptoms of eye and throat irritation, DNA damage, and acute and chronic lung effects.

Many emission processes generate stable air contaminants that do not change composition after emission from a source, such as solid particles from grinding or drilling. Material handling, such as moving, loading, or transferring bulk materials, coal, sand, flour, or chemicals, can produce airborne emissions with stable size distributions after the large particles (those larger than ~10 μm) settle out of suspension.

Spraying materials with a volatile solvent carrier produces droplets that may spread out over a surface and make a coating, but some will not contact the surface (overspray) and will remain as an aerosol. The overspray can also generate a smaller-sized solid aerosol when the solvent in the droplet evaporates. These processes apply to industrial painting and coating operations, as well as to hobbyist painting with an aerosol can or to someone cleaning an oven with spray-on cleaner. Many wet processes produce droplets, such as acid baths and spray dryers.

Environmental Transport of Emissions

Dispersion of airborne emissions has a profound effect on concentration and sometimes also on composition of an exposure. Factors that limit dispersion likewise have important effects on exposure. Placing a source in a room can severely limit dispersion. Three factors are important: the output rate or source strength, the volume of the room, and the ventilation of the room. Occupational hygienists are very concerned about exposures in confined spaces, especially in areas in which there is no ventilation, because they are a common cause of workplace fatalities. Even a small source, for example a quarter cup (200 ml) of solvent evaporating in a utility closet, can produce a very high exposure for a worker entering that closet.

Painting illustrates the importance of source strength: brush painting covers a small area with a limited amount of paint in an hour, roller painting can cover several times as much area with paint in an hour, and spray painting can cover large areas in minutes while also aerosolizing large quantities of paint and solvent. If the paint used in all three cases contains a toxic solvent, then the solvent exposure to the painter will be quite different: lowest for brush painting and highest for spray painting. It is important to have extra ventilation when painting indoors, and spray painting should be done in special partially enclosed areas with a high air flow, "spray booths."

Personal Contact with Emissions

One can use the source-receptor approach to analyze the basic features of an exposure situation to determine the nature of an exposure. Consider this example: a person spraying house plants using an aerosol can of pesticide purchased at the hardware store. The source is highly concentrated pesticide with potential health effects (hopefully noted by warnings on the can). The total quantity released is relatively small and localized. The volume of the affected area also will be small, and the dilution and dispersion of the emissions will be limited (most homes do not have high-volume ventilation). All of these factors taken together imply that there will be a high local concentration that will dissipate slowly.

Now look at the example of a worker driving a pesticide spray truck through an orchard or a small town to spray mosquitoes. He will plan to drive the truck with the wind blowing from the truck into the trees. If he is careful loading the pesticide into the sprayer tanks and if the wind is strong, he may have no more exposure than the homeowner spraying house plants, even though the quantities of pesticide used are orders of magnitude larger. However, if he is careless, or if the wind direction is variable and at times he drives the truck with the wind blowing against the sprayer, he may get a very large exposure. When identifying hazards, the same exposure situation analysis can be applied to indoor, outdoor, and occupational exposures to begin identifying sources, contaminants that are present, and factors that are likely to affect the intensity.

Another example is the asbestos exposure of people living in a town near an asbestos mine in Black Lake, Quebec. Airborne asbestos fibers can deposit deep in the lungs and cause cancer. An examination of the residential area and the area around it showed that there was a large tailings pile (waste rock) approximately 100 meters from the houses. The tailings are a waste rock produced after the asbestos has been extracted from the crushed ore. Asbestos fibers are formed by the crystallization of molten mineral forced through cracks in granite by a geological process. The asbestos extraction process is incomplete, and small amounts of asbestos remain with the waste rock. The waste rock leaves the crusher building in a steady stream that drops off a conveyor belt onto the tailings pile. Wind blowing through the airborne stream and over the tailings captures any loose asbestos and disperses it over the surrounding area. As a result, there is asbestos in the surface soil, on the sidewalks and streets in the residential area, and on surfaces inside homes. Surface deposits are made airborne by traffic and routine activities of the residents. "Dust bunnies," fist-size balls of loose chrysotile asbestos fiber, are sometimes seen rolling down the town streets near the tailings piles when the weather has been dry.

The question is: How much becomes airborne and transported into the breathing zones of the residents, where it can be inhaled? Only fibers with lengths greater than 5 μm are thought to cause carcinogenic effects, and particles must have a length-to-diameter ratio of at least 3:1 to be defined as a fiber. Most asbestos fibers are longer than 20 μm and have diameters less than 0.1 μm. The hazard must be measured by air sampling in the breathing zone and measurements made by a technique that can determine the number of airborne fibers meeting the size definition.

Multiple Emission Sources

When there are many sources distributed over an area, such as a city, then the emissions mix with each other, and wind-based dilution/dissipation processes may be less effective. Large-scale weather conditions can also be important, such as a temperature inversion (a layer of warm air above cold air near the ground) that limits the vertical mixing and dilution of contaminants. When elevated concentrations of emissions remain in the atmosphere for hours or more, photochemical smog may form, especially in warm southern climates. The accumulated emissions are irradiated with ultraviolet radiation from the sun, which breaks down some compounds and leads to reactions that form new ones, such as ozone and other highly reactive oxidants (Lippmann, Cohen et al. 2003; Manahan 2004). These reactive substances will interact with any suitable material they contact, including atmospheric particles and the cells lining the human respiratory tract.

Multiple source exposures are also common in occupational settings. The exposure of a worker to airborne emissions is a function of distance from each source, its output strength, room size, and air flow patterns within the workroom. Low-strength sources, such as soldering or hand wiping with solvents, will affect only small areas. High-strength sources, such as spray painting, can affect everyone in a large-volume room, whether they are working with the source or not.

Many people have difficulty understanding why their personal vehicles and other devices need to be controlled, because they are only small sources of emissions. California has generally led the United States in imposing air pollution controls on automobiles and industrial sources. An outcry resulted over plans for regulations to control emissions from charcoal grills and two-cycle engines on lawn mowers. Because these are small and used infrequently, it seemed inconceivable that they could contribute significantly to air pollution. Some are admittedly smoky and may affect the operator, but one lawn mower by itself is not a problem for general air pollution. However, when 10,000 lawn mowers are running at the same time on a Saturday morning, they can be a major contribution. Thus small sources cannot be ignored, especially when there are large numbers of them.

Exposure Controls

Several of the textbooks listed at the end of the chapter discuss methods for reducing hazardous exposures (Boleij, Buringh et al. 1995; Lippmann, Cohen et al. 2003; Manahan 2004; Gardiner and Harrington 2005), and the source-receptor model is also useful for identifying potential points for intervention (Figure 2.1). There are two primary points of intervention: at the source and at the person. The most effective and efficient intervention is to remove the source. Failing that, we can intervene at the source to capture emissions and prevent or reduce their release. Intervention to prevent or limit personal contact is much more difficult to do effectively after the chemical has been released and dispersed in the environment.

The most effective control is achieved by eliminating the emission source, such as changing from leaded to unleaded gasoline to eliminate community lead exposure from street traffic (Quinn, Kriebel et al. 1998). All other types of controls can achieve only a percentage reduction in emissions.

There are many types of control devices that can be applied to emission sources—for example, smokestack devices to capture selected gases and particles, or workplace controls such as local exhaust ventilation (LEV) that can modify personal exposures of workers, or even small personal air purifiers worn by an individual on an airplane. It is common in occupational settings to place partial enclosures ("hoods") around a source of hazardous materials and exhaust the contaminated air collected in them. Another strategy is to use suction from a flexible tube to directly capture emissions from a small point source, such as a grinder.

In occupational and some community settings it is sometimes necessary to control exposure at the person. Personal protective equipment is widely used in industrial workplaces to prevent inhalation or contact with toxic materials. Respirators, face masks with air-purifying cartridges to remove contaminants from inhaled air, are used where it is not feasible to reduce exposures to acceptable levels by intervening at the source. The cartridges come in a variety of types for capturing a broad range of pollutants, such as solvent vapors and toxic dusts. They are usually highly effective in purifying air inhaled through them, but significant reductions in air concentrations can be achieved only if there is a tight seal of the respirator to the wearer's face. Unfortunately, the quality of the seal is inversely proportional to the comfort of wearing the device. The full face mask units with their own air supply ("self-contained breathing apparatus"; SCBA), such as those worn by most firefighters, are highly effective, but they are most suited for emergency situations and cannot be used for regular or long duration exposures.

Homeowners may also use small air purifiers to reduce their exposure to outdoor air pollution and indoor contaminants, such as mold spores. Electrostatic and filter air cleaners can be purchased for this purpose, and they may reduce some types of airborne particles and gases, but they require regular maintenance. Poorly designed electrostatic units are a problem because they are inefficient particle collectors and can be significant sources of highly toxic ozone.

Water Contaminants

Contaminant Sources

Historically, streams and rivers have been used to carry away wastes—human, animal, and industrial—and treated and raw sewage are still major sources of water contaminants in rivers and the ocean. Industrial wastes entering municipal sewers are not generally removed by standard sewage treatment processes. Large volumes of animal wastes from concentrated feeding areas can run into nearby streams and rivers. Materials deposited on top of or within the soil, such as fertilizers and pesticides, can be carried into bodies of water by runoff of rain water. Air pollutants may settle out into waters or be scrubbed out of the air by precipitation processes (the source of acid rain), or pollutants deposited on the soil may be carried into waters by runoff.

Environmental Transport and Other Processes

Water behaves like air in many ways. Both are fluids and subject to the same fluid dynamics, but water is much more dense and viscous. Once added to flowing water, contaminants will be mixed into the bulk by turbulent mixing and diffusion, just as they are in air, but the processes are much slower. As a result, if the added material has some color, we may easily see it even a kilometer downstream when the mixing is incomplete. Stratification of water also inhibits mixing. For example, freshwater in rivers entering the ocean will disperse across the surface, because freshwater is less dense than seawater. Time and wave action are needed for saltwater to diffuse into the freshwater and to mix the surface layer into the seawater. Thermal stratification of standing water, in which temperature declines with depth, also inhibits mixing in lakes and seawater. Thus a stream of waste going into natural waters may not receive the dilution that might be expected.

Biological agents and dissolved toxic materials are probably the most common hazards in water. Standard water treatment processes are designed to remove suspended solids, especially bacteria, and to adjust pH. Drinking water is also chlorinated or treated in other ways to kill bacteria and viruses. Chlorination can also form small amounts of organic halogenated compounds, which are suspected human carcinogens (Manahan 2004). Water treatment is generally ineffective at removing dissolved substances such as the heavy metals lead and arsenic. Some organic materials with low water solubility can still dissolve at low concentrations and be a toxic risk, such as benzene and some solvents.

Water is present in nearly all soils, and wells are drilled to tap accumulated underground water. This groundwater has similar properties to water on the surface: it will flow from higher elevations to lower, and less dense liquid materials will float on groundwater. For example, gasoline leaking from old underground storage tanks is a serious pollution problem. Small amounts of some constituents of gasoline, such as benzene, will dissolve in the groundwater, but the bulk sits as a layer on top of the groundwater. In a few locations sufficient gasoline has been present in soil that it constitutes a fire hazard. This can happen when someone nearby is pumping out water, "drawing down" the groundwater table and making a depression in the water level. Gasoline will slowly flow into this depression, just as it would if it were spilled on the open ground. At some point, the water pump starts pulling out a mixture of gasoline and water. In the Cape Cod area of Massachusetts, long-term military operations at Otis Air Force Base spilled large quantities of jet fuel and solvents on the ground, which was not viewed as a problem at the time. This mass of solvents has formed several large underground plumes that are flowing toward the ocean along with the natural flow of rainwater deposited on the Cape. These plumes have contaminated household well water in a wide area around the base.

Sadly, it is very difficult to remediate these problems, and groundwater, once contaminated, is nearly impossible to thoroughly decontaminate. In these instances, the only option may be the less desirable strategy of intervening at the receptor rather than the source—removing water contamination at the faucet or supplying bottled drinking water from a remote source.

Bioavailability and Bioaccumulation

Bioavailability defines the quantity of a substance that is available in the body so that it may be absorbed. Lipid-soluble materials are those that are very soluble in fats, such as PCBs. On the other extreme, arsenic sulfide is extremely insoluble and, if ingested, will not dissolve under physiological conditions in the GI tract. Similarly, some metals and organic substances are tightly bound to soil constituents, so they cannot be absorbed by plants in contact with them or by animals that ingest them. Tightly bound materials do not present a hazard.

Bioaccumulation is the process whereby chemicals that are lipid soluble and minimally metabolized become concentrated as they move up the food chain. Initially they are taken up in plant lipids and then concentrated in herbivore lipids, because the herbivores eat relatively large amounts of plants and the stable chemicals are not excreted. Larger predator organisms consume relatively large amounts of smaller prey and further concentrate the chemicals from body fat of prey organisms. This concentration process can be repeated many times as the chemicals move up the food chain and can produce toxic levels in the top predators, including humans. Trace concentrations in the environment can be amplified by 100 to 10,000 compared with the initial water concentration.

The classic case of bioaccumulation is the insecticide dichloro-diphenyl-trichloroethane (DDT), which Rachel Carson highlighted in her seminal book, *Silent Spring* (Carson 1962). DDT has very low water solubility but high fat solubility, so that in water it diffuses into the cells of plants, where it is concentrated in plant lipids. Small animals eat the plants and are in turn consumed by larger animals, and so on up the food chain. At the top of the food chain, hawks eat fish contaminated with DDT concentrated several thousandfold. Similar problems have been seen for other stable lipid-soluble contaminants, such as methyl mercury in humans.

Food Contaminants

Sources of Contamination

Soil or groundwater contaminants can be taken up by plants, such as cadmium by corn. Contaminated plants can be eaten by animals, which in turn can be eaten by humans, or humans may eat the contaminated plants directly. If pesticide residues on food crops do not decompose before the food is prepared, and if the preparation processes do not remove the contamination, then the food can be a hazard. For example, some pesticides on the exterior of fruits and vegetables will decompose with exposure to sunlight and/or moisture, or they can be removed by washing. Foods may also be contaminated by the containers in which they are stored. Some pottery glazes contain lead, and when acidic foods, such as orange juice or red wine, are stored in these vessels, they will leach sufficient lead from the glaze to be a hazard.

Outbreaks of foodborne illnesses from contamination with pathogens such as salmonella or with toxic chemicals occur through failure to follow safety procedures or failure of government oversight. Unfortunately, government monitoring programs that feature limited and infrequent sampling of food can easily miss contaminated foods.

Personal Contact with Contaminated Food

Concerns about human toxic hazards from food have been spurred by instances of severe illness caused by eating contaminated foods. One of the most terrible was the poisoning of hundreds in Japan through eating fish contaminated with methyl mercury (discussed in more detail in Chapter 12). A more recent example is the apparently intentional addition of an inedible protein, melamine, to milk in China (Chan, Griffiths et al. 2008).

The methyl mercury case began when mining wastes containing mercury were dumped into a river which entered Minimata Bay in Japan. Once in the bay, mercury in the mine waste was methylated by microorganisms, rendering the mercury much more toxic. The methyl mercury bioaccumulated in fish, which were eaten by local fishermen and their families. A wide range of health problems occurred, including severe birth defects. Sadly, one of the lessons from this tragedy is that health hazards are often detected only through epidemiologic study, which means that substantial human suffering has already occurred before the problem is identified and action is taken to prevent it.

Skin Contamination

Sources of Contamination

Skin contamination results from several processes: (1) direct handling or use of chemicals in a way that leads to skin contact, usually with the hands and forearms; (2) contamination of clothing with liquids and fine dusts; (3) incidental contact with contaminated surfaces; and (4) deposition of liquids and dusts on the skin. Spraying or splashing liquids can deposit droplets on exposed skin and clothing. With liquids, clothing—even protective clothing—can play an important role in holding liquids in contact with the skin for longer than they would otherwise remain. A good example of this problem is seen when cytologists prepare microscope slides. They will wear gloves to protect their skin from microbiological contamination, as well as from the dyes applied to the slides. Xylene is used in the preparation of the slides, and it is highly soluble in latex. If latex gloves are worn, the xylene may penetrate and dissolve the glove material. Once the protective barrier of the gloves has been breached, the xylene dissolving in the latex is held next to the skin, increasing this route of exposure.

Personal Contact with Contaminants

Lipid-soluble materials are most likely to be absorbed through the skin because the skin is lipid in structure and generally not permeable to water-soluble (polar) materials. The most common problem materials are organic solvents and pesticides. For example, because xylene is readily absorbed through the skin (it is lipid soluble and has a low molecular weight), a day of spray painting by shipyard painters can lead to a high body burden despite low inhalation exposure due to use of supplied air masks, as shown by high urinary metabolite levels (Chang, Chen et al. 2007).

If respiratory intake is not protected, then a volatile water contaminant may be vaporized and inhaled. For example, showering with chloroform-contaminated water can lead to a significant inhalation exposure. Laboratory exposures in simulated baths containing < 100 ppb of chloroform studied by Corley and coworkers showed that skin absorption from the water for 30 minutes produced a low level of absorption that ranged from 1 to 28% of the amount that would have been absorbed in the GI tract from drinking water with the same chloroform content (Corley, Gordon et al. 2000). Thus, depending on the situation, measuring only one route of exposure may result in considerable misclassification of the exposure, as exposure through one route may be poorly correlated with exposure from another.

2.5. POTENTIAL EXPOSURE

In many epidemiological studies, subjects are distinguished by their *potential for exposure*, that is, by their probability of being exposed to an agent of interest. Although potential exposure may be a useful approach to exposure classification, it suffers from the lack of a clear definition. There are two types of questions to be answered. First, is the agent present in the environment, and can it come into contact with the subject? Second, given that the agent is present, are the concentration and duration of contact sufficient to be a hazard? We can usually answer the first question by an analysis of the sources and conditions present in the setting. However, some settings can be deceptive. In a petrochemical plant, hundreds of thousands of gallons of toxic materials may be processed daily. However, unless they escape from their containers—the pipes, reactors, and storage vessels—there is no exposure. Thus the presence of an agent is a necessary, but not sufficient, indicator of exposure. Potential for exposure depends on the likelihood of a release in the proximity of the subject. In a well-run, modern petrochemical refinery, workers may have no more exposure to the millions of gallons of gasoline they produce than you have when you fill your car at a self-service gasoline station. Because of the very wide range of potential for exposures, identifying the fact that a subject worked in a petrochemical refinery is insufficient information by itself to define high-exposure potential for gasoline. However, if the exposures of a small group are very intense and long-lasting when they occur, the increased risk of a rare disease may be evident even with considerable misclassification of exposure. Consider the campfire example again. A person sitting next to a campfire is potentially exposed to wood smoke from the fire. His or her intensity of exposure will depend on the size and smokiness of the fire and the wind direction. A small, clean-burning fire may produce minimal exposure intensity, even for people directly downwind from the fire. Fires of moderate size burning green wood will produce large amounts of highly irritating smoke. Around such a fire, few locations exist at which someone can sit relatively close to the fire and not be exposed to the smoke. Additionally, duration must also be considered. A person who sits all evening by a fire has a greater risk of smoke inhalation than someone who visits briefly. Given all of these factors, when we use a questionnaire to learn whether someone has "ever sat next to a campfire," it is difficult to assess the quality of a classification of subjects based on their answers. Some certainly have had substantial inhalation exposure to fire emissions. Many probably have had trivial exposures because they were sensitive to the smoke and they sat only upwind, or because the fire was not smoky. Consequently, "ever sat next to a campfire" is likely to be a poor classification variable to identify subjects with campfire smoke exposure. In fact, this question is based on the assumption (hypothesis) that people sitting next to a campfire *do* have exposure. As with the gasoline example, more information is needed to assess the potential for exposure. The implicit assumption underlying the "ever sat next to a campfire" classification is that at least some of the exposures were high enough to cause the effect of interest. Designing useful exposure questionnaires must be done with a clear understanding of the nature of the situation being probed (Armstrong, White et al. 1992; Nieuwenhuijsen 2003).

It has been common for epidemiologists to attempt to be "conservative" and count subjects with any conceivable nonzero probability of exposure as "exposed." As a result, anyone who has the slightest contact with the agent is sometimes counted as exposed. A person who occasionally walks through a busy street intersection with bus and truck traffic and the policeman who works for hours controlling the traffic at the intersection are both potentially exposed to diesel exhaust. However, there is probably more than an order of magnitude difference between the inhaled doses of those two individuals and probably a similar difference in their risk of being affected by the exposure. We see in the next chapter that exposure intensities and durations often have a highly skewed distribution. If very high exposures are rare, then many of the exposures will be trivial; trivial exposures often do not carry detectable risk. Thus the "conservative" approach can produce considerable misclassification and overestimation of the number of subjects exposed, which reduces the likelihood of detecting a causal relationship. We would like "exposed" to mean that there is *a reasonable probability of an exposure that may be sufficient to cause the effect* of interest. Under that definition, slight and very brief exposures would not count as "exposed."

When considering the use of a potential exposure classification strategy, it is important to consider the absolute magnitude of the exposures: a population with a large exposure gradient is needed to overcome misclassification. It is critical that a significant fraction of the subjects have identifiably "high" intensity exposures (usually through close proximity to high output sources). There are many historic examples in which the exposures met this requirement. For example, coke oven workers are a subgroup of steelworkers who have jobs in close proximity to large ovens in which carcinogenic volatile organic compounds are cooked out of coal. The ovens are difficult to seal, and large amounts of chemicals can be released into the environment. Lloyd (1971) did a retrospective mortality study of steelworkers and found that steelworkers as a class did not have an increased risk of lung cancer. However, those steelworkers who worked around the coke ovens had a substantial increase in risk. Jobs that were closer to oven emissions also carried a higher risk of lung cancer because the gradient in exposure was substantial. Thus the broad "potentially exposed" classification "ever a steelworker" was insufficiently precise to detect the effect of working near a coke oven, which was "hidden" inside the overly broad classification of steelworker.

In the next sections we review the basic methods and strategies for measuring exposures for an epidemiologic study.

2.6. MEASURING EXPOSURE

The first part of this chapter provided a framework for understanding the nature of exposure to environmental toxins. With this framework in place, we can now describe the methods for measuring exposures in a biologically relevant way, setting the stage for the application of these methods in epidemiologic studies. We will see that *how one measures* depends on *why one is measuring*—biologically based exposure assessment for epidemiology differs in important ways from exposure assessment approaches used to determine compliance with an environmental control regulation or to describe the movement of a contaminant through an ecosystem or a study of atmospheric photochemistry in smog formation.

This presentation of exposure measurement methods is an overview. It cannot provide detailed analytic methods; these will be found in references cited in the text. Our goal is to describe typical applications, common pitfalls, and key assumptions of exposure measurement.

Choosing What to Measure

Measuring every agent present in an exposure situation is rarely feasible. However, often we can use the characteristics of the observed health effects, their timing and target organs, to narrow the scope of the investigation to agents that could plausibly cause these effects—this is a key aspect of a biologically based exposure assessment. Consider the following example: We are asked to investigate complaints by family members doing spring window cleaning. They report acute irritation of the eyes, nose, and throat, followed within an hour by a severe dry cough. They reported using a newly mixed solution of ammonia and an aerosol spray glass cleaner with a strong odor. They also reported that the symptoms rapidly diminished when they finished cleaning.

Given the details of this case, how can we narrow the scope of possible agents and decide what to measure? The symptoms suggest an airborne agent that is deposited on moist membrane surfaces and upper airways. Our personal experience with ammonia and a review of toxicologic information on ammonia strongly supports the possibility that the agent is ammonia. Ammonia is gaseous and highly water soluble and is rapidly absorbed by the moist surfaces of the eyes and mucous membranes of the head and upper airways. Large-diameter (> 10 μm) particles that might be generated by the spray cleaner can also be deposited in the eyes, nose, and upper airways. As there were no symptoms of deep lung irritation, such as shortness of breath, the initial impression is that the irritant agent is not penetrating deep into the lungs. If the exposure were to a gas with low water solubility, such as nitrogen dioxide, it would be only weakly absorbed in the upper airway but highly absorbed in the lower airways and alveolar area of the deep lung. This would be expected to produce a combination of symptoms: limited upper airway irritation and slowly developing effects of irritation in the lower airways and alveolar area, such as shortness of breath from pulmonary edema.

The family's reports of rapid onset of eye and nose effects implies a direct route of entry and a rapid process for the development of the effects, such as direct contact with the target membranes. The rapid fading of irritant symptoms after exposure stopped clearly shows that the effects were reversible. Exposure to ammonia in the new cleaner is consistent with all of the reported symptoms. What about the spray cleaner? A review of the label showed that it contained "a surfactant, cleaning agents, fragrance, and water," which was not very helpful for this case, but there were no warnings about eye irritation or other irritant risks. Based on all of the data so far, our primary hypothesis about the agent is ammonia. If we were to conduct a formal epidemiologic investigation of all families in a neighborhood doing spring cleaning, we would want a method suitable for measuring acute ammonia exposure. A secondary hypothesis might be that the active agent is a component of the spray cleaner, but the product information has provided no chemical data with which to identify what to measure (this is a common problem in case investigations). Sometimes a call to the local poison control center can gain more detailed information on the product.

Although this simple example may be useful to illustrate the steps in identifying potential exposures for epidemiologic study, we want to note in passing that it would probably not be necessary to conduct a formal epidemiologic study in order to bring some relief to the window cleaners. The description of symptoms, their timing, and the fact that a new cleaner was just introduced provide sufficient evidence to justify some immediate action—probably changing the cleaning agent—to reduce the hazard.

This example shows how the nature of the health effects can be combined with preliminary data on the nature of the exposures and a list of materials known to be present in order to make a tentative identification of exposure components and their likely routes of entry. These possible agents lead to testable hypotheses—that each one is a contributor to the observed health effects. These hypotheses then guide the choice of exposure assessment methods, and possible groupings of subjects based on their potential contact with suspected agents.

Use of Surrogates and Markers of Exposure

Because we do not have a general measurement technique that will measure every chemical present, we must choose what we will measure and how to do it. It may not always be feasible to directly measure the hypothesized agent(s). Practical limitations in analytic methods, a lack of time or money, or insufficient knowledge with which to identify a specific agent often lead to the measurement of exposure surrogates or *markers*. A marker is a substance that can be more easily measured and that is known or hypothesized to be correlated with the true causal agent(s). But using an exposure marker can create as many problems as it solves. If the true agent is unknown, then how can we be confident that the marker is correlated with it? If a marker is used in an epidemiologic study and no association of risk is observed with the marker, it is tempting to conclude that the (real, unmeasured) exposure is not harmful. The reality may be that the marker is poorly correlated with a truly hazardous exposure. Finally, a positive finding of an association between the marker and disease risk may lead to the incorrect conclusion that the marker is the causal agent, "forgetting" that the marker is a surrogate. So markers should be used with care and with full disclosure of implied or explicit assumptions behind their use.

One common approach to choosing an exposure marker is to measure a broad, nonspecific category of contaminants in the hope that this summary quantity will be correlated with all of its component parts. For example, air pollution studies often measure the mass of all particles with diameters less than 2.5 μm ($PM_{2.5}$) under the assumption that this quantity is likely to correlate with the quantities of its various chemical components, one or more of which may be toxic. Concentrations of components that are a small percentage of the total may not be correlated with variations in the total if they come from different sources. However, this is not to say that using such a broad marker cannot be useful. An important argument for using crude, or overly broad, markers of exposure is that sometimes more detailed exposure-response information is not needed. If one can substantially limit exposure to a broad class of agents—$PM_{2.5}$, for example—then one has, by definition, limited exposure to all the components. Sometimes the most effective prevention strategies do not require the most detailed exposure-response information. Broad markers are most useful if they have a relatively stable composition over time and location, such as $PM_{2.5}$ as a marker of urban air pollution. Variations in disease risk with $PM_{2.5}$ exposure in different cities may be the result of different mixes of sources, which change the composition, making it more or less toxic.

Methods for Measuring Exposure

The complexity of environmental contaminant composition and physical forms and the nearly infinite variety of different media in which contaminants may be found lead to a wide array of measurement techniques (shown in Table 2.1). It is beyond the scope of this book to discuss these in detail; the interested reader is referred to one of the books listed at the end of the book. Also, chemical research and new sensor technologies continue to produce newer, faster, cheaper methods for measuring personal exposures, so that the latest exposure assessment literature on the agent(s) should always be reviewed to identify both the standard measurement methods and emerging new approaches. It is essential to understand the limitations of any method selected—such issues as the precision and accuracy of results and any known interferences (positive and negative). Epidemiologists and toxicologists working with exposure assessors need to have some familiarity with these ideas to use data produced by them. Here we briefly summarize basic principles that apply broadly and that are not always clearly explained in the reference books.

2.7. CHARACTERISTICS OF EXPOSURE METHODS

Measurement methods have several important basic characteristics that apply to all types of media being sampled or measured: a limit of detection, sensitivity, selectivity, accuracy, precision, averaging interval, and, for direct reading instruments, a response time. Unfortunately, some of these terms are also used in epidemiology, but with different meanings.

Limit of detection (LOD) is the concentration or quantity just large enough to be distinguishable from noise or random variability in the method. It is usually

Table 2.1 Common examples of measurement methods for environmental contaminants based on substance characteristics affecting entry, deposition, and uptake during exposure

Route of Entry	External Exposure	Sample Collection	Measurement Techniques	Direct Reading Instruments
1. *Inhalation*	*Airborne Gases*			
	water soluble	bubbler with binding agent	colorimetric analysis	specific for gas: SO_2, formaldehyde
	limited water solubility	badge with binding agent	colorimetric analysis	chlorine, ozone
	Airborne Vapors			
	organic materials	charcoal in glass tube	extract and GC	Nonspecific detector, FID; PID
	Airborne Particles			
	"inhalable" < 10 μm (EPA defines PM_{10})	collect > 10 μm by cyclone or impactor + filter	weigh filter; extract andAA for metals;GC, GC/MS, or HPLC for organics; IC for inorganic ions	light scattering photometer;deposit on quartz crystal; electron mobility analyzer
	"respirable" < 4 μm (defined by NIOSH)	collect > 4 μm by cyclone or impactor + filter		
	"$PM_{2.5}$" < 2.5 μm(EPA defines as fine particles)	collect > 2.5 μm by cyclone or impactor + filter		
	particle size distribution	10 stage impactor	weigh and extract same analyses as above	electron mobility analyzer
2. *Ingestion*	*Water & Food*			
	metals: lead, arsenic, cadmium	collect volume in container	digest and AA	metal and inorganic ions by specific electrodes
	organic compounds: pesticides, solvents, dioxin	collect in container	homogenize, extract and GC, GC/MS	none

Table 2.1 (continued) Common examples of measurement methods for environmental contaminants based on substance characteristics affecting entry, deposition, and uptake during exposure

Route of Entry	External Exposure	Sample Collection	Measurement Techniques	Direct Reading Instruments
3. *Skin Absorption*	*Liquid materials –*			
	lipid soluble, low molecular wt. (skin is barrier to polar compounds)	skin patches and wipes with solvent; fluorescent tracer added to liquid	extract; analyze by HPLC or GC; photograph skin under UV light for fluorescence	UV light inspection of skin if substance is fluorescent
	Dusty materials			
	lipid soluble, low molecular wt.	skin patches and wipes with solvent	extract; analyze by HPLC or GC	none
	Airborne to skin			
	v large >100 μmlipid soluble, low molecular wt.	Same	Same	none

Abbreviations: AA = atomic absorption spectroscopy; GC = gas chromatography; IC = ion chromatography; MS = mass spectrometry; HPLC = high-pressure liquid chromatography; FID = flame ionization detector; PID = photoionization detector; wt. = weight; $PM_{2.5}$ = particulate matter < 2.5 μm diameter.

defined as three times the baseline measurement variability or noise (Lippmann, Cohen et al. 2003). A measurement that is near the LOD indicates that the compound may be present, but the amount of material is highly uncertain. Such a measurement should be interpreted as having an error of at least ±100% of the measured value. However, even when there are a number of measurements below the LOD, the mean value of *all* the measurement values should estimate the true mean as long as the method is unbiased.

When designing a study we try to choose a method that has a low LOD relative to the minimum exposure we expect to be biologically relevant. If the minimum concentration needed for a biological response is not known, then we want the LOD to be generally smaller than the expected exposure levels. However, values at the LOD or below are common and do contain some information. Several data analysis strategies have been used to avoid losing this information. The value of the LOD divided by 2, the square root of 2, or a maximum likelihood estimate have all been used. Each approach has advantages and disadvantages (Rappaport and Selvin 1987; Helsel 2005; Newton and Rudel 2007).

Sensitivity is the response per unit of exposure of the measurement method, that is, the slope of the response-exposure calibration curve. A low sensitivity means that important changes in exposure may not be detected. This should not be confused with the sensitivity of a medical test, which is defined differently (Rothman and Greenland 1998).

Selectivity (or *specificity*) of a method defines how well the method can distinguish a specific material, such as benzene, from other possibly similar materials, such as toluene. Analytical chemists are fond of combining a highly specific separation method with a broadly nonspecific measuring device, the "detector." An example is the use of gas and liquid chromatographic (GC and LC) methods to separate a complex mixture into its components, which are then quantified separately by a nonspecific detector. For example, a GC column can separate a wide range of hydrocarbons, which can each be measured by electron emissions when they pass through a hydrogen flame. For example, nontoxic pentane can be easily separated from neurotoxic hexane by this approach. Carcinogenic polycyclic aromatic hydrocarbons (PAHs) can be separated by LC, and the PAHs measured by fluorescence under ultraviolet (UV) radiation. A more powerful combination is GC or LC together with mass spectrometry (MS), which yields a method that is both very sensitive and specific. Nonspecific measurement methods, such as photoionization detectors that respond to most gases and vapors in an air sample, can be very useful in situations in which a known mixture or only one contaminant is present.

Many analytic methods suffer from *interferences* by components of mixtures. Positive interferences cause an overestimate of the substance of interest, whereas negative interferences cause underestimates. Interferences can be especially problematic when using nonspecific measurements because many things can cause a response, but we cannot determine whether the signal represents the hazard. For example, measurements of $PM_{2.5}$ in a city located next to the ocean may contain varying amounts of nontoxic salt (a positive interference), which can make $PM_{2.5}$ in this case a poor indicator of air pollution. Often supplementary analyses are needed to ensure that interferences are not affecting nonspecific measurements. Alternatively, preseparation methods can be used to remove interferences before measurement.

Accuracy is the degree to which the method gives the true value for the exposure. Deviations from the true value are measured by the *bias*, which is the difference between the measured mean and the true mean value. Interferences lead to biases and problems with accuracy. Systematic problems with instruments and lab and field techniques can cause loss of accuracy. Note that methods with poor precision but no bias can be accurate on average.

Precision is the repeatability of a measurement and is described by the standard deviation (SD) of replicated measurements of a known or standard concentration. Precision of environmental methods is often indicated by the coefficient of variation (SD divided by the mean, expressed as a percentage: CV%). An epidemiologic study using highly precise measurements of a substance that is not the causal agent can provide seemingly strong evidence for the absence of an association, whereas imprecise measurements of the true causal agent can lead to lack of evidence for an association when one actually exists. When making measurements we often need to balance the requirement for precise measurements against the normal high variability of environmental conditions; the mean of a few precise measurements (high cost) may be less useful than the mean of many less precise measurements (moderate cost) because the latter gives us a more precise *mean exposure* at less total cost.

Averaging time interval: When integrated samples are collected, for example by pooling successive samples from a water supply or gathering aerosols on a filter, the sampling strategy must specify a sampling time interval over which the exposure concentration measurement will be averaged. We can use the air filtration example to illustrate. During the sampling interval, air passes through the collector at a fixed flow rate so that each minute contributes an amount to the collected material that is proportional to the air concentration. The total amount collected is the sum of all of the individual minutes' contributions during the sampling interval. When we divide the total collected by the total volume of air through the collector, we obtain the average concentration during the time interval sampled. As a result, when a fixed flow rate is used, the sample is called a *time-weighted average* (TWA) sample, and the duration of the sample is called the *averaging interval*. Typical sampling intervals for air pollution studies are 15 minutes and 1, 8, and 24 hours, but both longer and shorter intervals have been used. Similar intervals are used for water and other media sampling.

Temporal variability of concentrations observed in air, water, and soil during short averaging intervals tends to be higher than for longer sampling times because short-term variation is averaged out. It is also common to observe a correlation between adjacent temporal measurements (*autocorrelation*) because the physical processes generating and diluting or reducing the concentrations by turbulent mixing cannot change the concentration instantly, and so a high concentration at one moment is more likely to be followed by another high value than by a low one. The longer the time between observations, the less autocorrelation we will observe, unless there is some systematic process producing cyclic variations in emissions, such as diurnal weather variations or factory production cycles. There is also a similar spatial autocorrelation in which exposures at adjacent locations are correlated, and this correlation decreases as the distance between them increases. With flowing water in rivers and streams, temporal and spatial variations occur together.

Response time is a characteristic of direct reading instruments. It is defined as the time needed to go from baseline to 90% of the true concentration (other percentages may also be used). Sensor responses can range from fractions of seconds to minutes or longer. Even instruments using detection methods that are virtually instantaneous, such as absorption of UV light, have a response time because the external air or water must travel into the detection cell (causing a *lag* in response) and the volume of the cell must be flushed (causing a wash-in and wash-out process that broadens the time for a response). For epidemiologic studies, it is important that the response time be considerably shorter than the time scale on which the physiologic response occurs, so that all relevant variations in exposure can be measured. This is discussed further in Chapter 5.

2.8. OVERVIEW OF MEASUREMENT APPROACHES

There are two basic approaches for most measurements of exposure: "collect and analyze," in which we collect an integrated sample of environmental contaminants in the field, and "direct reading" or "real-time" measurements, in which we use a sensor-based instrument to make the measurements onsite (Table 2.1). In some settings, both collect-and-analyze and direct reading measurements may be used, because they provide somewhat different information on the nature of the exposure.

With the collect-and-analyze approach, you can bring to bear the full analytical power of the laboratory, including mixture separation and highly sensitive measurement methods, but you must collect sufficient material for the analyses. This approach necessarily integrates across time and perhaps locations, limiting your ability to resolve short-term differences in exposure that may be important. Resources are important determinants of what can be measured. Chemical composition of complex mixtures is difficult and expensive to assess. Highly reactive substances can be very challenging to measure in the lab because they are unstable, so some method must be found to stabilize or fix them at the time of sampling. Many established collect-and-analyze methods for known hazards have been developed by the environmental regulatory agencies. In the United States, the most important of these are the Environmental Protection Agency (EPA) (http://www.epa.gov/nerlcwww/methmans.html), the Food and Drug Administration (FDA) (http://www.fda.gov/ora/science_ref/lm/default.htm), the National Institute for Occupational Safety and Health (NIOSH) (http://www.cdc.gov/niosh/nmam/), and the Occupational Safety and Health Administration (OSHA) (http://www.osha.gov/dts/osta/otm/otm_toc.html). Similar institutions for the European Union, United Kingdom, and other countries have also developed methods or adopted them from elsewhere (http://www.eea.europa.eu/; http://www.environment-agency.gov.uk/). The standard analytic methods of these agencies should be examined before new methods are developed. The alternative approach uses direct reading instruments which can be taken into the field to measure exposure onsite. A number of instruments have been developed for this purpose, especially for gases and vapors in air, which include infrared and ultraviolet spectrometers, flame- and photo-ionization detectors, and even small mass spectrometers. The list of air contaminants that can be measured with these instruments is limited, and some methods are nonspecific, making them less useful for mixtures. Portable gas chromatographs with mass spectrometer detectors have been developed for field use. However, mixtures of contaminants are common in the environment and can be too challenging to separate and analyze in the field. The direct reading instruments available for water contaminants are much more limited. A combined strategy using both direct reading instruments and collector sampling with lab analysis can be an effective way to deal with the temporal variability and compositional complexity.

In addition to this distinction between collect-and-analyze and direct reading instruments, it is useful also to distinguish two general *locations* for environmental measurements: personal exposures at the portal of entry or ambient exposures at a fixed location selected to represent the exposure of nearby individuals. Both of these have advantages and disadvantages that must be weighed in the context of the aims and practical limitations of the study, which we discuss in the next chapter.

Airborne Exposures

Personal Sampling

Personal contaminant collectors have been most extensively developed for inhalation exposures. These sample collection devices are worn by individuals and carried around as they go about their lives to obtain a *personal sample*. There are two types: active samplers that have sample air pumps and passive samplers, also called "badge samplers," that use diffusion to move the contaminant into the collector. The passive samplers are the least intrusive for the individual wearing them. Both types are designed to have a steady air flow to obtain a time-weighted average exposure concentration for all of the activities and locations visited by the person wearing the sampler. When air sampling devices are attached to the individual's collar or other location close to the nose and mouth, they provide a *breathing zone* sample. To solve the problem that only a few activities or locations may be contributing the majority of the individual's exposures, some of the direct reading devices have been miniaturized and can also be worn to measure some air contaminants. This can be important if we suspect that a particular emission source is producing the overexposure and health problems. However, there are few of these devices, and they are expensive, so it is often not possible to use these for a comprehensive assessment of a large population.

Area or Fixed Location Sampling

Materials in air have long been sampled from fixed-location samplers. They might be placed in a workplace near a work station or set next to a home to collect periodic samples over an 8-hour, 24-hour, or longer period. High-volume samplers can collect large amounts of airborne material, sized particles, vapors, or gases suitable for extensive chemical analysis when we are concerned that composition may change with time or location. For example, the particle size distribution of ultrafine emissions from Los Angeles freeways changes with time and distance after emission, and this is suspected to alter their toxicity (Zhu, Hinds et al. 2002; Zhang, Wexler et al. 2004). Direct reading instruments are used to track these temporal changes. If the interest is in the exposures to vehicle occupants, samplers can be mounted inside a vehicle to estimate the conditions in the microenvironment of the passenger cabin (Fruin, Westerdahl et al. 2008). This is not strictly a personal sample, but it may be a good representation of personal exposures.

Water Contamination

A water sample is usually a spot or grab sample at a point in time—perhaps a glass bottle is filled with water from the faucet in the kitchen—but for an epidemiologic study, we are concerned with the exposure during the etiologic time interval when the exposure is relevant to some effect. For example, when studying risk of skin cancer from arsenic in drinking water, the long-term (years) total ingestion of arsenic would be relevant to the risk. Often we will not have continuous measurements to provide this information, and so we will have to use data from spot samples to represent a longer time frame by assuming that the concentration is stable, which may or may not be a good assumption. Supporting data from other sources would be important to check this critical assumption.

There are a few direct reading instruments that can be used in the field, such as pH, turbidity, and selected ion meters, but generally determination of chemical constituents requires collection of samples and laboratory analysis. Specialized sampling devices are available for collecting water samples at chosen depths from bodies of water, collecting bottom sediments, and collecting suspended particles. Although it might seem that water contaminants might be less variable in time than air contaminants, this is not necessarily true. The concentration of arsenic in the water of a specific well can vary substantially over time, for example in Bangladesh (Kile, Houseman et al. 2005). If the water in the well is drawing from different soil strata having different arsenic content, then the water concentration may vary depending on how rapidly water is taken out of the well and on the rate at which the well is replenished from different strata. Sampling tap water is also complicated by the tendency for some contaminants to settle out or be removed or accumulate while the water stands in the pipes, so that morning samples may be uncharacteristically low or high. Lead dissolving from plumbing in the home will usually be at its highest concentration after the water has stood in the pipes for a long time. Measurements are best made after an examination of the factors likely to affect the source(s) of the contaminants and their transport through the environment and of the temporal patterns that might result.

Measurement of groundwater contamination is generally costly and time-consuming because it must be collected by drilling test wells. A challenging sampling problem arises during the choice of locations for the test wells, a problem that is similar to the air pollution assessor's problem of where to place a fixed air monitoring station. Hydrological surveys and mathematical models are used to determine how groundwater is moving through the soil.

Ingestion and skin absorption are not the only routes of entry for water contamination. Water may also contain volatile contaminants, such as benzene and chlorinated hydrocarbon solvents. Those contaminants can volatilize if the water is sprayed through the air. A significant inhalation exposure can result from volatilization when a person takes a shower. The warm water temperature increases release into the air. Airborne releases from standing bodies of water are generally small, but skin absorption can be important.

Food Contamination

From the chemist's point of view, food contamination poses a particular challenge because it is often difficult to extract the contaminant from the organic "matrix" in which it is dissolved or bound. As with air and water, there is an important problem of understanding and accounting for temporal variability in foodborne exposures to toxins, but the approach is somewhat different. The major sources of exposure variability are two: day-to-day variability in food consumption patterns and variability in the contaminant level from one "batch" or meal to the next. Usually contamination is infrequent and only a very small fraction of the total food is affected. For example, pesticide levels on the skin of apples eaten by a family will vary over time, depending on when the apples were picked, how long they were stored, moisture levels in the storage areas, and other factors. Pesticide levels on apples within a batch also can vary because the application on the trees was uneven and because the time between pesticide application and consumption varies.

Known contaminants such as pesticides can be monitored through routine surveillance of samples of foods as they enter commerce. These surveillance approaches have important limitations, however: they can usually only sample a tiny fraction of the foods in circulation, and they measure only a small number of known hazards. Unless there is a reason to suspect problems with a specific type of food, its source, or its preparation, samples are usually collected in a "market basket" sample of a variety of foods from local retail sources. Post hoc sampling after someone has been affected is much more efficient and more likely to detect exposures than a limited routine surveillance program, but of course this is not a fully preventive strategy.

Food-frequency questionnaires are commonly used to determine food consumption patterns for the members of a cohort whose foodborne exposures are being studied (Rimm, Giovannucci et al. 1992). A large literature exists on the validation of these questionnaires with either direct observation of food habits or biological monitoring (to compare, for example, the level of vitamin A intake estimated from a questionnaire against the amount in serum). Measurements of food contamination can be linked with consumption to estimate intake. Negative findings from surveillance do not preclude the possibility of adverse effects from limited or episodic contamination.

Skin Exposure

Skin exposure is difficult to study. As noted earlier, the skin is both a target organ and a portal of entry for contaminants into the body. Skin permeability varies widely for different parts of the body, for different compounds, in different environmental conditions, and among individuals. Heat and humidity can increase permeability, whereas clothing either can act as a protective barrier or can increase uptake by holding a contaminant against the skin. One can wipe the skin and measure what is found there, but what remains on the skin during and after an exposure is only the material that was not absorbed or removed—it may bear little quantitative relation to the amount that either contacted the skin or passed through it. On the other hand, such methods can be useful qualitatively to determine whether dermal exposure is or is not occurring. Standard methods for skin wiping or washing to measure various types of contaminants have been developed by the U.S. Environmental Protection Agency (EPA) (http://www.epa.gov/compliance/monitoring/inspections/index.html) for monitoring compliance with pesticide regulations. These include detailed manuals for collecting samples, preparing them for analysis, and shipping them to the laboratory (e.g., see http://www.epa.gov/

compliance/resources/publications/monitoring/fifra/manuals/fifra/index.html). Fluorescent tracer dyes may be added to a chemical in industrial use to allow visual detection of skin contact (Black and Fenske 1996; Fenske and Birnbaum 1997). In another approach, sampling patches are attached to the skin before a work day to collect materials that would otherwise have deposited on the underlying skin (de Cock, Heederik et al. 1998). However, sampling patches do not work well on the hands, and particularly the fingers and palms, which are often the most likely areas to be exposed. Some investigators have tried rinsing the hands with solvent to measure the residues left on the hands. Because skin permeability varies with skin type, racial background, and the presence of abrasions and other injuries, knowing the total deposited may not well characterize the internal dose.

Because of the difficulties in quantitative assessment of skin uptake, occupational hygienists often recommend biological monitoring of blood, urine, or exhaled breath to measure contaminants absorbed through the skin (Kezic, Meuling et al. 2004; Semple 2004; Kim, Andersen et al. 2006). The feasibility of this approach depends on the particular contaminant being studied. There may be additional sources of variability, such as differences in subject skin characteristics or metabolic rates when using urinary or blood levels of metabolites to indicate internal levels and quantities absorbed. Also the internal levels will reflect contaminants entering by all routes, not just skin absorption. These issues are described in more detail in Chapter 5, which discusses the role of biomarkers in environmental assessment for epidemiology.

2.9. QUALITY CONTROL AND QUALITY ASSURANCE

A quality assurance program is critical for maintaining quality control (QC) on data produced by sampling and analytical procedures. Most measurements of environmental chemicals and other hazards involve trace analyses, which are difficult for even highly skilled analytical personnel. As a result, each of the agencies that develop and promulgate standard methods also include well-developed procedures for quality control. When the investigator sends samples out to commercial or academic labs for analysis, he or she should always include quality samples to determine whether the values are meaningful. A detailed presentation on quality control-quality assurance (QC-QA) protocols is beyond the scope of this book, but there are many good texts on this topic. Also, all standard and regulatory methods will have QC procedures specified.[2] The basic elements of a QC-QA program are:

1. Blind analyses: Sample media are sent to the laboratory without information on the history of the samples or on what the levels "should" be. Later follow-up is required to provide the background data so the lab can evaluate their performance.

2. Blank samples of two types:
   a. Field blanks: Sample media handled in every way the same as the samples, but no air, water, or other media actually contacted the collectors. Usually, one sample in 10 should be a blank. The limit of detection for the method can be calculated from the SD of the blanks.
   b. Laboratory blanks: Analysis of sampling media to determine laboratory contamination levels.
3. Replicate samples: Side-by-side samples collected under identical conditions are used to evaluate reproducibility. The SD of these will indicate the precision of the measurements.
4. Spiked samples of two types:
   a. Spiked blanks: A known amount of the analyte is added to the media to determine the percent recovery. These data can be plotted by date to detect systematic instrument changes and problems with the analytical instrument's function, such as reductions in responses, drifting, erratic responses, and so forth.
   b. Spiked field samples: A known amount of substance closely related to the analyte is added to the media containing a sample.
5. Instrument maintenance logs: Records keeping track of routine and other maintenance procedures must be systematic and organized.
6. Personnel training logs: Lab personnel should be required to attend skill-maintenance and development courses.
7. Participation in interlab quality sample exchanges: Some regulatory agencies and professional organizations have laboratory certification programs. As a component of these, regular circulation of quality samples is required to maintain certification. The U.S. National Institute of Standards and Technology (NIST) provides well-described environmental sample materials that labs can use as quality control materials.

Whenever you gather data on exposure conditions, you must have a quality assurance program that spells out how you will be able to verify that you have obtained meaningful findings. Even apparently simple things can be tricky. For example, we want a sample of the drinking water from someone's home to measure metals in the water. Using a new glass bottle, we place it under the faucet and run a half-liter of water into the bottle and cap it. When the lab results come back, we find that some of the metals have unreasonably low values and some are too high. Later we learn that some metals will be lost to the walls of the glass container. A small amount of acid must be added to the sample to prevent this. On the other hand, some types of glass contain metals that can be leached out and contaminate the sample. Some elements, such as boron, are very widespread in the environment, and extra care and special precautions must be taken to avoid contamination of samples. In general, we are always concerned about losses of sample components and additions of contaminants from extraneous sources. Whenever you receive data from a laboratory (including the in-house lab), you should carefully inspect the quality control data to convince yourself that the data are meaningful.

The field and laboratory technicians can explain what to look for and should identify any sets of samples that are not trustworthy because of QC problems.

NOTES

1. A good general discussion of atmospheric processes and air sampling devices is given in ACGIH's *Air Sampling Instruments*, 9th edition. A highly technical discussion of aerosol physics and behavior is given in W. Hinds's book, *Aerosol Technology*.
2. Governmental web sites for the Environmental Protection Agency (EPA) (http://www.epa.gov/nerlcwww/methmans.html), the National Institute for Occupational Safety and Health (NIOSH) (http://www.cdc.gov/niosh/nmam/), and the Occupational Safety and Health Administration (OSHA) (http://www.osha.gov/dts/osta/otm/otm_toc.html) in the United States provide access to extensive documentation on recommended or required methods for measuring contaminants and maintaining quality control.

3 Applications of Exposure Assessment in Epidemiology

If your experiment needs statistics, then you ought to have done a better experiment.

Ernest Rutherford (1st Baron Rutherford of Nelson)
(1871–1937) English physicist

Sir Rutherford obviously was not an epidemiologist!

The previous chapter summarized the characteristics and measurement of chemical exposures in various settings associated with different environmental media and routes of exposure. It also provided some guidance on how to develop sampling strategies. Given those tools now we consider the how to apply them in epidemiologic applications. Again, we will not address all of the important hazardous agents. Some will be omitted, such as viable organisms or physical hazards, but they can be approached with the same tools and concepts we develop here. Exposure assessment for epidemiologic studies has much in common with exposure evaluation for other purposes, including investigations of environmental chemistry, behavior of particles in air and water, determination of compliance with environmental and occupational regulations, and investigation of environmental disasters, such as chemical spills. The distinguishing feature of exposure assessment for epidemiologic studies is the biological considerations that must be addressed to characterize agents hypothesized to cause observed or suspected adverse health effects. This means that we must consider the hypothesized agent's concentration at the point of entry (or its suspected precursor), the agent's bioavailability for uptake under physiological conditions, and its temporal and spatial variations. In Chapter 2, we identified many of the ways that assessments must be structured to obtain meaningful data. One of the important features we only touched on there is the development of strategies for exposure characterization. In this chapter we explore these in more detail as they relate to epidemiologic studies and design issues.

One of the core issues of exposure in populations is how exposure varies among the individuals. When we design a study of environmental factors in a disease, we must be aware of exposure variation and its characteristics. Just as biostatistics is a major part of epidemiologic studies, statistical concepts and analysis are also critical for exposure assessment.

3.1. CHARACTERISTICS OF EXPOSURE VARIATION

When we talk about "exposure" we must be clear about which dimensions we are discussing. At an instant in time, an individual's exposure to chemicals has three dimensions: physical form, composition, and concentration (see Chapter 2). However, we are rarely interested in just an instant of exposure. Our earlier discussion of an individual's exposure has clearly shown that temporal and spatial variability are fundamental characteristics of environmental conditions. Understanding this variability is critical for the investigator of health effects associated with exposure. For an individual, temporal variation includes spatial variation, because people rarely stay in fixed locations. Because of its close linkage with epidemiologic data analysis strategies, understanding and quantifying temporal and spatial variability are very important aspects of exposure assessment for epidemiology. As we discuss in more detail later in this chapter, it is also critical to make our measurements on a time scale relevant to the biological responses.

Naturally occurring variations in exposure intensity over time, location, and personal activities are the primary sources of spatial and temporal variability. A street address often identifies an air pollution exposure, a domestic water supply, and a location relative to soil contamination by lead paint or toxic waste deposits, and hence the quality of the tap water in a home, its outdoor air quality, and the exposure of children playing in the yard. A job title identifies particular duties and work locations, which in turn may determine opportunities for contact with a toxic agent and its intensity. The existence of these kinds of natural differences in exposure is critical for environmental and occupational epidemiology.

For epidemiologic studies, we need to aggregate individuals into groups with differences in exposure and to estimate differences in disease risk among these groups. Thus exposure variation is important in two ways: we seek homogeneity of exposure within groups, so that the disease experience of a group can be clearly attributed to a particular level of exposure, but we also need heterogeneity in mean exposure *among* groups, so that the observed differences in relative risk between groups will be large enough that they are unlikely to be due to chance or to small undetected biases. Even moderate differences in group mean exposure may not be useful if the within-group variation is large. Thus variation in exposure is both an advantage and a disadvantage. We need variation to create contrasts among exposure groups, but that variation may make it hard to create homogeneous exposure groups.

When exposure groups are defined for an epidemiologic study of a chronic disease, subjects usually enter at different time points, so it is rarely possible to aggregate subjects in such a way that all the individuals in a group have identical exposure histories, even if they have the same exposure duration. Air, water, and occupational exposures over long periods in the United States and Europe often show a historic decline in concentration. As a result, 10 years' worth of exposure will have a different average depending on when it was experienced. This can cause within-group exposures based on duration to have a heterogeneous character. This is a source of error in epidemiologic models and is discussed further in Chapter 7.

Variation in Exposure Composition and Physical Form

Environmental or occupational exposure to single agents is unusual. Complex mixtures are much more common, such as cigarette smoke and industrial wastewater. These mixtures do not necessarily have fixed compositions, because the components in a source's output can vary over time. Physical form can also vary. Droplets evaporate to vapors, and vapors condense on particles. Dissolved materials in water can be deposited in sediments, and solids can dissolve. Temperature is an important determinant of phase changes: solid to liquid and liquid to vapor. These changes can be important for health risks.

For example, diesel exhaust has a variable composition of particles, inorganic gases, and organic vapors (HEI 1995). The composition of the emissions from a given diesel engine depend on the engine type, its age, its fuel composition, and how the engine is operated (Sjogren, Li et al. 1996). Older diesel engines release more emissions, and these emissions have a different composition from those of newer engines because of changes in engine design (Clark, Kern et al. 2002). Older high-sulfur fuels produce more particulate emissions than newer low-sulfur fuels. Idling engines emit moderate levels of particles but high levels of organic vapors, primarily unburned fuel, whereas engines under moderate load operated at steady highway speeds have low emissions for both particulate and vapor. Further, the emitted materials undergo changes with time after emissions: particles agglomerate with each other and with ambient particles, and vapors condense on or evaporate from the particles, depending on the specific hydrocarbons. As a result, exposures to emissions from trucks in stop-and-go traffic are qualitatively and quantitatively different from those during highway operations. Although all of these factors affect exposure significantly, if we have an exposure situation in which the engine type, age, fuel, and operating conditions are defined, then the composition and quantity of emissions will be reasonably consistent. The basic point is that "diesel exhaust particulate" is not a single entity, even though it is a widely used label, and, as a result, risks from diesel exposures may be different depending on the setting and historic period.

Systematic differences in composition, when they are known, can be used to form groups of subjects whose risk can be contrasted to determine whether variation in composition changes the risk. Unfortunately, it is more common that the variation in composition of a contaminant mixture is systematic with historic source changes, such as are seen in diesel engine modifications, but random for driving conditions in which engine operations vary with traffic density, speed, and terrain that produce haphazard variations across time. Thus "hours spent driving in traffic" may be a poor dose metric because of random variations in composition depending on those variables plus differences associated with the mix of car and truck traffic on different types of roads. In these situations, variations in composition introduce "random" variability (misclassification) that is difficult to characterize in the epidemiologic models.

Variation in Exposure Intensity

A large literature exists on statistical issues in assessing environmental variations (see *Statistical Tools for Environmental Quality Measurement*, (Ginevan and Splitstone 2003); also *Occupational Hygiene of Chemical and Biological Agents*, Chapter 4, (Boleij, Buringh et al. 1995)). We address the central issues and tools in this chapter. When we talk about variations in an individual's exposure intensity, we mean variations across time and location. Many of the sources of temporal and spatial variation are summarized in Table 3.1. Some of the variation consists of fixed differences across individuals, such as residence type and location, employer, and job title. Other variations in environmental exposure are associated with semiroutine personal activities (what one eats or drinks, regular commuting routes, etc.). Another set of variations consists of random changes in source output and a person's location relative to emission sources and changes in environmental fate and transport as emissions move away from the sources. Temporal variation occurs on a wide range of time scales, from seconds to years. Concentrations in different environmental media—air, water, surfaces, soil, and so forth—vary on different time scales. Mobile sources, such as vehicles in traffic in cities, are major sources of exposure and can be local (near a vehicle) or accumulated across wide areas (e.g., regional air pollution from traffic, industrial, commercial, and residential sources). As the number of sources and their locations vary during the day, so does the exposure to their emissions and photochemical reaction products. Similarly, wastewater releases (organic solvents and toxic metals) by many small businesses in a city will vary throughout the day. The sewer system collects these and mixes them at the sewage treatment plant; traditional primary and secondary sewage treatment

Table 3.1 Time scales of variation with examples

Time Scale	Environmental and Other Processes	Examples
Seconds/ minutes	Small-scale turbulent mixing in air and water; chemical reactions in air and water; source releases; soil interactions	Random variations in air and flowing water concentrations; reduction in groundwater concentrations as contaminants move in soil
Hours	Large-scale mixing; wind and water transport; photochemical processes; precipitation; variations in personal activities	Systematic changes in exposure with movement of air masses and development of photochemical smog; changes in personal exposures with changes in location, e.g., home, school, traveling, working, or recreation
Diurnal	Solar heating/nighttime cooling; atmospheric inversions; some business emissions to air and water; variations in personal activities; traffic variations	Day/night differences in air exposures, changes in chemical processes with temperature affecting exposures
Days	Long-range atmospheric and water transport; weather system variations; sedimentation in air and water	Systematic differences in area-wide exposures, weather, and exposures affected by activity patterns
Weeks	Weekday vs. weekend differences in personal and business activities	Weekday vs. weekend exposures to traffic emissions; metal content of municipal wastes from small business, e.g., plating operations
Months	Weather system variations; business production cycles	Monthly averages of air or water pollution; occupational exposures in business; also, outdoor weather affects indoor conditions
Seasons	Economic cycle variations; seasonal holidays; weather changes	Seasonal and holiday averages of air pollution; occupational exposures
Years	Long-term business trends affect pollution output; implementation of source control technology	Annual averages of community pollution; occupational exposures

accomplishes only modest removal of solvents and dissolved metals. Consequently, effluent going into a river and the subsequent exposure (composition and concentrations) downstream will vary with time and distance from the treatment plant.

What does this temporal variation look like? If we ask an individual to wear a direct reading air monitor, the resulting time profile of his or her airborne particle exposure would be something like that shown in Figure 3.1. We would expect a similar random time profile for water contaminants measured with regularly repeated water samples from a river if we sample downstream from a sewer outfall. However, the variations might be smaller from minute to minute.

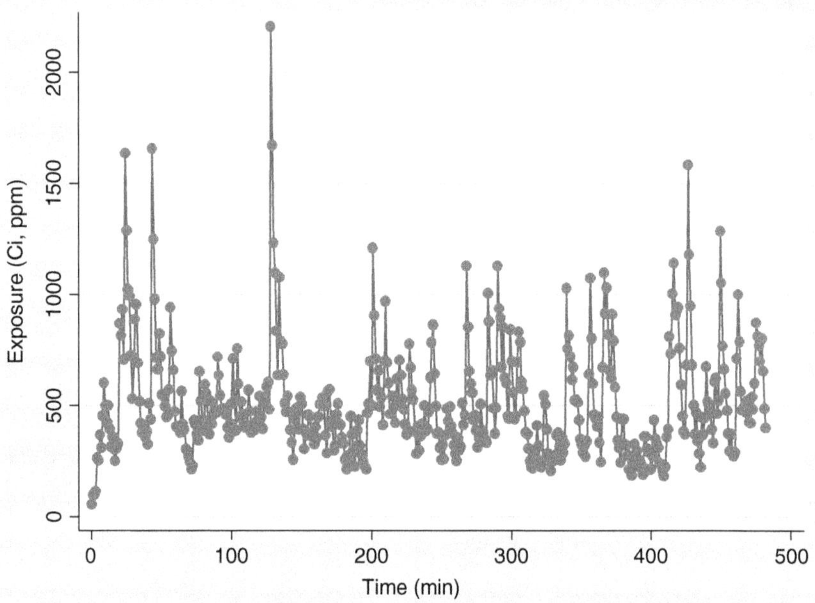

Figure 3.1 An example of a typical time profile of exposure intensity for an air or flowing water contaminant measured at a fixed location – once per min for 8 hr. Graphs with similar appearances but different scales could be obtained with time scales of seconds, hours, days, etc.

This variation in intensity over time has often been viewed primarily as a nuisance problem for epidemiology because it complicates the data analysis, introduces uncertainty into exposure assignments, and forces the collection of more data than would be needed if the conditions were less variable. This perspective often seems to be based on the unstated assumption that the average exposure is the primary determinant of risk, even when there is little or no evidence to support this idea. In fairness, it can be difficult to gather sufficient data to examine temporal variability. However, sometimes investigators have not taken advantage of time-course data when it was available, and we hope that the methods presented in this book will encourage more complete exploitation of information on temporal patterns of exposure intensity when they are available. Temporal variation in exposure can be used as a probe of biological processes and individual responses. Patterns of temporal variation in external concentrations often produce related variations in internal concentrations, which in turn can have important implications for the magnitude of risk. We explore these ideas further in Chapters 5 and 6.

One of the critical features of temporal variability is its time scale. When we are planning to link exposure data with health data, we must measure exposures on a time scale that is relevant to the disease processes producing the health responses. For example, acute respiratory symptoms can develop in the time scale of minutes to hours, which suggests that short-term variations in exposures may be relevant; or cancer risk from arsenic in drinking water may be associated with long-term water exposures over years. However, some chronic effects also can be caused by acute exposures, such as cancer from brief, intense ionizing radiation exposures.

Describing Exposure Variability

Given that exposure intensity is variable, how can we specify or define the exposure intensity[1] experienced by an individual or a population? It makes no sense to report the result of a single observation at a point in time and specific location, such as 3 mg/L of arsenic in a village well in Bangladesh, even if that was the actual concentration observed, because in a week or month or in another location, it would be different, perhaps considerably. When we wish to describe the exposure of a person, we must specify its composition and the statistical distribution of its concentration as a function of time and location (based on repeated measurements across time and location).

Random variation in measurements usually has a normal distribution. However, when we look at random variation in exposure intensity, commonly it is skewed to high values; that is, high exposures are more common than expected from a symmetric Gaussian distribution. Our source-receptor model of exposure implies that sources release concentrated emissions that are progressively reduced by repeated application of random factors, such as diffusion, mixing, and dilution, which each reduce the concentration; they define a series of stochastic (random) events. We show (following) that the distribution of concentrations resulting from multiplying together a series of random dilutions will tend to be right skewed.

Because skewed distributions are poorly fitted by the standard normal distribution, they require more complex mathematics than usual. This is important because there are many statistical questions we are interested in answering. For example, do two exposure groups have significantly different exposures? And what fraction of a group has exposures above some threshold level?

It simplifies our lives that environmental exposure concentrations tend to have approximately lognormal distributions (see Box 3.1), which means that if we take the logarithm of each observation we get a classical normal distribution. Because nonmathematicians are not comfortable working in logarithms that are hard to interpret, we take the antilogs of the mean and SD log values and obtain the geometric mean and geometric SD, which we discuss in detail later. Researchers have found theoretical reasons that the turbulent dilution processes that generate air and water contaminant distributions might generate lognormal concentration distributions across time (Rappaport and Selvin 1987; Day, Esmen et al. 1999). The theoretical fit of distribution is most likely for measurements made at a fixed location relative to a source emitting into a turbulent fluid flow (air or water). Because humans are rarely found in fixed locations, their exposures are better thought of as being drawn from a set of lognormal distributions, one for each location in which

Box 3.1 Lognormal Distribution

The key parameters of the lognormal distribution are its geometric mean (GM) and geometric standard deviation (GSD), which are log transformed mean and standard deviation. This transformed distribution can be described by the mean and SD of the $\ln(x_i)$ values. Because it is awkward to work on the log scale, we define the geometric mean (GM) as the antilogarithm of the mean of the log values:

$$GM = e^{\sum_{i=1}^{N} \ln(x_i)/N}$$

Unlike the normal (or "native scale" as it is sometimes called) SD, the *geometric standard deviation* (GSD) is a unitless multiplier because it is calculated from the square of each ln(xi) minus the mean of the ln(xi) values, which is identical to the log of the ratio of each observation to the GM, as shown:

$$GM = e^{\sum_{i=1}^{N} [\ln(x_i)-\ln(GM)]^2/(N-1)} = e^{\sum_{i=1}^{N} [\ln(x_i/GM)]^2/(N-1)}$$

The relations among the geometric parameters are multiplicative. To see this, note that a value of 1 standard deviation below the GM is calculated by dividing the GM by the GSD:

$$GM / GSD = e^{mean(\ln xi)} / e^{SD(\ln xi)} = e^{[mean(\ln xi)-SD(\ln xi)]}$$

Similarly, two standard deviations above the mean is calculated by multiplying the GM by the square of the GSD:

$$GM \times GSD^2 = e^{mean(\ln xi)} \times [e^{SD(\ln xi)}]^2 = e^{[mean(\ln xi)+2SD(\ln xi)]}$$

Because the GSD is a multiplier, the range of common GSD values for exposure is relatively narrow, approximately 1.4 to 5, and values are often close to 2. When the GSD is less than about 1.4, the distribution is nearly normal.

When environmental data are presented in tables as arithmetic mean and SD without showing the distribution, it is useful to examine the coefficient of variation (CV, which is equal to the SD divided by the arithmetic mean, expressed as a percentage). The normal distribution has a small CV that is usually < 40%, because of its approximately symmetric distribution, whereas the CV for lognormal exposure distributions is large, often > 100%, because of the asymmetric distribution. In this case, two standard deviations below the arithmetic mean is a meaningless negative number. A lognormal distribution with a GSD of about 2 will have arithmetic mean and SD that are about the same magnitude and a CV of about 100%. This is typical: the dispersion of exposure measurements for an individual across time. Note that the measurement error is usually normally distributed and has a much smaller CV ~ 20% based on replicate analyses for the same samples.

they spend time, which implies that a single lognormal distribution will probably fit only approximately. Because we rarely have enough data to statistically test the fit of the observed data to a true lognormal distribution, exposure distributions are often assumed to be *lognormal*. By convention, the logarithm to the base *e* is used, although there is no particularly compelling reason for this, aside from the fact that mathematicians like *e*.

Graphical Methods to Visualize Variation

Our goal in describing variability is to understand the likelihood, or the probability, of exposure intensities and composition. Graphical methods offer us a chance to visualize the distribution of sample values without making assumptions about the nature of the distribution, which is important because we rarely have a purely normal or lognormal distribution. The most direct description of intensities is a histogram that shows the frequency of different exposure values the fraction of time that we observe exposures within each of a series of nonoverlapping ranges of values. If we have made a sufficient number of observations, then a histogram can give an indication of the shape and type of distribution. Two histograms in Figure 3.2 were prepared from the time-course data in Figure 3.1 to indicate how frequently exposure intensities of different magnitudes were observed. The exposure histogram arrayed on a linear scale of intensity is shown in Figure 3.2A. From it we can see that the distribution is asymmetric, with a long "right" tail of the higher values. Whereas the log-scale distribution in Figure 3.2B is symmetrical and reasonably normal shaped. In other cases we may also be able to determine that the distribution has multiple modes, which is frequently seen for particle number-by-size distributions.

Another useful graphical approach for visualizing an exposure distribution's shape is to calculate a cumulative probability distribution, formally called the *quantile plot*. We like this approach because it gives us a quick visual assessment of how normal or lognormal a distribution is. The quantile plot is constructed by ordering the observed values from smallest to largest, and then, starting with the smallest, calculating the percent of values that fall below each observation. Then the concentration values are plotted against their percentiles on cumulative probability graph paper or using a statistical package. An example is shown in Figure 3.3 for the data in Figure 3.1. These plots are most useful when there are 50 or more data points.

There are two common types of plots: a normal quantile plot and a lognormal quantile plot. We can determine whether the data can be considered either normally or lognormally distributed by determining whether the data form a reasonably straight line on either of these graphs, respectively. In Figure 3.3, the data from Figure 3.1 have an approximately straight line in the log scale plot but a curved line with a linear scale. In practice, we find that most data sets will diverge

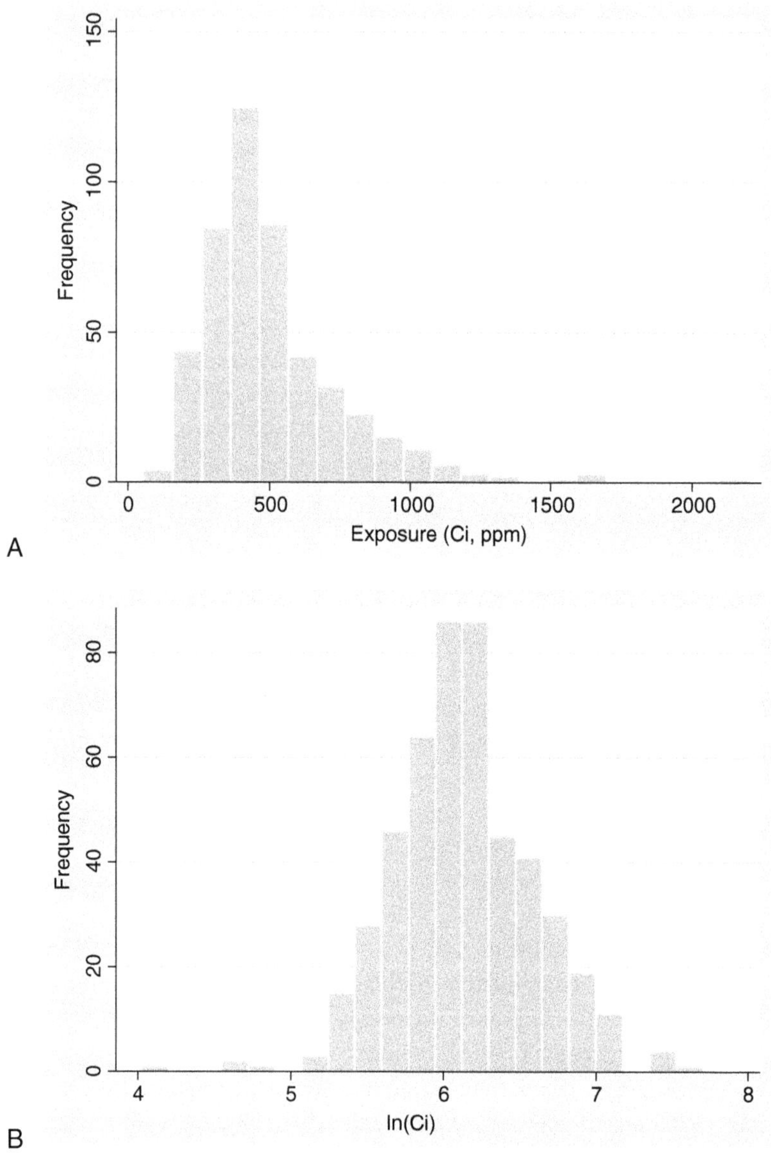

Figure 3.2 Frequency plots (histograms) of data from Figure 3.1. Part A: Frequency plot of concentration data from Figure 3.1 on a linear scale. Part B: Frequency plot of log-transformed concentration data from Figure 3.1.

from a straight line for both random and systematic reasons, especially at the ends of the distributions, such as shown in Figure 3.4. In these cases, it is harder to come to an unambiguous conclusion about the type of distribution. We also can use one of several formal tests to determine whether our data have a normal or a lognormal distribution.[2] However, the more data we have, the less likely they are

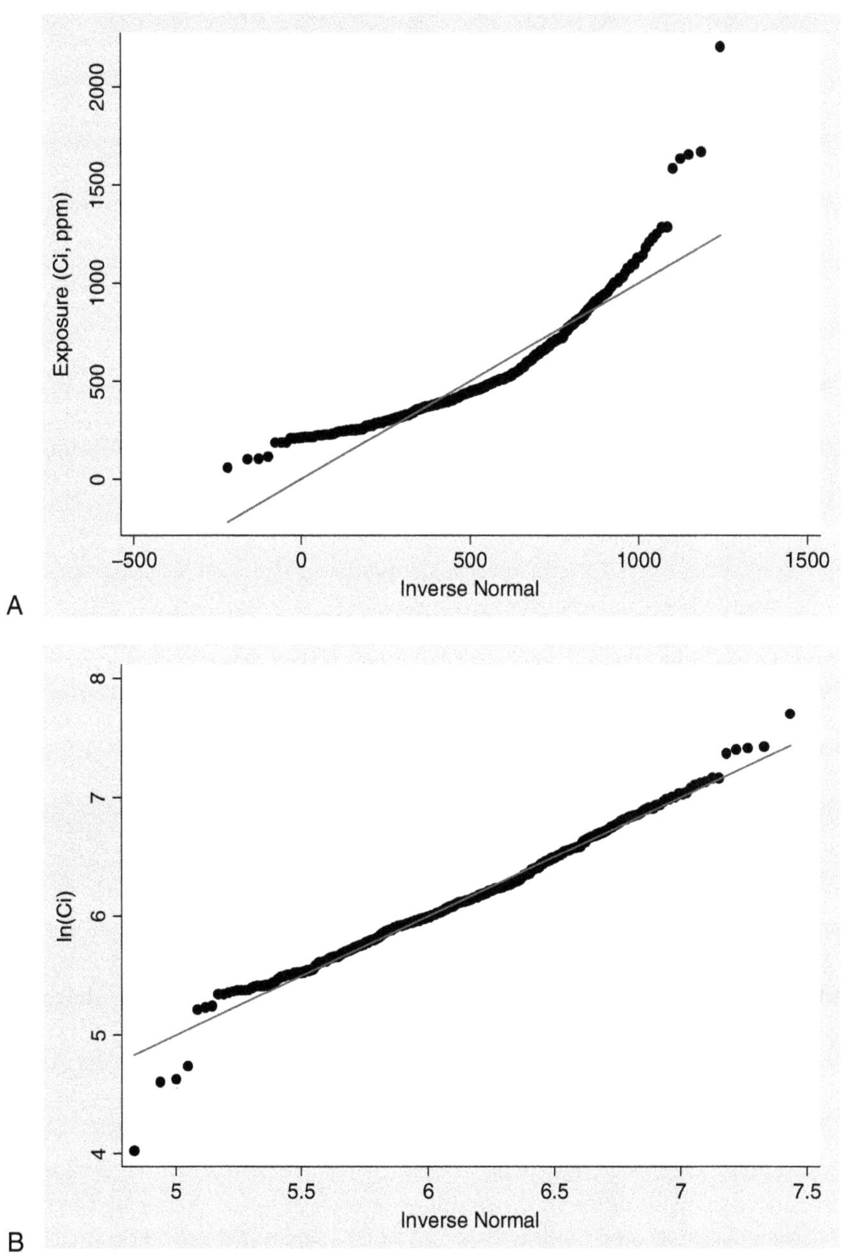

Figure 3.3 Quantile plots of concentration data from Figure 3.1. Part A: Normal quantile plot for data in Figure 3.1.
Part B: Quantile plots of log-transformed concentration data from Figure 3.1.

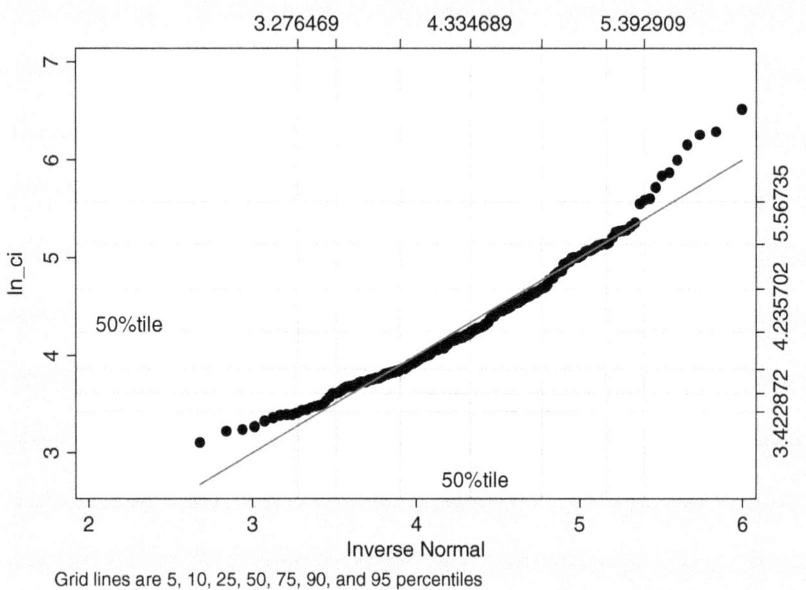

Figure 3.4 Quantile plot of combined distribution of data drawn equally from three underlying distributions (N = 100 for each) to simulate subject exposures. (Group 1 background: GM = 15, GSD = 1.5; Group 2: GM = 50, GSD = 2.5; Group 3: GM = 75, GSD = 2.5.)

to fit either the normal or lognormal distribution because the data are usually a composite sample of several underlying distributions. The most common approach is to assume that our sample distributions are *approximately* normal or lognormal. But this is generally sufficient, as we are rarely concerned with subtle differences in distributions.

The deviations of the shape from a straight line for the quantile plot can give us important clues to the nature of the exposure. One common shape is the bent line, as shown in Figure 3.4. This exposure distribution is a composite of three distributions, such as would result when a person spent equal amounts of time in three locations with different distributions of intensity—in which case the overall frequency distribution would look curved on the lognormal plot. We discuss why that may happen when we discuss time-weighted averaging later. Another common form is an apparent "baseline" of low values, as shown in Figure 3.4, in which a third of the values were drawn from a distribution that had a low value and little variation (Group 1: GM = 15, GSD = 1.5). In other words, the line is initially less steep and then rises approximately linearly in a quantile plot. When beginning an investigation of a new exposure situation, it can be very instructive to make a cumulative-probability quantile plot of the data.

Parametric Methods to Characterize Variation

When our exposure data have distributional shapes that indicate they are approximately normal, or when we can reasonably assume a normal distribution (when the number of observations is too small to show it), then we can employ the powerful parametric methods developed for data with a normal distribution, such as using the mean and SD to calculate the probability of values above a threshold, or testing for differences. If the data are reasonably lognormal, then the logarithms of the values can be treated as if they were normally distributed, and we can use all of the analytical tools that assume normality. Many fewer data are needed to calculate the mean and standard deviation (SD) than are needed to determine the distribution's characteristics from its frequency distribution by graphical methods. We like parametric tests because they are more powerful for a given number of data points than nonparameteric methods. The mean and SD or the geometric mean (GM) and geometric standard deviation (GSD) (formerly defined) are particularly useful because they can be used to estimate the probability of occurrence of any given value, assuming normality.

Cumulative Dose and the Arithmetic Mean

One of the most important parameters we want to estimate is the total amount of a chemical taken into the body during an exposure. For example, each day we take in approximately the same total volume of air or water. The quantity of air contaminants deposited in our lungs over a long period of time, N days, will be the sum of the amount inhaled each day, which is given by the inhalation rate, I m^3/day, times the fraction retained, f, times the average air concentration, $C_{expo}[i]$, each day. This is mathematically the same as the constants I and f times the arithmetic mean amount inhaled multiplied by the number of days.

$$Mass\ Deposited = \sum_{i=1}^{N} IfC_{expo}[i]\Delta t = If(\overline{C}_{expo}N\Delta t)$$

Equation 3.1 Inhaled dose relationship

Thus the total quantity of contaminants inhaled is proportional to the arithmetic mean times the duration of the exposure ($N\Delta t$). This biological relationship is true regardless of the frequency distribution of the exposure. Even though the exposure data are often lognormal, we want to calculate the arithmetic mean to estimate total contaminants inhaled, not the geometric mean. This is sometimes confusing because we often want the best estimator of the central tendency, but not in this case; here we want the arithmetic mean to calculate the dose metric.

Unfortunately, directly calculating the arithmetic mean of lognormal data may be biased. First, the skew to large values of the lognormal distribution makes the arithmetic mean calculated from a modest sample ($N < 10$) biased relative to the true population mean; it tends to overestimate the population arithmetic mean.[3] There is a likelihood estimator approach that we can use to calculate an unbiased estimate of the arithmetic mean and its confidence interval for a small sample when the data have a lognormal distribution (Rappaport and Selvin 1987; Armstrong 1992). Second, another situation that can result in bias of the arithmetic mean is left truncation of a sample value distribution due to values less than the LOD. Some researchers have proposed that, depending on the skewness, values based on the LOD be substituted for the missing values, such that LOD/2 might be used with large skew (GSD > 3) or LOD/$\sqrt{2}$ with modest skew (Hornung and Reed 1990; Nieuwenhuijsen 2003). It is critical to recognize that sufficient samples may be needed to characterize the full distribution, not just the mean.

Time-Weighted Average Exposures

This statistic is widely used to describe average exposure intensities during a period of time: 15 min, 1 hr, 8 hr, 24 hr, or 1 year. As described earlier, the *time-weighted average* (TWA) is calculated by dividing a time period, T, into increments and measuring the exposure intensity in each increment (i); then we sum the product of the exposure concentration ($C_{\text{expo}}[i]$) times duration (t_i) of each increment (note that the intervals do not need to be of the same duration) and divide by the total time (T), as shown in Equation 3.2.

$$\overline{C}_{TWA} = \frac{\sum_{i=1}^{N} C_{expo}[i]t_i}{\sum_{i=1}^{N} t_i} = \frac{\sum_{i=1}^{N} C_{expo}[i]t_i}{T}$$

Equation 3.2 Formula for the time-weighted average

This concept was developed for air pollution and occupational studies because it was recognized that people have relatively even breathing rates during each minute, and the amount inhaled is proportional to the air concentration. The total amount deposited in the lungs during a period of exposure is proportional to the TWA of breathing zone concentration. We could develop a similar expression for intake of water contaminants if we had the concentration and volume ingested of a series of water volumes. Later we will see that use of a TWA measurement in a health study implies that there is a uniform risk per unit of exposure, e.g. one min of 1ppm, so that high-intensity, short-duration exposures ("peaks" e.g. 0.1 min of 10ppm) do not have disproportionately higher risk. The TWA smooths out the short-term variations occurring within the averaging time period. TWA is relatively easy to do with an environmental sampler by automatically collecting an equal amount of air or water during each minute so that a TWA sample is obtained. The TWA is one of the more practical and universal forms for health studies, but its use carries some implicit assumptions about the nature of the exposure-response relationship—that is, that peak exposures do not produce disproportionate responses.

As noted earlier, environmental levels often vary by location because of proximity to different sources. A person moving through a series of locations during a day experiences a TWA of the location-specific exposures. The TWA of personal exposure can be obtained by having an individual wear a sampling device to collect air from his or her breathing zone, but this can be intrusive, and some devices are difficult to wear for periods of more than a few hours. Alternatively, an individual's exposure profile over time may be estimated by measuring exposures in each of the locations in which a person spends time and then using a *time-activity diary* to determine how long he or she is in each location so that a TWA can be calculated. In a hypothetical case, an individual spends 10 hr per day in the bedroom, 4 hr in the kitchen, 2 hr commuting, and 8 hr in a office job. The air contaminants and concentrations can be substantially different in some, and possibly all, of these locations. For example, exposures can range from very low in the bedroom to very intense while trapped in stop-and-go traffic. If we had data on the arithmetic mean exposure for a chosen pollutant in each of those locations, we could calculate the TWA, as shown:

$$\overline{C}_{TWA} = (10hr \times \overline{C}_{bedroom} + 4hr \times \overline{C}_{kitchen} + 2hr \times \overline{C}_{traffic} + 8hr \times \overline{C}_{work}) / 24hr$$

Remember that we are calculating the TWA for its relevance to biological dose, not because it is the best estimate of central tendency, so even though the data may be lognormal, we want the arithmetic mean and not the GM. It is not necessary that the pollutant be present in all of the microenvironments listed; in some locations, the concentration may be zero. This approach can lead to some surprising findings. For example, if an individual's home is in a highly polluted urban area in central Los Angeles but he works the night shift as a janitor in an office building in a seaside community north of Los Angeles, the majority of the integrated inhalation exposure may take place during the time spent at home from pollutants that infiltrate into the house.

The statistical characteristics of the TWA can be very complex. At the extremes, if one high exposure accounts for most of the time, then the distribution of TWA values will follow very closely that of the high-exposure values. On the other hand, if two or more substantially different exposures contribute equally, then the TWA exposure will be a blend of the distributions. Another common situation is one in which different days come from different distributions because of work at different sites, resulting in wider variation in the TWA. Thus a person's exposure through time will depend on the particular activities and locations in which he or she may come in contact with emissions.

Between-Subject Variability

We know that an individual's exposure varies, so what does it mean that a group of individuals have the "same" exposure? A "commonsense" response might be that the measured personal exposures of these individuals are essentially identical day after day (or over some other relevant time scale). But because of both the inherent moment-to-moment variability at a location and the additional variation of time-activity patterns, no two persons' exposures are identical. More important, we rarely have the detailed exposure data, such as shown in Figure 3.1, for all subjects. Thus the commonsense definition is not very useful.

A more practical definition of "the same exposure" is to say that all the exposures of individuals have been drawn from the same exposure *probability distribution* and that that distribution is stable over time; that is, it is a "stationary" distribution (see section 3.2). Therefore, we expect that there will be some differences among the measured exposure values, but taken as a set, these differences will be small enough that the observations will appear to have been randomly drawn from the same underlying distribution. When our objective is estimating exposures for an epidemiologic study, we add the additional requirement that groups whose risks are compared should have exposures drawn from *different* underlying distributions so they have different group mean exposures. Although this general principle may be clear enough, it is another thing to specify how similar the within-group data must be and how different the between-group data should be.

Analysis of variance (ANOVA) is the tool to determine whether groups of individuals are statistically different from each other, given some amount of within-group variability. Kromhout, Rappaport, and others have used an analysis of variance to look at the relative amounts of between-subject and within-subject (across time) variability (Kromhout, Symanski et al. 1993; Rappaport, Kromhout et al. 1993). Job title, which has been commonly used alone for exposure assignments, was shown to be a poor classifier in many occupations. Jobs such as welding, in which an individual's work habits or work location can strongly influence his or her average exposure, can produce large between-subject variability within groups. Consequently, even large differences in mean exposure (e.g., 50% difference in the means) between groups of welders may be meaningless, because the groups have large GSDs and overlap in the exposures of their subjects, as shown for groups 2 and 3 in Figure 3.5. On the other hand, production workers, such as machine operators in different departments at an automobile manufacturing plant, may have small GSD variations in exposures because their conditions of exposure are highly constrained. In that situation, small differences in group mean exposure (e.g., 20%) may be statistically significant and potentially useful for detecting differences in risk. Thus, when we are using group means to define differences in exposure for epidemiologic comparisons, large between-subject variation within groups can invalidate group comparisons, even though the group means are significantly different. This is especially true for large groups because the large number

The raw exposure data (one observation per individual)

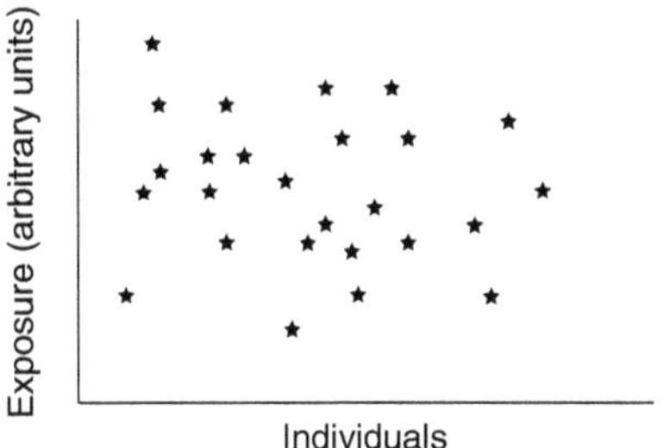

The same data organized for an epidemiologic study

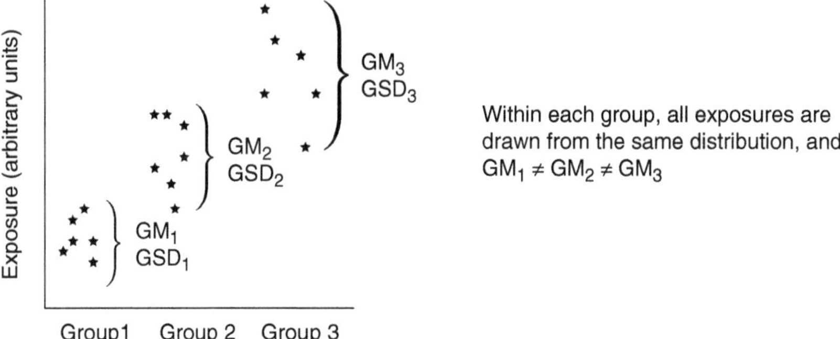

Figure 3.5 Exposure variability in epidemiologic studies: within-group exposures should be drawn from the same distributions, but the groups should have different underlying exposure distributions.

of values enables small differences in the arithmetic means to be statistically significant.

An ANOVA evaluation assuming a random-effects model can determine whether between-group differences are significant in the presence of different levels of within-group variation in individual mean exposures. Where the exposure variation is lognormal for individual exposures, it will likely be lognormal for personal mean exposures, too. Therefore the ANOVA is calculated based on the $\ln(C_{expo}[i])$ values. An example of an ANOVA analysis is shown in the Box 3.2. More data on personal exposures are needed for this calculation than are usually collected. Power calculations can be performed with standard packages using the log values. The findings will indicate how large the ratio of group means must be to be significantly different and suitable for an epidemiologic analysis.

Peak Exposures: Upper Tail of the Exposure Distribution

It is often assumed that the overall contribution of exposure to risk will be proportional to the magnitude of the mean exposure. The degree to which this general

Box 3.2 Analysis of Variance for Between- and Within-Subject Variation

One of the basic features of groups of subjects in an environmental epidemiologic study is their variation in exposure, which must be considered in the design of sampling and analysis of data. A nested ANOVA is the right test for this. The random-effects statistical model is $Y_{ijk} = \mu + \alpha_i + \beta_{ij} + \varepsilon_{ijk}$, where α_i is the group effect, β_{ij} is the subject effect, and ε_{ijk} is the within-subject day-to-day variation. The ANOVA table for this model is shown here.

Source	Degrees of Freedom	Mean Square (MS)	F-Value	Expected(MS)
Groups	a-1	$MS_a = SS_a/(a-1)$	MS_a/MS_b	$\sigma_e^2 + n\sigma_b^2 + nb\sigma_a^2$
Subjects	a(b-1)	$MS_b = SS_b/a(b-1)$	MS_b/MS_e	$\sigma_e^2 + n\sigma_b^2$
Error (within)	ab(n-1)	$MS_e = SS_e/ab(n-1)$		σ_e^2

Note: SS is the sum of squares for the indicated source level, and n is the number of observations (obs) per subject, b is the number of subjects per group, and a is the number of groups. More complex forms are available for uneven designs.

An example is shown here, which uses the data in Figure 3.4.

Number of obs = 42 R-squared = 0.8738
Root MSE = 0.342593 Adj R-squared = 0.8010

Source	Partial SS	df	MS	F Prob >	F
Model	21.128530	15	1.40856868	12.00	0.0000
Group	9.247170	1	9.24717057	11.06	0.0050
Subject|Group	11.708837	14	0.83634556		
Subject|Group	11.708837	14	0.83634556	7.13	0.0000
Residual	3.051624	26	0.11737017		
Total	24.180154	41	0.58975987		

Subj|Group represents the variance of subjects within groups.

Using the preceding model we can calculate the individual variances shown:

$$GSD_{within} = 1.41, GSD_{subj} = 1.90, \text{and } GSD_{grp} = 1.88.$$

Even though there is large between-subject variation, there are also large group mean differences.

principle holds will depend on the underlying disease mechanism, and it is not hard to think of pathophysiologic processes that will lead to quite different effects. For example, when there is a threshold for a response to exposure below which people do not respond to the toxin, then the extent to which a group will show a toxic response depends on the fraction of exposures above the threshold.

Note that the value of the mean and the fraction of exposures above some value are not independent. However, when the fraction is small, it may have little impact on the mean. There are also conditions under which a very high tissue concentration produces an entirely different type of response than lower concentrations, such as deep lung respiratory irritants. These phenomena focus attention on so-called peak exposures—those in the upper tail of the exposure distribution—rather than simply on the mean exposure. This topic is be discussed in more detail in Chapters 13 and 14.

Determination of the frequency of very high exposures is often difficult because they are usually infrequent. For example, if the upper 1% of exposures are defined as peaks and are thought to produce disproportionate responses, then we must collect hundreds of samples to observe a small number of peaks. In the work environment, peaks may be associated with specific tasks or working conditions, which can be sampled and the distribution characterized. In a similar way, peak environmental exposures may be more likely to occur during certain meteorological conditions that can be anticipated, such as temperature inversions, and monitored more intensely. Peak exposures become important as a result of the biology. If the damage processes are all linear (simple proportionality, effect$[i] = K_{eff}C_{exp}[i]$), then we do not need to characterize peaks; the mean exposure is good enough.

3.2. EXPOSURE STABILITY

A key feature of epidemiologic analysis is the comparison of groups with defined differences in exposure to determine whether they have exposure-related differences in disease risk. For this to work, we must either have measured all of everyone's exposures or have stable exposures within groups, such as residence or job groups. However, given all of the variability noted earlier, does it make any sense to talk about stable exposures? It does, but only in a dynamic sense. If the environmental conditions are "stable"—that is, the source output has a regular temporal pattern, the environmental transport and mixing processes have stable characteristics, and the exposed person has a defined set of activity relative to contact with the source releases—then they will define a stable exposure distribution through time. That is, the central tendency and dispersion of the distribution will remain stable. When those conditions occur, the exposure is said to have a *stationary distribution*. When we make comparisons, we must make certain that we are comparing groups within similar time frames using averaging intervals that will smooth out shorter term irrelevant variations, which come from many sources, as shown in Table 3.1.

Unfortunately, environmental conditions are rarely precisely stable, as noted earlier. There are diurnal and seasonal cycles of temperature, humidity, rainfall, and wind speed, but they often vary with some reasonable regularity. Cyclic variations of diurnal and seasonal differences can be sources of systematic variation in exposure. As a result, we need to compare groups within the same season in the same year.

Other regular changes in environmental conditions can produce trends in exposure, such as during the period moving from summer to winter or vice versa. Some businesses and the general community experience variations in activities associated with seasonal and holiday economic factors, such as the trucking industry and retail sales, that can have large effects on emissions and pollution. One of our key concepts is that the time scale of the exposure variation is very important, and whenever possible it must be considered and matched to the time scale of the variations in health effects. This is discussed in more detail in the next chapter.

3.3. SAMPLING STRATEGIES

Up to this point we have discussed the range of techniques for measuring the exposures of individuals and locations and the nature of exposure variation. Next we need to consider integrating our measurements with the dynamics of health effects and outcomes in epidemiologic studies. A *sampling strategy* is a detailed, specific plan stating who, where, when, and how we will sample and make our measurements. Because our goal is to relate the exposures to some adverse health effect(s) or epidemiological outcomes, we also must define the relevant biological features to guide the development of the plan. Note that the strategy used for epidemiologic studies is not the same as would be used for determining compliance with legal limits for allowable exposures. First, from the nature of the effects, we can define a target organ or tissues. Toxicokinetics can be used to model the target tissue concentrations given the temporal pattern of exposures. We will use this to define the time scale of averaging time for the exposures. Second, we must define the approximate time course of the effects, the etiologic time interval (ETI; minutes, hours, days, etc.). This will define the total time we must measure into the past, beginning from the outcome and working back, in order to determine the relevant exposure. Rapidly developing acute effects suggest that the exposures immediately preceding the effect are most relevant, although this might not be true for an immunologic effect. Slowly developing effects tend to have slow recoveries when exposure stops, which suggests that exposures over longer time scales may be important. Irreversible effects suggest that the whole period of exposure is important, perhaps offset by a lag to allow time for the effect to develop. These considerations will guide setting the ETI. Third, the nature of exposure variability is important for the measurement strategy. In some cases, the exposure is limited to specific locations, such as a workroom or a neighborhood, or to specific activities, such as commuting in traffic. For others, exposure is diffuse and extensive, such as with some types of community air and water pollution.

There are four types of questions to be answered in the development of any environmental measurement strategy:

1. What should be measured, and in what concentration range? What methods (specificity, precision, accuracy, LOD, suitability for personal sampling,

equipment needed, practicality, cost) are available? Can we use surrogates for suspected agents?
2. What time periods (ETI, averaging intervals) are important? Can measurement techniques characterize those time periods? If necessary, how can we estimate past unmeasured exposures?
3. Who should be measured? Where? When? Do we measure specific individuals over time or random individuals to assess group exposures? Should we do time-activity pattern estimates or TWA sampling?
4. What statistical data (estimates of the mean and SD, or GM and GSD, or frequency of "high" exposures for individuals or groups) do we need? How many samples are needed to estimate relevant parameters for individuals and groups with suitable precision?

Answers to those strategy questions will depend on the design of the epidemiologic study. Table 3.2 lists the three common epidemiologic designs and the exposure characterization strategies commonly used for each. A small, prospective cohort study can take advantage of current sophisticated measurement and monitoring tools to characterize each subject's personal exposure in extensive detail. At the other extreme, a large, retrospective case-control study is a powerful design

Table 3.2 Exposure strategies for common epidemiologic study designs and responses

Epidemiologic Design	Acute Response, Short ETI (min to hr)	Chronic Disease, Long ETI (yr)
Prospective cohort	*Small study population:* Personal sampling, all subjects; real-time measurements to identify peak exposures during ETI period *Large study population:* Personal sampling to identify exposure determinants[a] and calibrate questionnaire; apply questionnaire to all subjects	Personal sampling to identify groups and estimate by main exposure determinants[a]; repeated measurements/questionnaires over ETI period
Retrospective cohort	Cross-sectional measurements to define groups and estimate exposures by main determinants; assume no change in recent past or extrapolate with determinants	Cross-sectional personal and location measurements to define groups and estimate exposures by main determinants; assess retrospective changes in determinants to make past exposure estimates
Case-control	(design not used)	Define exposure groups by main exposure determinants; retrospective exposure estimates based on changes in determinants

a. Exposure determinants are those factors, such as job and work area characteristics, residence location and characteristics, commuting pattern, drinking water or food source, etc., that are associated with major differences in mean or extreme exposures.

from the epidemiologic perspective but a very difficult one from an exposure assessment perspective, because we cannot make personal measurements retrospectively and the subjects come from a very wide range of backgrounds, so methods for developing estimates of past exposures are very limited.

A sampling strategy usually includes a combination of personal and fixed-location measurements designed to describe the exposure components of a study population, which will be supplemented with background data from questionnaires, records, and various databases. The population size, access to the subjects, and the type of epidemiologic study design will define the needs for exposure measurements. An exposure study performed in an epidemiological setting has the added constraints that the data must be relevant to the route of entry and uptake (bioavailability) and to the need to estimate relevant exposures for population subgroups or individuals during selected periods of time. In the next section we look in more detail at the application of exposure assessment to epidemiologic studies.[4]

3.4. ESTIMATION OF UNMEASURED EXPOSURES

When conducting a large-scale epidemiologic study, we are often faced with the problem of estimating unmeasured exposures. When we collect existing data in records from company occupational surveillance programs or pollution monitoring data from city and state health departments, we find frequent gaps or missing data. Alternatively, when we conduct a cross-sectional sampling study, we may conserve resources by minimally monitoring groups with little or no evident exposure. If we want to relate an environmental exposure to a chronic disease, such as cancer or chronic kidney disease, we will need estimates of exposure over very long time intervals: 20, 30, even 40 years. And, because the systematic collection and archiving of exposure data are relatively recent phenomena, the historical exposure data are often not available. This problem has been most extensively addressed in occupational health studies, but there have been some environmental studies, too. These situations leave us with a problem of how to estimate the missed exposures. We must estimate two types of unmeasured exposures: unmeasured exposures of groups within a cohort, which were not measured at the time of cross-sectional sampling, and unmeasured past exposures.

Several approaches for estimating past exposures have been developed (see the chapter by Smith et al. in (Gardiner and Harrington 2005). They are briefly summarized here.

The Basic Approach

The estimation of long-term past exposures poses challenges, both because of the frequent absence of all types of data and because of the many components of temporal variability in exposures that were discussed earlier. To confidently estimate exposures, we must determine that exposure conditions have been stable or have changed over time in a defined way, which requires knowledge of the characteristics of the emission sources, the environmental transport processes, and the ways in which individuals come in contact with emissions. In the United States, direct measurements of personal exposures are generally unavailable for exposures occurring before the 1980s. Extensive occupational exposure monitoring was very unusual prior to the mid-1970s; air and water monitoring networks were developed in the 1960s, but many areas have had only spotty monitoring. Thus, in cases in which the subjects in an epidemiologic study began their exposures prior to the period for which data are available, it is necessary to develop a strategy to extrapolate the estimates in order to have complete exposure histories. Even when exposures can be extrapolated, it may be impossible to determine their accuracy and precision.

Assigning quantitative or semiquantitative exposure estimates to individuals for time periods or locations without individual measurement data generally proceeds from two types of data: (1) knowledge of the physical locations of the individuals in the past (residential street addresses, job locations, etc.); and (2) estimates of the exposures in those physical locations. For example, to estimate each individual's drinking water exposure history, records of residential histories may be combined with historical reconstructions of the concentrations of a contaminant in the drinking water supply of a region, as was done at a toxic waste site in Woburn, Massachusetts (Lagakos, Wessen et al. 1986; Durant, Chen et al. 1995).

Many variations on the theme of combining these two types of data have been reported in the literature. As always, the details vary somewhat depending on the particular study population and exposure. Key principles for exposure reconstruction are summarized here, with references to the literature. It is useful to discuss separately those aspects that relate to the personal "behavioral" data, such things as residential or occupational histories, for example, and those that relate to the environmental exposure estimates, for example, solvent concentrations in drinking water, levels of urban air pollution, exposures in various locations, or jobs in a factory.

Personal data

Our gold standard for exposure estimates is personal data and samples on each member of our study group, which can capture all the personal idiosyncrasies of behavior, diet, job, and ambient exposures. Unfortunately, such data are rarely available outside of small, prospective studies. When we have personal measurement data, they are usually available only for a small sample of our subjects, so they must be used to define the exposure distribution of individuals in groups having other common factors, such as a job. As noted earlier in this chapter, personal information about things such as residential and occupational histories can sometimes be gathered without direct interviews of the subjects. Occupational histories are often available from work records, and current residence is often available from medical records. Historical residential data usually require an interview, but investigators often assume that this history, although limited by recall, will not be seriously biased by the subject's beliefs about exposure-disease associations, as the interview does not directly address exposure but the more "neutral" topic of residences. Food-frequency questionnaires and diet diaries are used extensively to determine nutrient intake, as well as foodborne contaminant exposures.

Occupational studies have relied heavily on personal job histories, usually obtained from employers. Collection of historic data and formulation of exposure models are needed to estimate the exposure composition and intensities associated with job titles, work areas or organizational units (departments), or even specific tasks. However, considerable investigation of the plant layout, raw materials, operations, job definitions, and changes over time may be needed to make meaningful exposure assignments. Job titles can be arbitrary, with meanings that change over time. For example, in the gasoline distribution business during the 1940s and 1950s, the job of "chauffeur" was not the driver of the boss's limousine; he was a gasoline truck driver (Smith, Hammond et al. 1993). Once this information is obtained, an individual's job history can be used to make a semipersonal TWA estimate of exposure. This type of occupational exposure study is very labor intensive and costly and can require years to complete.

Ambient Environmental Data

Most of the historic data on exposures consist of data from centrally located monitors: central-site air and water system data for cities and towns, area monitors used in work sites and factories, and so forth. These data can be used with time-activity data to approximate personal exposures. However, there are often systematic differences between personal and area or location samples collected at the same time because of heterogeneous spatial distributions of emission sources and environmental contaminants; proximity of an individual to sources varies, whereas area samplers have fixed locations. Exposure avoidance behavior can reduce personal exposures relative to area measurements, whereas personal activities close to sources can lead to higher exposures. Indoor and outdoor air contaminants differ because of proximity to sources. Indoor sources produce internal contaminants, for example, $PM_{2.5}$ and NO_2 from gas-burning stoves, but those can also come from outdoor sources, and some contaminants with outdoor sources do not penetrate indoors, for example, ozone. Area data are more common because they are easier to collect, but we want to know personal exposures. Available concurrent personal and area data and time-activity data can sometimes be used to determine the relationship between personal and area exposures, and area data can be adjusted to make reasonable personal estimates. For example, water supply data are often based on periodic monitoring from fixed locations in the supply system, and these must be extrapolated to link them to the residential histories of study subjects (see Lagakos et al. 1986). Extrapolation of ambient exposures can be done in several basic ways. First, one can assume that exposures have remained constant or did not vary spatially in the past. Although this makes exposure extrapolation simple, it is generally a poor assumption, unless there are data to support this stability. Environmental studies of water quality using historic records have seen some stability in basic parameters, such as microorganisms and biological oxygen demand, but chemical parameters have changed as local business and industries have changed. In some long-running industries with simple processes, it has been appropriate to assume no change in occupational exposures because the raw materials, operations, jobs, and work settings have not changed. However, since the 1980s there has been a pattern of increasing change in exposures across the industrial landscape in the United States and other Western countries (Symanski, Kupper et al. 1996). As countries across the world have developed their manufacturing bases, they also have made changes. For example, the People's Republic of China had a long period of stable heavy industry operations, but since roughly the early 1990s they have been modernizing their industrial plants, resulting in lower exposures. Consequently, the overall long-term pattern of occupational and environmental exposures has been one of change. The assumption of stability should be made only when there are supporting data to verify the lack of changes.

A second approach to extrapolating exposure data is to develop a statistical exposure model based on the available data (Boleij, Buringh et al. 1995). These models, although "naïve," may perform well at capturing sources of variability in exposures. For example, when exposure data on airborne formaldehyde exposure in a mortality study of embalmers collected across time were fit to a regression model that included job, task, workroom characteristics (size, ventilation), process type, and other descriptors to predict missing data, it was found that the year in which the samples were collected was an important determinant of the measured formaldehyde levels (Bennett, Feigley et al. 1996; Hornung, Herrick et al. 1996). The year was probably capturing the effects of changes in activities, raw materials, and/or job activities that were not explicitly documented. As with any purely empirical prediction model, one must be careful when extrapolating outside the range (time period, for example) of the data. In this example, the causal factors associated with "year" may easily have been different in earlier time periods. This is a fundamental limitation of using empirical models—they provide only a weak basis to support the resulting predictions in conditions or locations beyond those for which the model was formulated. That said, these models are probably often better than the default assumption of constant exposures, and they are increasingly applied in historic exposure reconstruction (Nieuwenhuijsen 2003).

The third approach to environmental exposure extrapolation is the use of source-receptor models described in Chapter 2. These models have similarities to the empirical models, but they differ in one important aspect: they rely on an understanding of the physical and chemical behavior of the underlying environmental processes to inform estimates of unmeasured exposures. This makes them potentially more reliable than the empirical models if sufficient information is available about the exposure setting. For example, in Chapter 2, we discussed the atmospheric processes that dilute and disperse emissions from a campfire. Models of those processes are applicable to all campfires. If we know the amount of emissions, the wind speed and direction, and the topographical features of the surrounding landscape, then we can make meaningful predictions about the concentrations of emissions at locations around the fire. Even without knowing the amount of emissions, given the wind and landscape data, we can make good estimates of the average relative reduction in concentration with distance and direction from the fire. This modeling approach might allow us to confidently predict exposure levels. Even without environmental measurement data, we might predict the *relative* exposure concentrations for different residential locations with a high degree of certainty. For example, we could predict that residences in one area of a city had much less exposure to a power plant's emissions than residences in another area if we had knowledge of wind direction and wind speed. If, in addition, we had some limited exposure data for one area, we might combine wind direction and speed data with the limited exposure data in a source-receptor model to estimate exposure levels in other unmeasured areas. The model developed by Held and coworkers is a good example of using a source-receptor model for regional air pollution, petrochemicals, and so forth (Held, Ying et al. 2005).

Regardless of the approach used, extrapolation models may be biased if the available environmental data is gathered in a nonrandom way. In fact, this is often the case. In surveillance programs, to conserve resources, low exposures without a perceived hazard are frequently not measured. If exposure data are gathered to assess compliance with environmental rules, then the sampling strategy may be quite consciously biased toward looking for the high values. For example, the monitoring of urban beaches for bacterial contamination may be organized to sample more frequently after heavy rains, when runoff may increase pollution. For health surveillance, this is an appropriate sampling strategy, but for epidemiology, care must be taken to compensate for this selective sampling strategy. This type of bias has been found in the large database developed by the U.S. Occupational Safety and Health Administration (OSHA) for data collected during worksite inspections (Middendorf 2004).

Unfortunately, we have come to recognize that hazards for some agents are present at levels previously thought to be safe. Research projects assessing exposures need to verify that subjects who are assumed to have low exposure or who are thought to be unexposed are indeed unexposed. Some epidemiologic studies have found no difference in risk because control subjects had exposures that were not much different from those of the exposed subjects; the investigators failed to verify the assumption of no exposure for the controls.

Extrapolation of Past Occupational Exposures

Occupational hygienists have developed a wide array of methodologies for estimating past exposures in the workplace.[5] A good example of these approaches is the retrospective exposure assessment for the man-made vitreous fiber industry (Quinn, Smith et al. 2001; Smith, Quinn et al. 2001). This was a large cohort study of workers drawn from 20 different plant sites across the United States. The link for the exposure estimates with the subjects was their personal job histories obtained from the companies. The work histories were chronological listings of job and department names with starting dates. For each plant a job-department dictionary was developed, which collapsed the thousands of verbatim job-department titles appearing in the histories into short lists of unique titles. Long-term employees from each plant carefully reviewed all dictionary assignments for accuracy (a very tedious job). Extensive plant histories of job activities, departmental operations, materials used, and operational configurations for work areas were also obtained for each plant. Those data were used to develop time lines for specific qualitative exposures (yes/no for PAHs, arsenic, silica, formaldehyde, etc.) for each unique job-department pair. Where sufficient quantitative data were available, exposure intensities, such as fibers/cc and mg/m^3 formaldehyde, were estimated. These data were assembled into a job-department exposure matrix for each agent for each plant.

The agent matrices were prepared for each unique job-department pair, identifying temporal periods in which exposures were approximately constant because the operating conditions and jobs did not change. The matrices were constructed by working backward from recent exposure measurements and operational data. When there were significant changes in conditions known to affect exposures (automation added, change in fiber generation, modified fiber binder, ventilation added, etc.), a multiplier based on available data was applied to the measured exposures to extrapolate past exposures. A small amount of historic data were available to check the estimates. In general, they were within a factor of two of the measurements. The defining feature of this approach is exposure models based on the characteristics of each workplace. Algorithms were developed for the defining characteristics that quantitatively indicated how exposures had been affected by historical changes in work activities, workplace layout, operations, addition of exposure controls, and raw materials. It cannot be done without access to highly detailed job history records for the subjects, good historical data on production operations and plant structure, and baseline measurement data that can be the starting point for backward extrapolation of exposures.

Available exposure data can be used to calibrate an exposure model for particular locations. Then historical data on the operations, jobs, work activities, and so forth can be used to identify relative changes in the past, which can then be used to extrapolate changes in exposure by backward extrapolation. For example, measured airborne fiber concentrations were 0.1 fibers/cc in work locations close to a fiber production operation but only 0.03 fibers/cc at a similar operation in another plant. The rooms containing these two operations were of different sizes and had different ventilation characteristics and somewhat different local ventilation near the emission sources. However, in both cases, workers at locations distant from the sources had half the exposure levels of those working close to them. This pattern was seen across most of the plants sampled. Other data indicated that, historically, one plant had fewer production units (sources of emissions) in one of the work locations. This led to an extrapolated exposure that was proportionally lower prior to the date that the production units were removed. Sampling data were later found that showed that the extrapolated exposure was close to the measured values. The deterministic relationship, fewer emission sources leads to lower exposures, is true without measurements to prove it in every case. One note of caution: this approach focuses on the major factors that affect exposure, but in some cases the smaller factors can combine to affect changes leading to both differences in magnitude and direction of the exposure changes. As a result, the extrapolated values may differ by more than a factor of two from the measured values where there are data to test this, but usually they differ by less than a factor of two. A rich literature has developed on exposure modeling that the interested reader is recommended to investigate.[6]

A good example of exposure modeling is a study by M. E. Davis and colleagues (Davis, Smith et al. 2006). Elemental carbon (EC) exposures of workers in the U.S. trucking industry were measured cross-sectionally with the goal of predicting unmeasured exposure using features of the workplace that would represent sources of exposure and modifying environmental factors for both indoors and outdoors. A three-tiered model was developed based on concurrent measurements: outdoor background levels (upwind); indoor work area levels; and personal exposures indoors. As these were all measured concurrently, it could be observed how background and work area concentrations contributed to personal exposures. Data analysis showed that outdoor contaminants from upwind contributed substantially to indoor concentrations of EC, 50–70%.

3.5. EXPOSURE ASSESSMENT FOR EPIDEMIOLOGY

Up to this point we have been discussing the nature of exposure as it is generally understood by those whose business is the assessment of chemical hazards in the environment. However, epidemiologists generally do not think about exposure in the same way, because their goal is to use "exposure" as an explanatory variable in a health study. This usually means forming groups of subjects in which "exposure" is the same within group and differs among groups. We put the word *exposure* in quotes here to emphasize that the term is often used very loosely to describe one of many different possible characteristics of a group of people, including some that may not have much to do with the concepts presented in Chapter 2 (psychosocial "exposures" such as sexual harassment, for example).

The introductory chapter described the uneasy tension between exposure assessors and epidemiologists that was one of the motivations for writing this book. In the course of designing and conducting an epidemiologic study, it is often the choice of exposure definitions that brings out these differences. The epidemiologist is often much more comfortable than the exposure assessor with very simple, categorical exposure definitions.

Living at a street address within a certain distance from a toxic waste site may be used to define exposure to one or more components of the wastes dumped there, whereas residents of distant communities may be considered unexposed. In the same way, job titles are often used to define exposure in occupational epidemiology. The title "truck driver" may be presumed to have something to do with exposure to diesel exhaust, whereas the job title "bookkeeper" may be considered unexposed to diesel exhaust. It is probably already clear to the reader that there are at least two problems with these exposure definitions. First, street address or job title may not really be associated with the intensity of exposure to the substance of interest (solvents in the waste site or diesel exhaust from a truck, for example). Second, there may be other differences between the "exposed" and "unexposed" groups—other hazards or gradients in underlying health status that account for any observed differences in disease risk. If either of these two situations occurs, then we will make a mistake if we attribute to "exposure" the patterns of disease risk we observe among the groups. Simple definitions can be useful, but we need to verify that they mean what we think they mean.

Despite these limitations, a substantial amount of useful information about environmental disease risks has been gained using qualitative definitions of exposure and surrogates for exposure, such as address, duration of residence at an address, job title, duration in job, or source of drinking water and duration of use.[7] These methods also benefit from simplicity and low cost. A job-title approach to diesel exhaust epidemiology can be accomplished with little or no direct measurements of individual diesel exposures but may rely instead on assumptions about average differences in exposure levels; these may, in fact, be reasonably correct but untested assumptions. If there *are* quantitative data on exposure differences, then the study findings will probably be more interpretable and more convincing. One goal of this book is to provide a rationale and systematic approach for moving beyond studies in which "exposure" is defined qualitatively by the presence of individual attributes such as address or job title and in which exposure is assumed.

Qualitative Exposure Assignments

Questionnaires and interviews are the primary subjective means of characterizing personal attributes and factors affecting population exposures, such as age, gender, race, smoking history, occupational and residential histories, and other personal characteristics. Personal interviews and review of records may be the only means to estimate past exposures. At its simplest, we identify distinct groups that are or are not exposed, that is, that do or do not have some potential contact with an agent ("exposure" = yes or no). Dichotomous yes/no exposure assignments are probably the most widely used approach in epidemiology and are particularly suitable when questionnaires are used to define exposures.

There are, of course, many types of exposures about which study subjects are not aware, and so the simple "yes/no" question cannot be used or will not provide useful information in those cases. The contaminants in our drinking water are not generally known to us. We do not know the pesticides in the bug spray we used for roaches in the kitchen cabinets, but we do know how often we sprayed. Workers rarely know the names of the chemicals they handle. If asked, they may say they used the stuff in the blue can, or they may know the product by a slang term unfamiliar to investigators, but they will know whether they did a task in which a particular chemical was used. A fisherman will not know whether the fish he catches are contaminated with mercury or polychlorinated biphenyls (PCBs), but he can tell you what kind of fish he caught.

Sometimes we are not actually interested in whether or not subjects had *any* exposure but rather in whether they were exposed to a "sufficient" level of exposure to increase risk. What this level might be is often unclear in advance of the study. Imagine, for example, an epidemiologic study of environmental tobacco smoke (ETS). It would be very difficult to find a truly unexposed group in any country in which smoking is widespread, but because it is unlikely that very small amounts of incidental exposure have serious health risks, it probably does not introduce serious bias to allow the "unexposed" group to include people with low levels of casual contact. But at the same time, this easing of the definition of the unexposed group creates another problem: how to set the threshold for what will be considered "sufficient" exposure. This is not an easy problem to solve because it depends on the strength and "shape" of the true underlying exposure-risk relation being investigated, as well as on the exposure distribution in the population being studied.

Questionnaires are developed by including questions that obtain data on factors or situations that are assumed to modify the likelihood of exposure and exposure intensity. However, they are rarely calibrated against measurement data to define the relevance of the information collected. Hammond and coworkers explored how ETS exposure might be documented by questionnaire and how that compared with two biomarkers of ETS exposure: a urinary nicotine metabolite and a hemoglobin adduct (Coghlin, Hammond et al. 1989; Hammond, Coghlin et al. 1993). The investigators did not ask "Are you exposed to ETS?" but instead asked a series of questions about whether the subject lived with a smoker, worked with a smoker, spent time in smoky clubs, and so on (Coghlin, Hammond et al. 1989). An analysis of variance was conducted that identified three questions that explained the most variance in the biomarkers and showed that many of the questions contributed little to resolving differences. Various combinations of those three (yes/no) calibrated questions could identify a range of semiquantitative exposure intensities. They also provide the ability to investigate the effect of altering the yes/no exposure cutoff on the epidemiologic result. It is desirable to be able to test different definitions during data analysis.

There is another issue that complicates the seemly straightforward yes/no exposure definition. The subjects who are determined to be "exposed," by whatever definition, will often display a wide range in actual exposure intensities. If our ETS exposure definition is based on detailed questions of the sort noted earlier, we might decide that our exposed group would include everyone who had experienced at least 1,000 hours of "moderate" exposure to ETS. Setting aside issues of accuracy in measuring this, there will be the additional limitation that this group may include subjects with widely different levels of exposure. There could easily be some subjects with levels of exposure an order of magnitude higher. If ETS is truly a cause of the disease being studied, then it is likely that there will be substantial differences in risk *within* this "exposed" group. As a result, we would expect that another study using the same exposure definition but a different distribution of exposures within the exposed group would have a different risk.

Semiquantitative Measures of Exposure

There are often situations in which it is possible to gather a variety of different types of data that will reflect exposure intensities, even though direct quantification is not possible. In questionnaires, for example, one can ask questions about the frequency of use of a product or the amount of time spent in different environments or activities. These types of data can provide information about important *determinants of exposure* that can either be used directly in epidemiologic studies or used to estimate levels of exposure. Using again the example of ETS, a questionnaire can ask whether a housemate smokes, how frequently she or he smokes, and so on. The choice of questions is based on the investigator's understanding of what factors and situations may lead to exposure and which ones are likely to be most important in the study population. Pilot studies can be very important for weeding out questions that concern activities or behaviors that turn out to be uncommon or that have vague answers in the study population. A pilot study might also include a direct validation in which some quantitative measurements are made, and these are used to formally evaluate an exposure definition based on the questionnaire. As if it were not already difficult to use a questionnaire to accurately assess an exposure, the investigator is almost always limited in the number of questions that can be asked, because long questionnaires are much less likely to be completed than short ones.

Questionnaires that have been calibrated against measurement data are a very powerful tool for inexpensive, semiquantitative personal exposure evaluations.

A related approach is to use measurement data to determine the effects of specific factors on exposure levels and then use this knowledge to sort the population into groups with different exposures. Thus well-calibrated questionnaires can be powerful tools for assessing exposures. A study by Coughlin et al. found that the highest personal exposures to ETS were found in response to the question "Do you commute with someone who smokes?" (Coghlin, Hammond et al. 1989). The small enclosed volume of a car tended to retain the ETS and produce very high personal exposures for the riders. On the other hand, several questions that seemed relevant, such as "being with a few smokers in a social situation," contributed little to defining the exposure. As a general rule, be careful of questions with "reasonable" but untested assumptions because they represent hypotheses about exposure that may be false, and calibration of the questionnaire will discover that fact.

Quantitative Exposure Estimates

In this chapter we have summarized methods to characterize environmental and occupational exposures in a way that is clearly related to our understanding and hypotheses about the biology of the adverse effects under study. Much more will be said about the use of quantitative exposure data in epidemiology in section B of this book, and particularly in Chapter 8. Here we want to make one basic point that is often overlooked in presentations on quantitative exposure assessment: after exposure data have been gathered, they must be summarized in some way to create the exposure variable that will be used in the epidemiologic model. In this book our general term for these summary measures is *dose metrics*. This step, creating the dose metrics as *summary measures of exposure*, also should be guided by a biologically based hypothesis about the form of the true underlying exposure-risk relation. In later parts of the book, we propose a systematic way to formulate these dose metrics.

Consider a simple example: if one is studying electromagnetic field (EMF) exposure and its association with brain cancer, one must decide what properties (frequency, amplitude, peaks, or combination measures) of EMF increase brain cancer risk and then produce dose metrics to represent those particular aspects of the total EMF exposure pattern. This step is often done without much thought or consideration of the possible underlying biology, and one often finds epidemiologic studies in which there is no explicit discussion of why the exposure data have been represented by their mean, or their duration, or the cumulative exposure, to cite three common examples of dose metrics. We show that all dose or summary metrics have implicit biological assumptions about the nature of the adverse effects. Also, using the wrong summary measure of exposure can introduce exposure misclassification that is just as serious a threat to study validity as is using a very crude method of gathering exposure data in the field.

Exposure Measurement Error in Epidemiology

Epidemiologists are very concerned about "error" in the assignment and estimation of exposures for health studies (Zeger, Thomas et al. 2000). Exposure assessors also are concerned about the accuracy and precision of environmental measurements, but there are some important differences in the concerns of these two professions. Epidemiologists are concerned with errors that will reduce the precision or validity of the measure of association between exposure (however measured or estimated) and disease, whereas exposure assessors generally define error more narrowly, as any difference between the "true" exposure and what is measured. This difference is nicely illustrated by a study by Peters and colleagues of adverse respiratory health effects from air pollution among children in 10 communities in the Los Angeles area (Peters, Avol et al. 1999a; Peters, Avol et al. 1999b). Daily exposure to mass of particles <2.5 μm was measured with a central monitoring station in each of the 10 communities. At the outset, there was concern about the adequacy of these central "community" measures (Z_{cent}) for representing the true personal exposure of each child (X_{pers}). A calibration study was performed. Three sources of variability were identified: (1) the difference between Z_{cent} and a local measurement at the neighborhood school (Z_{local}); (2) the difference between Z_{local} and a personal measurement (Z_{pers}); and (3) the difference between Z_{pers} and the true exposure (X_{pers}). Thus the investigators decomposed the error in the central measurement like this:

$$X_{pers} = Z_{cent} + (Z_{local} - Z_{cent}) + (Z_{pers} - Z_{local}) + (X_{pers} - Z_{pers}).$$

Each of these error terms has a mean and variance that were determined separately in the calibration study. If any of the mean differences were not zero, then there would be bias in the next level of measurement. For example, if one of the central stations was located in an area of the community with a high traffic density, then its measurements might be systematically higher than local measurements, with nonzero mean difference ($Z_{local} - Z_{cent}$). All three of these sources of variation would be potentially important sources of bias—error in the epidemiologically measured exposure-risk association—if subjects were to be assigned exposure based on the central measurement. However, the exposure assessor might define only the difference between the measured and the true exposure ($X_{pers} - Z_{pers}$) as measurement error; the others would be sources of variability to be determined by concurrent measurements in each area or setting.

The calibration study found that the magnitude of differences between communities was much larger than the other sources of variation between the measurements within a community (Navidi and Lurmann 1995; Berhane, Gauderman et al. 2004). Thus the investigators concluded that the central station data captured the dominant differences in personal exposure between communities and thus were suitable for the detection of differences in air pollution effects associated with community exposures, even though they might be poor estimates of some individuals' exposure.

Another illustration of the different perspectives on measurement error in exposure assessment and epidemiology is the treatment of complex mixtures. Take, for example, the study of the health effects of particulate matter in the size range less than 2.5 μm in diameter ($PM_{2.5}$) (Su 2003). Exposure assessors and many epidemiologists are aware that there is no reason to assume that $PM_{2.5}$ has a fixed composition wherever it is measured.

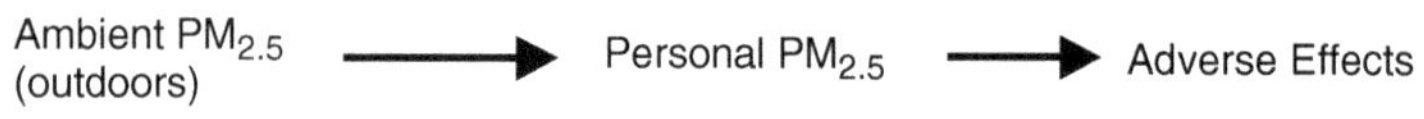

Ambient $PM_{2.5}$ (outdoors) → Personal $PM_{2.5}$ → Adverse Effects

However, epidemiologists and risk assessors often discuss the effects of $PM_{2.5}$ as if "it" were a single, well-characterized environmental contaminant and agent of the effects. When discussing acute mortality effects of daily $PM_{2.5}$ level, a common implicit conceptual model might be represented by the assumption that the observed relationship between ambient $PM_{2.5}$ and acute health effects is directly mediated through personal exposures to the same agent (Ostro, Feng et al. 2008). The correlation between ambient and personal $PM_{2.5}$ levels depends on the amount of time spent outdoors and the degree of penetration of ambient particles into the subject's indoor environment. It is also true that some indoor particles come from additional indoor sources (cooking, mold spores, pet dander) that may not have the same toxicity as the ambient particles. Thus unless the ambient $PM_{2.5}$ is a major fraction of personal $PM_{2.5}$, exposures may have different associations with the health effects because the composition and toxicity of all $PM_{2.5}$ are not the same. Some heterogeneity has been seen in the association of PM_{10} with daily cardiovascular mortality across cities, which has been attributed to the variation in composition (Samet, Dominici et al. 2000).

3.6. SUMMARY OF KEY FEATURES OF EXPOSURE CHARACTERIZATION FOR EPIDEMIOLOGY

This chapter has covered a broad range of topics concerning the characterization of environmental chemical exposures. Our aim was to provide a general, systematic overview and guidance for the design of exposure assessment studies. Following is a summary of the main points of the chapter.

Source-receptor models can usefully describe the relationship between sources of environmental releases and exposure. Sources have defined characteristics, which include composition and concentration of emissions and release rate into environmental media (air, water, surfaces, food/drink, and soil).

Once released, there are environmental transport, removal, and transformation processes that dilute and disperse emissions and modify their concentration, composition, and even their physical form. They also add to the temporal and location variability of concentrations.

Exposure variability applies to both composition (the agents) and an agent's concentration (intensity). An individual's exposure varies over time, often on several different time scales, and usually depends on the individual's pattern of activities and locations. Exposure intensity is quantified by defining the frequency distribution of concentrations at the point of entry across time for an agent or marker. A stable exposure has a stationary concentration distribution across time.

A sampling or measurement strategy is the plan for selecting who, what, where, and when measurements will be collected and should in part be determined by an understanding of the pathophysiologic processes linking exposure to the disease of interest.

Past exposures can be estimated with source-receptor models and historical data on changes in sources and other conditions over time.

If the reader wishes to learn more about these topics, several good textbooks are listed at the end of the book that address them. These will be the basis for the developments in later chapters, especially the modeling work.

NOTES

1. In a well-defined exposure situation, in which the agents are known, the term *exposure* is often used interchangeably with *exposure intensity*. For example, we might say "exposure increased with decreasing distance to the source."
2. A frequency distribution can be tested for normality using the Kolmogorov-Smirnov, G, or chi-square statistics (e.g., see Rosner, B. (2006). *Fundamentals of Biostatistics*, Brooks/Cole Cengage Learning.).
3. Several investigators have examined the problem of estimating the arithmetic mean of a lognormal distribution; for example, see Rappaport, S. M. and S. Selvin (1987). "A method for evaluating the mean exposure from a lognormal distribution." *American Industrial Hygiene Association Journal* **48**(4): 374-379, Seixas, N. S., T. G. Robins, et al. (1988). "The use of geometric and arithmetic mean exposures in occupational epidemiology." *American Journal of Industrial Medicine* **14**(4): 465-77.
4. Basic textbooks on air and water pollution and industrial hygiene have extensive discussions of sampling strategy and the associated statistical issues. For example, see Lippmann, M., B. S. Cohen, et al. (2003). *Environmental Health Science*. New York, Oxford University Press.
5. Retrospective exposure studies include: Heederik, D., H. Pouwels, et al. (1989). "Chronic non-specific lung disease and occupational exposures estimated by means of a job exposure matrix: the Zutphen Study." *International Journal of Epidemiology* **18**(2): 382-9, Loomis, D. P., L. A. Peipins, et al. (1994). "Organization and classification of work history data in industry-wide studies: an application to the electric power industry [see comments]." *American Journal of*

Industrial Medicine **26**(3): 413-25, Kromhout, H., D. P. Loomis, et al. (1995). "Assessment and grouping of occupational magnetic field exposure in five electric utility companies." *Scandinavian Journal of Work, Environment & Health* **21**(1): 43-50, Seixas, N. S. and H. Checkoway (1995). "Exposure assessment in industry specific retrospective occupational epidemiology studies." *Occupational & Environmental Medicine* **52**(10): 625-33, Seixas, N. S., N. J. Heyer, et al. (1997). "Quantification of historical dust exposures in the diatomaceous earth industry." *Annals of Occupational Hygiene* **41**(5): 591-604, Burstyn, I., H. Kromhout, et al. (2000). "Statistical modelling of the determinants of historical exposure to bitumen and polycyclic aromatic hydrocarbons among paving workers." *Annals of Occupational Hygiene* **44**(1): 43-56.

6. Good references for exposure modeling are Boleij, J. S. M., E. Buringh, et al. (1995). *Occupational Hygiene of Chemical and Biological Agents*. Amsterdam, Elsevier, Nieuwenhuijsen, M. (2003). *Exposure Assessment in Occupational and Environmental Epidemiology*. New York, Oxford University Press, Inc.
7. The use of questionnaires and other semiquantitative exposure assessment methods is well presented in Armstrong et al. (1992).

4 Personal Exposure-Tissue Concentration Relationships

Toxic chemicals in the environment do not cause ill health; rather, it is a toxic chemical or its activated form in a target tissue that causes the adverse effects. The chemical has to get into the body and reach the target in an active form to be dangerous. Even for tissues in direct contact with the environment, such as the eyes and skin, the agent must enter the tissue to have an effect. You can swim in water full of toxic metal ions, but as long as you avoid ingesting the water and getting it in your eyes, nose, and ears and you have no breaks in your skin, there will be no absorption and no effects. The processes by which materials enter the body, are distributed to the tissues, and remain there across time have important effects on the risk of tissue damage and disease. It follows that there must be a stronger relationship between the time course of tissue concentration and disease risk than between the environmental exposure and the risk. Investigations of exposure-disease relationships can be more focused by taking advantage of this truism. In this chapter, we address the following questions: (1) What biologic processes determine the time course of a toxic material in the target tissues? (2) How do these processes define the temporal relationships between exposure and tissue concentration? We show how to construct mathematical models as a means of representing the temporal relationships between external exposure and tissue dose.

Although internal physiologic and biochemical processes are complex, you will see that it is possible to assemble useful models from simple components. You may have been introduced to some of these concepts in a toxicology or physiology course; indeed, a highly evolved discipline of physiologically based toxicokinetic (PBTK) modeling has this same objective and meets it admirably (Andersen 2003).

However, we believe that exposure-dose models (toxicokinetic models) that are adequate for epidemiology can be considerably simpler than those used by pharmacologists to study drug metabolism and other clinical applications (Roach 1966; Roach 1977). In some cases in which considerable existing information is available, the simpler approach that we propose can borrow the essential features of the fancier models. We believe that our simpler approach can work for epidemiologic applications because we are not attempting to directly estimate tissue concentrations but to represent their time course, which is somewhat easier. Even simple representations can allow us to describe physiologic relationships and see how they interact across time—which is often difficult to conceptualize—and useful for developing dose metrics.

At the start of an investigation of a human exposure-dose-response relationship for a toxic agent, there is often limited information with which to build a toxicokinetic model. We show later in the chapter that even in that situation, useful simple representations can be constructed through reasoning by analogy and making educated guesses (hypotheses) about likely broad mechanisms of action. Armed with a good guess, we can then use the extensive physiologic data available on the human body and data on the properties of similar chemicals to create an adequate toxicokinetic model. Ah, but, you ask, what if the toxicokinetic model is wrong? Using the wrong dose metric will introduce misclassification and greater error in our epidemiologic analysis. We must test a model against some real data to see how well it fits. Without testing, the model remains no more than a guess or hypothesis about what might be happening.

We begin with an example illustrating how we get from knowledge of an exposure and the physiologic process underlying the body's reaction to this exposure to a toxicokinetic model.

4.1. LEAD POISONING BY SHOTGUN: AN INTRODUCTION TO TOXICOKINETICS

When you take a pill or inhale a toxic material, the drug or toxin is not instantaneously distributed through your body to the target sites. Pharmacologists and toxicologists have long studied the time pattern of blood and tissue concentrations produced when a drug or toxin is introduced into a patient or test animal. This time pattern of changing concentrations is called the time course, which reflects the toxicokinetics of the substance (often called the pharmacokinetics). We can observe the time course by collecting timed samples of blood, exhaled breath, urine, or other biological media and measuring the amount of chemical or of a metabolite in them. These biological samples are often used as biomarkers (see Chapter 5). These time-course data contain important information about the internal processes of drug and toxic material uptake, distribution, biotransformation, tissue storage, and excretion. This is especially true if the input concentration changes with time. A useful mathematical model that mimics the temporal behavior of a toxin in one or more body fluids or tissues can be constructed using simple formulae combined with what is known (or reasonably hypothesized) about the physiological behavior of the agent under study or a chemically similar but better studied material.

Consider what happened when an unfortunate individual was accidentally shot in the shoulder with lead pellets from an associate's shotgun (Gerhardsson, Dahlin et al. 2002). Immediately after the accident, emergency treatment dealt effectively with the wound and risk of infection. Fifty days later, a surgeon performed reconstructive surgery and removed as much of the lead shot as possible, but some was left in place. The victim's physician collected periodic blood samples, which were analyzed for lead (Figure 4.1). Metallic lead (the kind in shotgun pellets but not in paint chips) has a low solubility under physiological conditions, but it will slowly dissolve and enter the bloodstream. The victim's blood levels show a time pattern characteristic of drugs and toxic materials that enter the body in one instant of time (and a shotgun blast to the shoulder is a pretty good example of such an event!). Immediately after the accident there may have been a delay in the increase in blood concentration, but no data were gathered in this period. After 23 days, there was a rapid rise in blood concentration, which slowed briefly after the operation at 51 days but later continued to increase until it peaked at 81 days after the event. Then followed a long, slow decline in blood levels until 1 year after the accident, when blood lead concentration was the same as when first measured. This pattern is typical of an acute exposure, but the timing of the peak and the "washout" phase will vary depending on the personal characteristics of the individual and the specific agent and its behavior under physiologic conditions.

The kinetics of lead transport, excretion, and storage are complex. In this case, additional complexity was introduced by the body's reaction to the pellets—they were probably at least partially sequestered in fibrous tissue, thus reducing the "normal" rate of lead solubilization. Despite this complexity, the temporal change in blood lead over time after the peak can be represented by a simple mathematical model, an exponential decay function, which is typical. This type of removal kinetics is produced when a fixed percentage of the lead is removed per time unit, for example, 20% per month. Eventually the level will stop declining when nearly all of the lead shot has been removed and the amount entering the blood is small relative to other sources, such as dietary lead. Thus, even though the internal processes are complex, the observed time course often can be described with a simple toxicokinetic model.

Why can complex, multistep processes be described with simple models? Often the reason is that the slowest processes determine the overall rate. Nearly everyone has experienced the effect of one slow driver on a narrow highway. Unless you can pass the slow car, you are forced to go at that car's speed, as is everyone else. The same effect occurs for internal processes that must occur in a sequence: A follows B, which follows C, and so forth. For example, some steps in

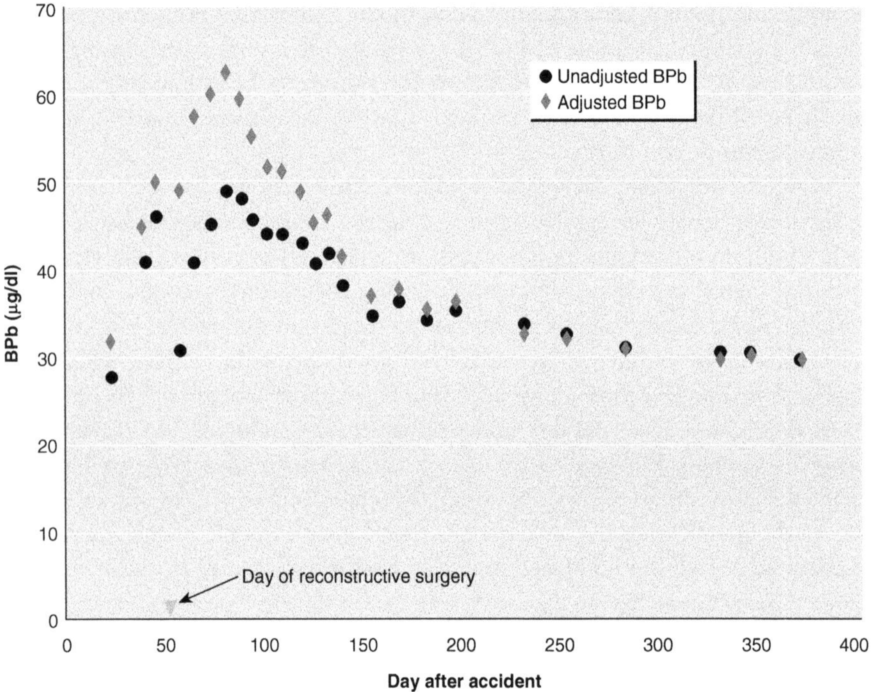

Figure 4.1 Time course of lead in blood following shotgun wound to shoulder (Figure 1 in Gerhardsson, 2002).

metabolic processes in the liver are mediated by efficient enzymes and are very fast, but overall metabolism can proceed only as fast as the slowest enzyme step. This means that we are observing the slowest step when we see a simple time course for concentrations coming from a complex process. When a material is metabolized in the liver, even if the process is fast, the material cannot be metabolized any faster than the rate at which the blood brings it into the liver. In that case, we talk about flow-limited metabolism because the rate is limited by the blood flow and not the enzyme kinetics. Many processes are limited by blood flow to the tissue—excretion into the urine, oxygen delivery to tissues, and skin absorption—and so they may be key rate-limiting steps. The time course of blood concentrations will indicate when there is a limitation; we do not need to determine what it is. We will take advantage of this kind of physiologic knowledge in developing our toxicokinetic dose models.

What specific information about your body's internal processes could we obtain using time-course data if you had a small number of lead pellets implanted under your skin? (Our university institutional review board will be very unhappy with us if you try this experiment at home!) The lead pellets would be solubilized by dissolving into your interstitial fluids, and the dissolved lead would pass into your bloodstream, where it would rapidly bind with components of red blood cells and plasma. The beginning of the solubilization process appears to take some time, reflected in a delay in the rise of the blood concentration. Once the solubilization begins, your blood lead level would begin to rise. Lead in general circulation is removed by urinary excretion and storage: short-term storage in the soft tissues and long-term storage in the bones. During the solubilization process, the blood lead concentration would rise, with the rate of increase depending on the balance between the rates of absorption and removal. The highest blood concentration would occur when the rate of lead entering the blood equaled the rate of removal. Redistribution of lead from soft tissues to bone would also contribute to lead in the blood and its removal. When solubilization neared completion, the input rate would fall to zero, and redistribution and removal would dominate. At this point, the blood lead concentration would decline at a rate characteristic of the combined redistribution and removal processes, continuing its decline until nearly all of the circulating lead was removed. The overall rate of removal could be estimated from the slope of the removal phase (a straight line on logarithmic concentration vs. time plot). Given early time-course data (which was not measured in Gerhardsson et al.'s [2002] study), we could estimate the rate of uptake from the slope of the rising part of the time course, after we have subtracted out the removal processes. From this basic description, you can see that the time-course curve would provide considerable information about your body's processes of lead uptake, distribution, and removal.

What if we repeated this experiment and implanted the same number of lead pellets in 10 other people? You would see that each of them had a similar *shape* of time course, but the highest blood concentration and the time it occurred would, in all likelihood, not be the same, nor would the lag time until some lead was observed in the blood, nor the rate of decline after the highest concentration (the peak). For all of these parameters, there would be some variation among the subjects; not necessarily a large amount, but some variation. The overall similarity in the time trend would occur because of similarities in basic anatomy and the internal physiologic processes across individuals. The variations occur because the detailed processes that combine to produce the broad pattern differ from person to person for reasons of genetic variability and environmental differences such as diet. These individual factors lead to differences in absorption, distribution, and urinary excretion kinetics, for example. But we stress that for epidemiologic studies, it is the common patterns of temporal behavior of a toxin in the body that are most useful; the individual variations on the theme are less important, although when we can define them it can sometimes remove a significant additional amount of variation.

4.2. IT'S BASICALLY JUST PLUMBING: A PRIMER OF TOXICOKINETIC PROCESSES

We want to use toxicokinetic modeling to estimate the time course of internal concentrations, such as blood concentrations. Inside the body there are a hundreds of distinct tissues (compartments) connected by blood flow, and within the tissues many complex processes occur. Because of this complexity the body cannot be truly modeled. However, as we have already noted, there are some important simplifications that we can use to make useful mathematical models. We can lump "similar" tissues together and only distinguish processes and tissues that need to be directly considered to describe what we see in the time course of blood or key tissue concentrations. For example, when considering a substance with low lipid solubility (high water solubility) we do not need to have the body fat as a separate compartment. If the substance of interest is not metabolized, we do not need to have a separate liver compartment. To make the model work, we must include the sum of all tissue volumes, and we must account for the total blood flow (mass balance). We discuss the mathematical models in more detail later in the chapter.

As shown in Figures 4.2–4.5 and 4.7, we can represent the body as a series of tissue compartments (boxes for groups of similar tissues) connected in parallel by blood flow (arrows). Although this may not seem like a simplification looking at the figures, it is. As a result, when we use these mathematical models, we must always check them to be sure that they are giving meaningful estimates for internal levels we can measure, such as blood, exhaled breath, or urine.

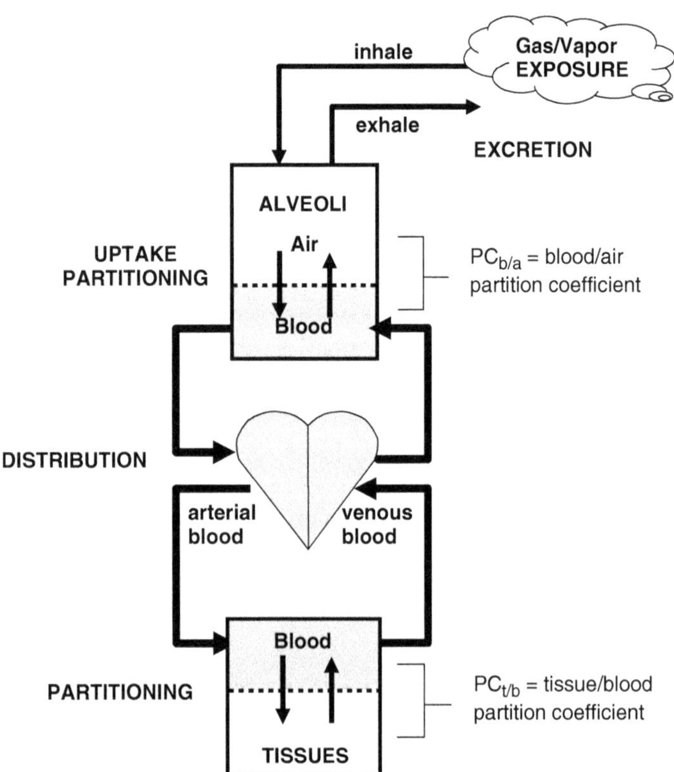

Figure 4.2 Descriptive compartmental model for an inhalation exposure to a gas or vapor with low water solubility, which is not metabolized.

In the next section we begin by describing simple models for the important features of different routes of entry. When making our models, we need to include only the tissues and processes that affect the kinetics. For example, if we are concerned about a respiratory route of entry, we usually do not need to also consider the GI system; its tissues can be lumped together with the other high-blood-flow organs, as shown in Figure 4.3. Different representations are used for different purposes, which are discussed later.

As shown in Figure 4.2, there are a set of five basic processes that affect substances entering the body: uptake, distribution, metabolism, excretion, and storage. We discuss each in turn and then look at some models. If you have little background in physiology, you will find it useful to read about basic physiologic processes in a textbook on human physiology. Alternatively, if you would like to do modeling and have a good background in physiology and math, you may want to delve into a textbook on physiologically based toxicokinetic modeling (see suggested references at the end of the book).

Getting in the Door: Routes of Entry

The route of entry of a toxin is defined largely by chemical and physical characteristics of the agent and the route of exposure, as we noted in the previous chapter. The principal routes are breathing for air contaminants, ingestion for contaminants in food and/or water, and skin absorption for many solids, liquids, and even some gases. Multiple concurrent routes of entry are not unusual. For example, both skin contact and breathing can be routes of entry when a volatile, lipid soluble solvent, such as kerosene, is used on a rag to hand clean bicycle parts. As noted earlier in Chapter 2, an important part of an exposure assessment is a determination of the ways that people in a setting are, or might be, contacting environmental contaminants. The goal of environmental sampling is to determine the composition and concentrations of contaminants at the points of entry and how they vary across time. In some settings, there also can be more exotic routes of entry. For example, toxic materials may be mechanically injected through the skin by high-pressure jets of liquids (or shotgun blasts). Each route of entry has distinctive characteristics that affect the possibility of entry and rate of uptake for chemical hazards and where they enter the bloodstream. The three most common routes of entry, inhalation, ingestion, and skin absorption, are discussed next.

Inhalation

Breathing efficiently transfers air into the lungs, where oxygen moves into arterial blood and carbon dioxide moves out of venous blood. Many toxic gases, vapors, and airborne particles also can efficiently enter the body by this route. Airborne materials have long been a primary concern of industrial hygienists and environmental scientists because of their direct access to arterial blood. However, not every air contaminant can easily enter the bloodstream by this route. The mouth, nose, and airways are lined with watery fluid, and the latter two are also coated with mucus. As a result, low concentrations of water-soluble air contaminants, such as ammonia, will be removed by dissolving in the fluid lining the mouth, nose, and upper airways, where they may irritate tissues in those locations. At high concentrations and during mouth breathing, these gases may penetrate more deeply into the lungs, where they can also damage alveolar tissues and produce irritation and pulmonary edema. Less water-soluble gases and vapors, such as nitrogen dioxide and organic solvent vapors, will penetrate into the alveoli, where they can produce deep lung irritation without upper airway effects. Some gases and vapors, such as ammonia and ozone, will directly attack the respiratory system before entering the bloodstream, whereas others must be transported to their sites of action elsewhere in the body. Gases and vapors absorbed into arterial blood are transported to the tissues. As a result, the organs with rich blood supplies have rapid increases in concentration of the potential toxicant at the onset of exposure. If the chemical is also rapidly cleared by exhalation or biotransformation, then the blood concentration will rapidly decline when exposure stops. A rapid rise and fall is a characteristic time course for slightly water-soluble gases and organic vapors. As we see later, this time course can lead to the rapid onset of some effects, such as anesthesia. Water-soluble gases and vapors, such as ethyl alcohol, are readily absorbed into the blood, but little is exhaled, so they are cleared much more slowly by metabolism.

Particle interactions with the respiratory system are more complex (Figure 4.3). Beginning with the trachea, the airways have a large number of branching points, places where the air must change direction. When rapid changes in air flow direction occur, particles (solid and liquid) that are too large and heavy to follow the air flow will impact on the airway surface. The deposited particles in the airways land on the mucus-covered surface, which is moved by cilia up to the back of the throat, where it and its particle burden may be swallowed (usually unconsciously).

Particle deposition in the airways and alveoli is a function of the aerodynamic diameter of the particles, which depends on their size, density, and shape. Larger spherical particles settle (or fall) through the air faster than small ones. Density is clearly important (spherical lead and Styrofoam particles of the same size will not fall at the same rate); so is shape (feathers fall more slowly than the lead pellets of the same weight). Aerodynamic diameter is defined as the diameter of a spherical particle with density of 1.0 g/mL that settles through the air at the same rate as the particle of interest. A second way to conceptualize aerodynamic diameter is to use a looser but approximately correct definition by asking the question "What is the diameter of a unit density spherical particle that behaves in the same way as the particle I am looking at?"

The largest particles (aerodynamic diameter 10–100 μm) mostly impact in the nose, mouth, and throat. Particles in the range of about 4 to 10 μm are deposited

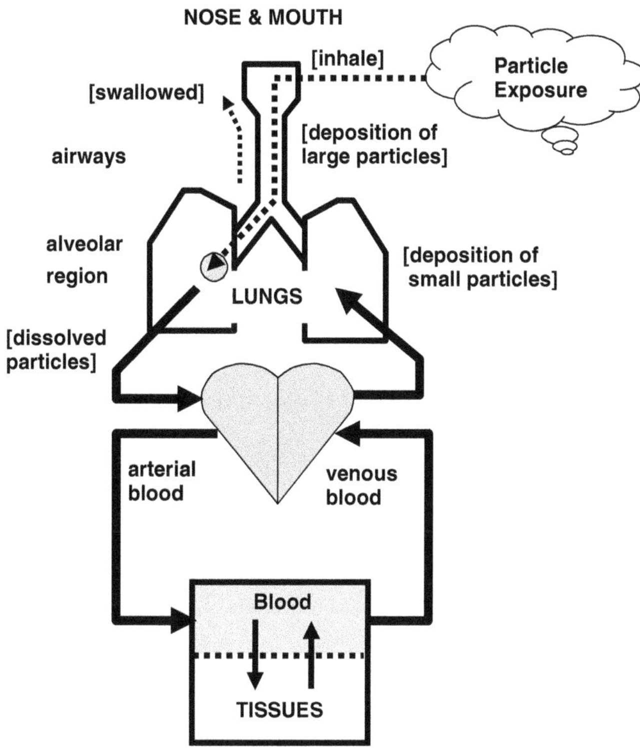

Figure 4.3 Descriptive compartmental model for inhalation route of entry for particulate air contaminants showing the different pathways for deposition and removal by particle size.

in the upper airways (nose, mouth, large bronchi, and trachea). Soluble materials in this range may be absorbed by tissues at the site of deposition, or they may be moved to enter the GI tract, where they may be absorbed. Finally, particles less than about 4 μm are deposited in the small airways (bronchioles) and alveoli, where soluble substances may pass rapidly into the bloodstream. Less soluble particles are taken up by alveolar macrophages, where they are exposed to digestive enzymes and acidic conditions in the lysosomes. Acid-soluble materials, such as some metal oxides like lead oxide, can be solubilized and released to the bloodstream. Insoluble particles are also usually taken up by macrophages and removed from the lungs to lymph nodes or into the airways and removed by the mucus transport process. However, some insoluble particles may penetrate through the alveolar epithelium and be lodged in the interstitial area of the lung tissues where they can cause chronic effects, such as silicosis.

In summary, only certain types of gases, vapors, and particles can penetrate into the respiratory system, and they may be deposited in different locations depending on their chemical and physical characteristics. The location of an adverse

Table 4.1 Health effects in different locations of the respiratory tract tend to be caused by different types of particles and gases

Respiratory Tract Region	Particles	Gases
Nose, throat	Very large >10 μm	Water soluble
Upper airways	Large 4-10 μm	Water soluble
Lower airways & alveoli	Small < 4 μm	Low water solubility

Note that high concentrations and heavy breathing by mouth (vigorous exercise) may allow some large particles and water-soluble gases into the lower respiratory tract.

respiratory system response can be a useful clue about the type of air contaminants that may be causing the problem (Table 4.1).

Ingestion

There are more barriers to prevent ingested materials from entering the bloodstream than there are to prevent inhaled materials. Toxic materials must be solubilized and brought into the blood or tissues to be a hazard; otherwise, they pass through into the feces without effects. Materials with very low solubility under GI tract conditions are unlikely to be a hazard by ingestion. Arsenic sulfide is a good example of this. Despite the dramatic toxicity of arsenicals in general, the sulfide is quite insoluble under physiologic conditions, and so it is relatively nontoxic. However, some materials, such as toxic heavy metals and pesticides, are readily taken up and enter the circulation by this route. Ingested toxic materials must be stable when exposed to strong acids in the stomach, and they must be solubilized if they are in solid form, although nanoparticles can be taken up by cells lining the GI tract. Once absorbed into the bloodstream, these materials travel by the portal circulation to the liver before entering general circulation, which provides an opportunity for attack by metabolizing enzymes in the liver, where some substances are activated (e.g. converted to epoxides) or detoxified (Figure 4.4). Ingested materials that are detoxified in the liver will be more rapidly broken down than the same toxins entering the circulation from the lungs or skin would be, because only about 25% of the peripheral blood flow passes through the liver, which means there is more time for target tissue contact. After passing through the liver, the material can enter the venous circulation, where it is diluted by mixing with venous blood from the other tissues. Then it passes through the lungs, where it may be exhaled or further metabolized before it passes through the tissues. Ingested toxic materials will enter arterial circulation more slowly than inhaled materials, and ingestion produces lower blood levels of toxic material with the same absorbed dose.

We noted earlier that inhaled large particles (approximately 4–10 μm diameter) are deposited in the airways and cleared by mucociliary transport into the

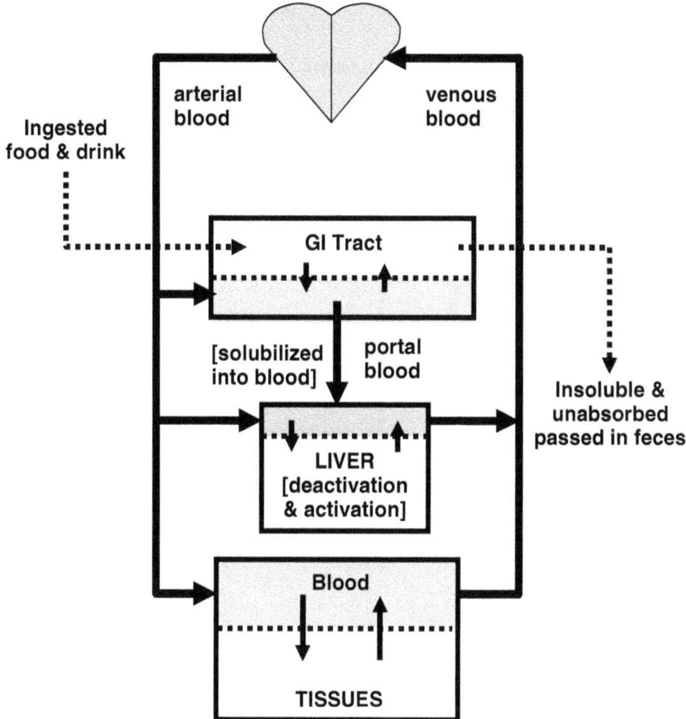

Figure 4.4 Descriptive compartmental model for entry through ingestion. This model also applies to large particles deposited in the airways that are brought to the back of the throat and swallowed.

GI tract (Figure 4.4). If the toxic materials can be solubilized or absorbed within the GI tract, then a toxic hazard may exist. For example, this route of entry is important for airborne lead exposures. A person removing lead paint in an old house with a grinder can produce high dust concentrations.

Skin Absorption

The types of materials that may enter by skin absorption are even more limited than those that enter by ingestion. However, there are a number of important lipid-soluble toxic materials that are hazardous by this route.

Materials may be deposited on the skin by direct contact with liquids or solids or by absorption of gases or vapors. Airborne materials, such as liquid droplets splashed or sprayed into the air, can be deposited on skin. Materials entering by skin absorption move by diffusion through the stratum corneum, which is the principal barrier (Figure 4.5). Very small particles (< 0.05 μm dia.), called nanoparticles, can also enter the skin by mechanical penetration. The rate of absorption

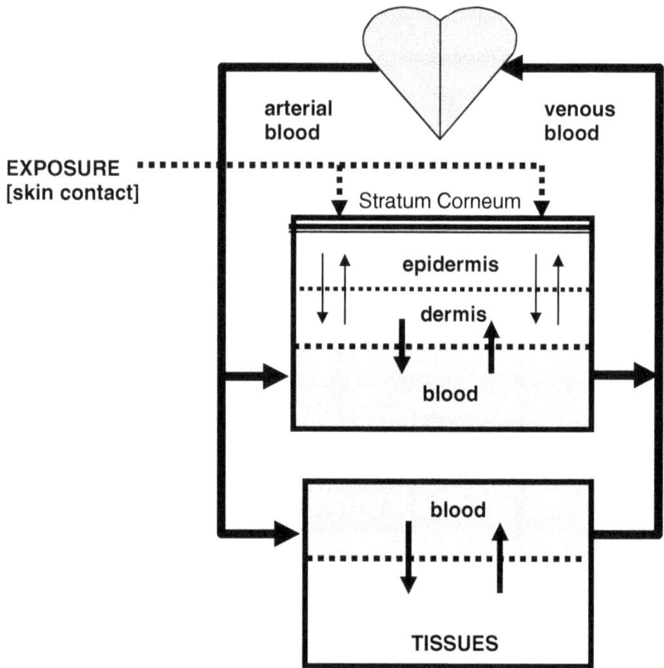

Figure 4.5 Descriptive compartmental model for skin exposure to a gas, liquid, or solid.

depends on the condition and thickness of the stratum corneum and the blood supply to the underlying tissues, and it may be facilitated by the presence of some solvents, such as dimethyl sulfoxide (DMSO). Small lipid-soluble molecules such as benzene and some pesticides penetrate most easily. Fat in the skin can be a reservoir of absorbed lipid-soluble material. Some types of water-soluble materials will also diffuse through the skin, especially if the skin is well hydrated, where hydration is the process of absorbing water. You may have noticed how your skin swells and becomes wrinkly when you have your hands in water for a long time; those are the effects of hydration. If the water contained suitable toxic chemicals, after hydration your skin would absorb much more than when your unhydrated skin was newly immersed. Physical damage such as cuts and abrasions will increase penetration and may allow materials in via the skin that would not normally be hazards. Allergenic materials penetrating the outer layers can interact with dendritic cells in the skin to produce sensitization and dermatitis (allergenic responses are discussed in more detail in Chapter 13).

After passing through the skin, absorbed materials enter the venous circulation, where they are diluted by mixing with blood from other tissues. They must also pass through the lungs before entering arterial circulation, where some may be exhaled. Because of the skin's barrier properties, uptake from skin exposure is usually much slower than uptake from inhalation exposure. However, contamination of the skin may prolong exposure even after the offending activity has ceased, as can occur when people do not shower after applying pesticides and continue wearing contaminated work clothes, for example.

Combined Routes of Entry

Sometimes toxic materials enter the body by more than one route. Some common examples are:

1. Skin contamination of the hands can contaminate food and lead to entry by ingestion, in addition to skin absorption. People who do not wash their hands after handling toxic materials may be exposed by this route. A television commercial about changing one's own motor oil shows an individual watching television with a sandwich made of white bread on the table next to him. He picks up the sandwich and takes a bite. When he replaces the sandwich on the plate, there is a black handprint on the bread. The announcer says, "A job well done." But as used motor oil contains carcinogens and other toxic materials, there could be significant risk from eating that contaminated sandwich (it would also probably not taste very good).
2. Transfer of hand contamination onto cigarettes or cigars can lead to significant respiratory contact with substances unaffected by combustion temperatures, for metals such as lead and cadmium and some plastics. For example, cadmium workers who smoke cigarettes have been shown to have higher urine cadmium levels than nonsmokers in the same jobs.
3. Skin contamination with volatile substances can lead to a secondary inhalation exposure in addition to skin absorption. This commonly occurs when using solvents for manual cleaning, such as cleaning off tar or paint with solvents. The relative quantities of material entering by skin versus inhalation routes depend on the relative rate of skin uptake compared with the concentration inhaled. For highly volatile materials the inhalation exposure may substantially exceed the skin uptake.

Moving Around Inside: Internal Transport Processes

Once a toxic substance passes through a route of entry and enters the blood, internal processes lead to its distribution, biotransformation, storage, and excretion and produce a target tissue concentration.

Blood and Lymph Transportation

Blood circulation is the primary means of transportation among the tissues for most toxic materials. Materials that enter through the alveoli of the lungs go directly into the arterial blood, whereas materials that enter through other routes of entry go into the venous blood. Lipid-soluble materials are transported in blood lipids. In the lungs and the GI tract, some particles and lipid-soluble chemicals are also transported through the lymphatic system. Particles can be trapped in lymph nodes.

Biotransformation: Activation and Detoxification

Biotransformation processes enzymatically convert a chemical from one form to another by adding, modifying, or removing functional groups, such as hydroxyl (-OH), amine (-NH_3), sulfhydryl (-SH), or acid (-COOH) groups. Some biotransformations produce more reactive compounds that are more toxic than the parent, such as the conversion of alcohols to aldehydes, and are called "activating." Others modify toxic groups, such as hydrolyzing epoxides into alcohols, making them less toxic, and those are called "detoxifying". These processes are often complex and can differ widely for different toxic materials (interested readers should consult standard texts on toxicology, such as *Casarett and Doull's Toxicology*; (Klaassen 2001)). Metabolism of toxic materials is often a two-step process: first, depending on their structure, chemicals may be oxidized (hydrogens removed; e.g., alcohol dehydrogenase: -CH_2OH $\rightarrow$ -CHO), reduced (hydrogens added; e.g., disulfide reduction: -CH_2SSCH_2- $\rightarrow$ two -CH_2SH) or hydrolyzed (water added; e.g., epoxide hydrolase: -CH-O-CH_2 $\rightarrow$ -$CHOHCH_2OH$). Second, they are conjugated (reacted) with glutathione or sulfate or other water-soluble materials so they may leave the body more readily. Cytochrome P-450 enzymes are the most important family of first step biotransformation enzymes. Glutathione-S-transferases are an important family of second-step detoxification enzymes. If the toxic agent is formed by activation of an environmental contaminant, there is a characteristic time course during which the concentration of the activated form in the blood lags behind that of the parent compound. The relative rates of activation and detoxification are both very important, especially relative to each other, because they determine the concentrations that may be achieved. The worst case for an individual is to have rapid activation enzymes and slow detoxification enzymes, which can lead to the formation of large amounts of activated material that are only slowly removed by detoxification.

Storage

Substances that bind tightly to tissues, or are incorporated into bone, or are highly lipid soluble and stored in body fat may be retained in the body for long periods. For example, cadmium is tightly bound to cell proteins in the kidneys, where it will accumulate until there is cell damage. Bone-seeking elements, such as lead and radium, are integrated into the matrix of the bone's structure. Because the rates of bone turnover are normally very low, little is released unless there is some increase in bone remodeling such as occurs during pregnancy and some disease states. Storage in body fat is passive (dissolved in the fat without binding) and strongly determined by fat solubility and metabolism. Long-term body fat storage is most important for substances with low biotransformation rates that make them more water soluble (like the PCBs mentioned earlier) and that limit the rate of removal. High fat solubility alone is not sufficient to define long-term retention. Fat-stored materials may be mobilized and reintroduced into the bloodstream at an elevated rate during illness or weight loss which mobilizes the body fat.

Excretion

Toxic materials and their metabolites can leave the body in a variety of ways. The more water-soluble materials can be readily eliminated in urine. Materials that can be inhaled can often be eliminated by exhalation if their blood solubility is not too high, because highly soluble materials produce low breath concentrations, so little is lost by exhalation. Excretion by sweat glands and biliary excretion to the feces are also important routes of removal for some substances. Small amounts of some materials, such as arsenic and mercury or reactive chemicals, are removed by binding with sulfur groups in proteins in hair and in fingernails and toenails, which eventually separate from the body. This is not important for removal of these toxicants from the body because only a small fraction of the circulating substance is bound, but it does provide a readily accessible long-term biomarker of the amount that was circulating in the blood.

These are admittedly quick sketches of the many complexities of the physiologic processing of foreign substances. A basic textbook on toxicology and human physiology can fill out the picture considerably, and we recommend that you have such a book in hand before embarking on your first effort to understand how chemicals are processed within the body. Useful help can also be found in reference texts of physiologic data. In addition, the International Commission on Radiological Protection (ICRP) has developed extensive data on the characteristics of the human body: highly detailed estimates of tissue blood flow rates; respiratory gas exchange, kidney clearance, bile production, and sweat and urine flow rates under various conditions of temperature and exercise; and water and food intake rates (ICRP 1975). Other data have been developed by the U.S. Environmental Protection Agency (EPA; 1994) and other government agencies (EPA 1994; ILSI 1994).

4.3. BOXES AND ARROWS: HOW TO BUILD YOUR OWN DESCRIPTIVE PHYSIOLOGIC MODELS

Now that we have reviewed the basic toxicokinetic processes, we can take the next step in modeling by making simple descriptive physiologic models. These are composed of boxes and arrows, representing tissues and blood flows, respectively. Later, we turn these descriptive models into mathematical ones that can be tested for their fit to time-course data, which will allow us to decide how useful the box-and-arrow version is.

A useful representation of the body is to think of it as discrete organs and tissues connected by the circulatory system, which provides oxygen and nutrients to tissues and concurrently takes away carbon dioxide and wastes. The key elements of this exchange and transport system are shown in Figure 4.2. For example, it shows how blood picks up toxic gases and vapors in the lungs and transports them to and from the tissues. In physiologic modeling we identify groups of similar tissues as "compartments" that are connected by the circulatory system. Our goal with descriptive modeling is to represent the physiologic relationship between the target tissue and the external exposure by initially diagramming how materials enter and are taken up, transported to the tissues, metabolized, stored, and excreted, emphasizing the target tissues in which primary effects caused by the agent occur. We want to capture all of the features and processes that may significantly affect the time course of the agent's concentration in the target tissues, and ignore those that have only small or no effects.[1] Although there are some common features, usually these models are specific to the agent, its metabolites, and the target tissues. If the internal processes are not known, we begin with one or more plausible descriptions that can be the basis of hypotheses about what happens.

The basic components of a model are (1) one or more routes of entry into the body, (2) tissue compartments, and (3) one or more routes of removal (very few substances are completely retained). We can diagram the basic physiological relationships among the tissues by indicating the tissues as a set of compartments (boxes) with blood flow shown by arrows that connect the boxes and entry and removal processes shown by arrows entering or leaving boxes, as shown in Figure 4.2. The level of detail in a model will depend on the application and information available about the exposure, the chemical, the target tissue, and the level of detail in any time-course data. This descriptive model will be later converted into a mathematical model; we are setting the form through this descriptive process.

Chemical Partitioning between Media

One of the important basic physiologic processes noted earlier for defining tissue concentrations is the process of partitioning, by which a portion of a chemical in the blood or tissues diffuses, or is transported between one medium and the other. Two common partitioning situations are shown in Figure 4.2: agents moving between inhaled air and blood and agents moving between blood and tissues. Chemicals will diffuse from regions with higher concentrations to regions with lower concentrations unless there is an active transport process to move them against the gradient. For example, when air containing ammonia comes into contact with your eyes,

Box 4.1 Basic Physiologic and Toxicokinetic Background

As shown in Figures 4.2–4.5 and 4.7, we can represent the body as a series of tissue compartments (boxes) connected by blood flow (arrows) that carry materials between them. The basic principle of these models is conservation of mass (volume) and blood flow. A person's weight can be subdivided into compartments (volume = weight × density). There are scaling functions that define tissue size as a fraction of the total body weight, such as shown in Tables 4.2 and 4.3. Similarly, total cardiac blood flow can be subdivided by each compartment's relative share, also shown in Tables 4.2 and 4.3. Different representations are used for different purposes. For example, Figures 4.3–4.5 show representations of the three point-of-entry tissues, which can be used when the point of entry is defined. Models are constructed based on the lipid solubility of the toxic substance, the target tissue, and point-of-entry tissue.

Compartments can be single organs, such as the liver, lungs, or groups of tissues. A compartment will have three important characteristics: tissue volume, blood flow, and lipid content. Separate tissues can be grouped together if they have similar perfusion (blood flow/volume) and lipid content. The combined tissue groups will have a volume and blood flow each equal to the total of all components. For example, all of the organs can be grouped into a well-perfused (WP) compartment. The muscles and skin can be grouped into a poorly perfused (PP) compartment. Body fat is not a well-defined tissue but is the remainder when you subtract WP, PP, and bone weights from the total body weight.

As shown in Figure 4.2, there are a set of basic processes that affect substances entering the body: uptake, distribution, metabolism, excretion, and storage.

The initial step is uptake, in which the substance is actively or passively (by diffusion) taken into the route-of-entry tissues. Uptake by tissues from blood is a function of relative solubility in the tissues versus the blood, indicated by the partition coefficient, and sometimes active transport, such as heavy metals transported out of the gut. From the entry tissue, the substance may enter the bloodstream to be transported to other tissues in the body. Some substances will reversibly bind to proteins or other blood components, which will slow the movement (partitioning) out of the blood into tissues.

In the liver, but also in some other tissues, there can be metabolic enzymes that will activate nontoxic substances (Phase I enzymes), such as P450 enzymes oxidize PAHs to reactive epoxides, and others (Phase II enzymes) that will deactivate reactive substances, such as glutathione-S-transferases and epoxide hydroxylases. In some cases, one set of enzymes activates while another deactivates the products of the first set. For many of these enzymes, individuals with different genetic backgrounds have different forms (the enzymes are polymorphic) that have different intrinsic activities, which epidemiologic studies have shown to affect susceptibility to an adverse effect, such as cancer risk.

Figures 4.3–4.5 show simplified tissue and transport representations of the three main routes of entry. The figures show that materials entering by the different routes vary in the amount of dilution and metabolism in the liver they may experience before being transported by the arterial blood throughout the body: entering by inhalation, substances go directly into arterial blood; entering by ingestion, substances have to go through the liver, where they may be metabolized (activated or deactivated) and then mixed into venous blood; entering by skin absorption, substances go into venous blood, leaving the skin and mixing with the venous blood from other parts of the body before entering the arterial blood. These differences affect the timing of how rapidly the target tissue concentration rises and the production and removal of active substances.

There are several routes of excretion/removal from the body: exhaled breath, urinary, biliary/fecal, and minor routes. Each of these has its own characteristics and types of substances it will process. Exhaled breath is important for small, lipid-soluble molecules, such as solvents like benzene and trichloroethylene. Urinary excretion is important for small to medium molecular weight, water-soluble substances, such as metals, and Phase II metabolites of many organic substances. Biliary/fecal excretion is important for high molecular weight, lipid-soluble materials such as PAHs and PCBs.

Storage is important for tissues with slow turnover, such as fat and bones. High molecular weight, lipid-soluble materials, such as PAHs and PCBs with limited metabolism, can be sequestered in body fat for very long time periods. Similarly, substances that are retained in the bone matrix, such as lead and radioactive elements, can also be retained for long periods. The substances stored in these locations are not bound in place; there is slow exchange with the blood that may lead to losses from the body. As a result, changes in a person's physiology, such as pregnancy or weight loss, can mobilize these stored materials and produce relatively high blood levels.

The overall toxicokinetics of a substance, such as its retention and metabolism, are a function of its properties and interactions in the body. Therefore, it is very important that you determine what is known about a substance to understand how it is handled in the body. Additional, more detailed background can be obtained from textbooks on toxicology, such as *Casarett and Doull's Toxicology* (Klaassen 2001), and specific information on hazards and properties of specific substances can be obtained from comprehensive databases on substance toxicity, such as the EPA's Integrated Risk Information System (IRIS) and many other on-line databases.

the ammonia partitions between the air and the watery fluid on the surface of your eye (Figure 4.6). The ammonia very rapidly diffuses into the surface fluid until it reaches a steady state in which the diffusion rate into the fluid equals the diffusion rate out. Because ammonia is very water soluble and it ionizes when in water (the ionized form is not volatile), it preferentially goes into the eye fluid. The fluid has only a small volume, and as a result the concentration of ammonia in the eye fluid can become quite high very rapidly. This high concentration is what activates irritant receptors on the surface of the cornea, leading to pain, tearing, and avoidance behavior. Consider a simple thought experiment: if we had a small cup that we could seal over the eye and introduce a low concentration of ammonia into the air inside the cup, we could measure the partitioning process between the eye surface and the air (please do *not* try this—your human subjects committee will never approve it!). We would observe that the ammonia was very rapidly absorbed such that the remaining concentration in the air would be much less than that in the eye fluid. The fluid on the eye also has some buffering capacity (to neutralize the strong base, ammonium hydroxide), which will also increase the amount of ammonia dissolved. If we did a new experiment and put a small amount of ozone in the eye cup, we would observe that after a short time period, the ozone concentration in the air would be nearly unchanged, and the concentration of ozone in the eye fluid would be much less than in the air because ozone has a low water solubility, so little dissolves. The steady-state concentration ratio between eye fluid and air is called the *eye fluid-air partition coefficient* ($PC_{eye/air}$). Partition coefficients can be calculated for any two distinct media, such as air and water, or tissue and blood, or adjacent tissues. They are expressed as ratios of the steady-state concentrations in the two media. The ammonia has a large $PC_{eye/air}$, whereas the ozone has a small $PC_{eye/air}$, with a value less than 1. Thus even if ammonia and ozone were equally irritating per µg/L in eye fluid, a much higher air concentration of ozone would be needed to produce the same degree of eye irritation as a given air level of ammonia. Figure 4.6 also shows how we can simplify the complexity of the eye into a one-compartment model that captures the kinetics of the exposure situation, assuming the irritant sensors are on the surface of the eye. If we hypothesize sensors within the eye, then we need a more complex model, and the simple model will not fit the data.

The partition coefficients that we use most frequently are the blood-air partition coefficient ($PC_{b/a}$) for chemicals that reach the gas exchange area of the lungs and the tissue-blood partition coefficient ($PC_{t/b}$) for a material in the blood passing through a given tissue (Figure 4.2). Some examples of partition coefficients for common gases and vapors are shown in Table 4.2. If there is reversible protein or other binding in the blood or tissues, it can have a dramatic effect on the partitioning process. For example, oxygen has only a limited solubility in water, so without hemoglobin blood would not be sufficient to support tissue needs for oxygen. Oxygen reversibly binds to hemoglobin, and the hemoglobin capacity depends on the oxygen concentration. Large amounts of oxygen can be taken up by the blood in the lungs, where the concentration is high, but upon reaching the tissues where the oxygen concentration is low the oxygen is released. Carbon monoxide (CO) binds to hemoglobin more tightly than oxygen does, so it blocks oxygen uptake.

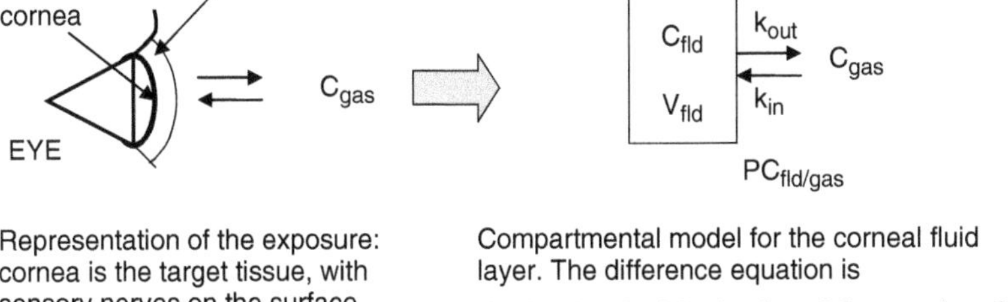

Representation of the exposure: cornea is the target tissue, with sensory nerves on the surface.

Compartmental model for the corneal fluid layer. The difference equation is

$$C_{fld}[i+1] = C_{fld}[i] + k_{in}\, C_{gas}\, PC_{fld/gas} - k_{out}\, C_{fld}[i]$$

where k_{in} and k_{out} are gas exchange constants.

Figure 4.6 Simple one-compartment model for tissues in contact with the external environment. This illustrates the application for eye exposures to gases with varying water solubility, such as ammonia and chlorine, or even low-solubility gases, such as ozone. A similar model can be used for mucous membranes in the nose or mouth or for skin effects from direct exposure.

Table 4.2 General physiologic parameters for the standard man.

Physiologic Component	Physiologic Processes	Volume (V, Male)	Air and Blood Flow (Q, Male)	Perfusion(Q/V)	Partition Coefficient (~lipid content)	
					Toluene	Butadiene
Body Height (Ht)		170 cm				
Body Weight (BW)		70 kg				
Lean Body Mass[a] (without body fat)		55 kg				
Total Blood Flow (cardiac output, Q_{art})	Distribution		**6.0 L/min at rest** ($0.024*BW^{0.425}Ht^{0.725}$)		$PC_{b/a}$ **15**	**1.5**
Minute Ventilation	Total air flow		**6.0 L/min** (~1.25 of alv vent.)			
Aveolar Ventilation	Gas exchange		**4.8 L/min at rest** ($0.019*BW^{0.425}Ht^{0.725}$)			
Well-Perfused Group (Target organs)	Metabolism, excretion, digestion, reproduction, mental work, endocrine, etc.	Percentages of LBM[a,b] **20%**	Percentage of Q_{art} **70%**	Fraction flushed per minute **~0.4 min^{-1}**	$PC_{t/b}$ **3** Eff. Perfusion[c] 0.13	 **0.3** 1.3
Poorly Perfused Group (Muscles & skin)	Movement, work Temperature control	Percentages of LBM[a] **49%**	Percentage of Q_{art} **20%**	Fraction flushed per minute **~0.04 min^{-1}**	$PC_{t/b}$ **~1.4** Eff. Perfusion 0.03	 **0.25** 0.16

Very Poorly Perfused Group (Fat)	Energy storage Some metabolism	(BW-LBM) Percentage of BW **25%**	Percentage of Q_{art} **5%**	Fraction flushed per minute **~0.02 min^{-1}**	$PC_{t/b}$ **64** **7.2** Eff. Perfusion 0.0003 0.003

Notes:

a. Lean body mass (LBM) is calculated based on gender, height, and weight. For young adult males: LBM = (-12.98 +0.1757*Ht +0.3331*BW)/0.732. Clearly, body build varies across individuals. Obese (BMI>30) and heavily muscled individuals will be poorly fit by this equation.

b. The density of most tissues, excluding bone, is ~1 kg/L; volumes can be estimated directly from the tissue mass. Organ size and tissue compartment size are variable fractions of BW because of variations in body fat. Therefore, estimating organ volume using an estimate of lean body mass (LBM) gives more accurate estimates than using body weight (BW; Fiserova-Bergerova 1995).

c. Eff. Perfusion = the effective perfusion which is estimated by dividing the perfusion (Q/V) by $PC_{t/b}$. This index indicates how much the apparent volume is changed by the solubility of the agent in the tissues relative to the blood.

CO that is bound to hemoglobin also changes the hemoglobin dissociation curve for oxygen's release in the tissues. Consequently, a relatively small amount of CO exposure can have deadly effects on oxygen delivery to the tissues. Acidic or basic gases and vapors that are dissociated at physiologic pH also have high partition coefficients from air into blood because the blood is buffered, which limits the change in pH and maintains the dissociated ionic forms of acids and bases that are not volatile. Partitioning is very important for the uptake and internal physiologic processes affecting environmental chemicals.

To give you a better feeling for how the partitioning process links exposure and a tissue concentration, look again at the simple model in Figure 4.2. You can see that at the start of exposure, an inhaled gas would partition into arterial blood, which is then pumped to the tissues by the heart. On reaching the tissues, the gas partitions out of the blood by diffusion because the tissue concentration of gas is low. After a sufficiently long period of exposure, the rate of transport of gas into and out of the tissues would be the same, because both the blood and tissue concentrations would have reached their *steady-state* values. Note that although the concentrations do not change, diffusion is still occurring, because steady state is a dynamic condition. Also the air, blood, and tissue concentrations would not be equal unless the partition coefficients were all equal to 1. If we stop exposure, then venous blood will be stripped of most of its gas by partitioning back into clean inhaled air, arterial blood will contain little of the gas, and in the tissues the retained gas will be removed by partitioning back into venous blood. The direction of movement and the transfer rate for an air contaminant gas into or out of the blood and tissues depends on the concentration gradient.

All cells in an organ or tissue also are not equally affected by a toxic agent in the blood. Even when a group of cells is uniformly exposed to a lethal concentration in a petri dish, there is variability in the time to cell death. In the same way, when cells in a tissue are exposed to a moderate toxic concentration of an agent, all of them are not equally affected because of microscopic variations in concentration and local variations in their sensitivity. Given a steady blood concentration, there is initially a concentration gradient; the concentration is highest for the cells closest to a capillary entering the tissue and lowest for those most distant from a capillary exiting the tissue. Only after the tissue has come to steady state will all cells experience the same concentration. Because environmental exposures are rarely steady, the incoming blood concentration will generally fluctuate with variations in exposure, and the tissues will not reach steady-state stability, so the cells do not have the same probability of effects.

Compartment Characteristics: Inside the Boxes

A compartment (a box in our cartoon models) may be a single organ, such as the brain or liver, or a group of tissues with similar properties, such as the vessel-rich group that includes all tissues with large blood flows per volume of tissue. Tissues have three general properties and two important characteristics relative to a specific toxic chemical:

1. General physiologic characteristics
   a. Volume (L)
   b. Blood flow (L/min)
   c. Perfusion (blood flow/volume; min^{-1})
2. Tissue-specific characteristics
   a. Partition coefficient, $PC_{t/b}$ (relative solubility [and binding] of the chemical in tissue versus blood)
   b. Susceptibility to the chemical or tissue-specific potency of the toxin.

Additionally, tissues may also have specific physiologic processes, such as gas exchange, excretion, biotransformation, and so forth, that affect response to an agent.

A large value of $PC_{t/b}$ means that at steady state, the tissue concentration is large relative to the blood. For example, lipid-soluble materials are highly soluble in fatty tissues relative to blood and have large tissue-blood partition coefficients for fatty tissues; for example, styrene has a $PC_{t/b}$ of 60 for fat relative to blood. The partition coefficient is independent of the concentration level of the environmental chemical because it represents the *ratio* of concentrations, which will be the same whether the blood concentration is high or low. Chemical solubility, binding, and tissue volume define the total quantity of chemical that can be retained in a tissue. The effective volume of a tissue, the product of $PC_{t/b}$ and the volume, is the volume the tissue would have if the partition coefficient was 1.0. We will see later that this rather abstract quantity is useful to simplify calculations.

Diffusion of small molecules over short distances between tissues and blood in capillaries is generally faster than the blood flow rate through the tissues. As a result, the rate at which a chemical's concentration can change in a tissue depends primarily on the tissue's relative blood flow—its perfusion—which is determined by the blood flow rate divided by the tissue volume. The perfusion represents the fraction of the tissue's volume flushed with blood per minute. Critical organs that are often target tissues, such as the brain, kidneys, lungs, and liver, have high blood flow per volume and high perfusions, whereas storage tissues such as body fat and bone matrix have low blood flow per volume and low perfusions. In poorly perfused tissues such as body fat, diffusion may be as fast as or faster than perfusion. As a result, it is sometimes useful to divide the body fat into two or more compartments (very slow and very, very slow perfusion) or to use a different form of model. In tissues with high perfusions, the concentration can change rapidly after a change in blood concentration, whereas in low perfusion tissues, concentration change takes relatively much longer because there is only a small blood flow relative to the total volume. The low perfusion tissues generally have little effect on the time course of concentrations in the organs (often target tissues), which is important when we want to simplify our model.

Do tissues with high capacity for a toxin always clear slowly? Not necessarily. Both the brain and body fat have a high capacity for storing lipid-soluble compounds such as anesthetic gases. However, given the same arterial concentration entering both, when administration of an anesthetic is started, the brain will rapidly approach its steady state with the large flow of arterial blood per unit of tissue, whereas the body fat will only slowly accumulate anesthetic because of its much lower perfusion. As noted earlier, even though all of the anesthetic is stripped out the blood flowing through the fat, the amount of blood flow per unit of fat is small, and it is distributed throughout the relatively large fat compartment so its concentration in the fat is low. When an anesthetic exposure has a long duration and a large amount has accumulated in body fat, will this produce a high blood concentration after exposure stops? No, because the low fraction of total blood flow passing through the fat is highly diluted when it mixes with the venous blood from other tissues that have more rapidly cleared; the anesthetic in the body fat will leave the same way it entered, a little at a time. A useful rule of thumb for tissues is "fast in-fast out" and "slow in-slow out," which means that the speed with which a tissue takes up or clears a chemical depends on its perfusion, as long as there are no reactions or binding.

4.4. TOXICOKINETIC MODELING FOR EPIDEMIOLOGY: WHAT DO WE REALLY NEED? THE GOLDILOCKS PRINCIPLE

Our goal in using a simplified form of PBTK modeling is to be able to develop dose metrics in which the basic dose metric has two important characteristics: (1) its magnitude is proportional to the target tissue dose and (2) it has the same temporal characteristics as the tissue concentration, that is, it will rise and fall on the same time scale as the tissue concentration. In addition, a dose metric covers the time period relevant for the etiology of the adverse effect, which is discussed in detail in Chapter 6. Therefore, we want to develop models that are complex enough to capture the essence of tissue kinetics and that will give us the desired proportionality, but we do not wish to invest in developing fully detailed PBTK models with more complexity than needed; we want the complexity and model to be "just right," like Goldilocks's porridge, recognizing that there is no single model that will work in all situations.

Researchers, such as Andersen and Moolgavkar, have developed a variety of highly detailed complex models for a range of substances (Andersen 2003; Moolgavkar and Luebeck 2003). However, early work by Roach, Rappaport, and Droz used simple one-compartment models to look at the effects of simple approximations for estimating body burden at near steady-state conditions (Roach 1966; Roach 1977; Rappaport and Spear 1988; Droz 1992). They found that they could make reasonable estimates of tissue and blood or urine concentrations, validated by measurements, using a simple one-compartment model. The key conclusion from this is that we do not necessarily need to use complex PBTK models, but we can use models like that shown in Figure 4.6 for dosimetric estimates under common situations: low to moderate exposures and relatively long exposure intervals, such as hours. Often, we are studying chronic effects of fairly low-level long-term exposures, and under those conditions the internal concentrations tend to be directly proportional to the exposures. As a result, simple dose metrics calculated directly from exposure will often be proportional to tissue dose and should be suitable for defining dose-response relationships (Rappaport 1985). This conclusion is consistent with the findings from many epidemiologic studies that have seen good relationships between irreversible effects and cumulative exposure and other simple exposure metrics (Blair and Stewart 1992). However, we also have found that there are a growing number of situations in which those low-intensity and long-term conditions do not apply. We are interested in reversible effects or the effects of short-term intense exposures, in which temporal relationships are more complex. In this section, we first discuss the components of PBTK models and then consider when we need to do such modeling in epidemiology.

What Defines the Kinetics of an Agent in a Tissue?

Basic Relationship

The basic model can work for one compartment or a number of compartments. For each compartment, change in concentration of a foreign substance in a generic tissue is given by Equation 4.1. This difference equation represents the mass balance of material moving into and

$$C_{tis}[i+1] = C_{tis}[i] + \frac{Q_{tis}}{V_{tis}} \Delta t (C_{art}[i] - C_{tis}[i] / PC_{t/b})$$

Equation 4.1 Basic compartment equation

out of a compartment. It divides the time line into intervals Δt wide, each is numbered by i, C_{tis} is the tissue concentration, Q_{tis} is its blood flow in and out, V_{tis} is its volume, and $PC_{t/b}$ is its tissue-to-blood partition coefficient. Each tissue or tissue group in a multicompartment model has an equation like this in a blood-flow-limited model, that is, one in which the blood flow through the tissue defines how fast the toxic agent can be moved into or out of the tissue; the other common alternative is diffusion limited. It also may have additional terms that define input and removal from the compartment by a route of entry or metabolism, respectively. This equation has three key parts that define a compartment's kinetic behavior in a model:

1. Q_{tis}/V_{tis}: This multiplier is the perfusion rate constant (units are min^{-1}). It defines the fraction of tissue that will be flushed with blood each minute. The larger this value is, the faster the tissue concentration will change over time.

When the perfusion rate becomes very small, such as for body fat in an obese person, then diffusion may become the rate-limiting step.

2. ($C_{art}[i]$ - $C_{tis}[i]/PC_{t/b}$): This term is the difference between the arterial blood concentration coming into the tissue and the venous blood concentration leaving the tissue, assuming that the blood leaving reaches local steady state before it leaves. Dividing the tissue concentration by the tissue-blood partition coefficient converts it into the equivalent venous blood concentration. When these two blood concentrations are equal, the tissue is at steady state. The difference can be either positive or negative. Positive values occur because arterial blood is bringing more agent into the tissues, and negative values occur because the venous blood is removing out more agent than the arterial blood is bringing into the tissue. If the agent is highly soluble in the tissue, such as dioxin in fat, then it will take a relatively long time for fat to reach steady state. The relationship in Equation 4.1 assumes that the overall rate of exchange between tissues and blood is limited by the blood flow, but with large molecules, such as dioxin, in poorly perfused tissues the rate of diffusion may be the limiting rate, which introduces a nonlinear relationship. If the agent binds to blood proteins or tissue components, then the relationship becomes more complex.

3. Tissue specific processes: Additional terms are needed to account for input, output, and storage processes operating in some tissues, such as ingestion and skin absorption, and for removal, such as urinary output or metabolism or reactions with local materials, such as glutathione. Most commonly, we represent metabolism as occurring in the liver and use nonlinear Michaelis-Menten kinetics (Klaassen 2001). Equation 4.2 shows the term we would add to the equation for the concentration in the liver compartment to represent enzyme removal of the parent compound by metabolism. In a related model for the metabolite, this term would be positive representing its formation.

$$Parent\ Metabolism\ (mg\,/\,\mathrm{min}) = -\frac{V_{\mathrm{max}} C_{Liv}[i]}{K_m + C_{Liv}[i]}$$

Equation 4.2 Michaelis-Menten relationship for enzyme kinetics

The metabolic rate is determined by two constants characteristic of the enzyme's action on the substance: V_{max}, the maximum rate, and K_m, the concentration when the rate is half of V_{max}. At low exposures, $K_m >> C_{Liv}[i]$, and the metabolism is approximately linear $\cong \frac{V_{\mathrm{max}} C_{liv}[i]}{K_m}$ and proportional to liver concentration. At high exposures, $K_m << C_{Liv}[i]$, and the rate becomes approximately constant, V_{max}. Other capacity-limited biological processes have similar relationships. Note that we do not need to do formal enzyme kinetics to estimate the rate constants; we can use the blood time course for the metabolite to estimate the apparent linear rate or the maximum rate depending on the shape of the time course.

Successful physiologically based models have been developed with a variety of structures. There is a broadly useful model for lipid-soluble substances, which requires three compartments composed of tissue groups with similar perfusion: the well-perfused tissues, poorly perfused tissues, and body fat. If we are studying a water-soluble substance, then we may need only two compartments. The selection of these three principal groups is based on the observation that tissues with approximately the same perfusion will have similar kinetic characteristics. The well-perfused tissue group contains nearly all of the organs that are potential target tissues: brain, lungs, kidneys, liver, reproductive organs, endocrine glands, and GI tract. In some cases, a separate compartment for the liver is needed to account for metabolism of toxic substances or for the kidneys when modeling urinary excretion; then we may pull those and their blood flow out of the larger organ group. The use of a separate compartment for these processes can make concentration estimates more accurate on short time scales, but in many situations we do not need the improved accuracy; we only need proportionality and approximately accurate temporal behavior, so it may not be necessary to add these compartments when constructing a model for use in an epidemiologic study. The relatively poorly perfused (PP) tissues, the skin and muscles, are rarely targets, but they account for a large fraction of the total tissues and modify the temporal behavior of the blood concentration. An important consideration is that the PP tissues have variable perfusion depending on exercise activity and ambient temperature (heat and cold). The third group is the body fat, which is an ill-defined group of adipose tissues and interstitial fat deposits.

An example of this model is diagrammed in Figure 4.7, and its parameters are given in Table 4.4. The spreadsheet showing the parameters and the equations for each of the model's two compartments is given in Figure 4.8. (Note that the lungs and heart are assumed to be at equilibrium in each interval.) The model and its parameters are scaled to the height, weight, age, and gender of the subject. Gender is important because body fat and some other tissue proportions change with gender. Likewise, some proportions change with age. Those differences can cause differences in time course and tissue concentration that systematically affect tissue dose. Even though we are not attempting to do fully detailed PBTK models, we still wish to take advantage of the PBTK structure because it allows us to better represent the effects of body size, gender, and age differences that are present in study populations.

Defining Model Structure

We have not specified the necessary number of compartments needed to define a useful model because it is not a fixed number but depends on the situation. We can gain some insight about how many might be needed to describe temporal data

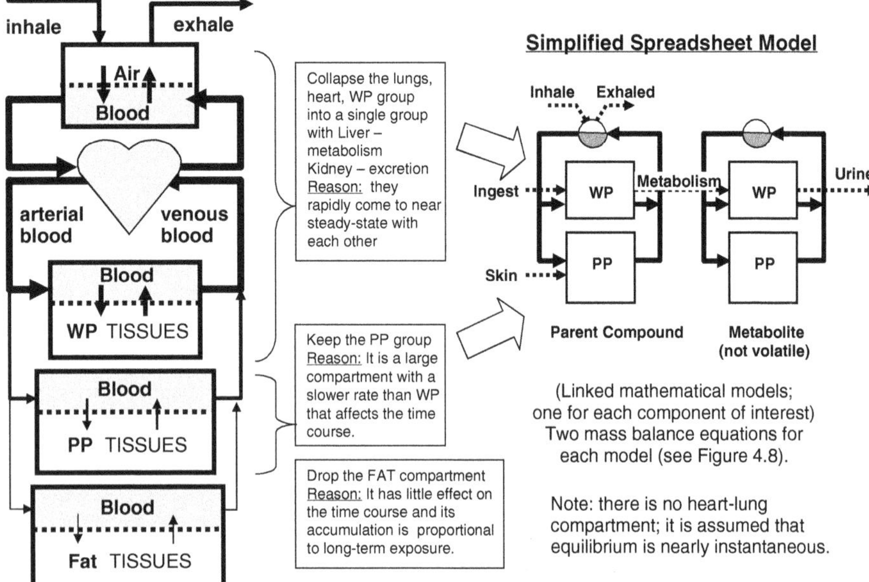

Figure 4.7 Derivation of the simplified physicologically-based toxicokinetic (PBTK) model for preliminary modeling studies.

by examining the time courses of venous blood measurements made after an injection or after a brief period of acute exposure (Pratt and Taylor 1990). When those data are plotted as the log(concentration) against time, very commonly we see two linear segments with different slopes; the first segment has a steep slope (fast clearance) and the second a flatter slope (slower clearance). Figure 4.9 shows the calculated time course with concentration on a linear scale, but you can see that the well-perfused tissues (C_{wp}) after 180-min exposure shows a steep drop when exposure stops, followed by a slower decline. This behavior has been observed in many drug and toxicologic studies with animals and humans. Analyses have shown that usually a two-compartment empirical model can fit the data and that sometimes only a one-compartment model is needed, as seen for the model in Figure 4.6. Occasionally, with highly precise and extended measurements, a third, even slower phase can be detected, implying a third compartment. This strongly suggests that in most short- or moderate-duration exposure situations (less than a few days) we could use a two-compartment model to describe blood levels, even though we know that the internal system is much more complex than this. For chronic exposures, in which there is accumulation in a long-term storage tissue (fat or bone), the three-compartment model may be a more accurate model. That finding is very useful, but how do we decide on one, two, or three compartments? Part of the answer comes from defining what time scale and tissue we are interested in representing.

We need to do PBTK modeling under two sets of conditions: (1) when there are nonlinearities in toxicokinetic processes, such as metabolism, protein binding,

Table 4.4 Simplified spreadsheet parameters for general PBPT model with effects of gender, exercise and metabolism

Tissues	Volume		Air and Blood Flow		Metabolism and Exercise Effects
	Male	Female	Male	Female	
Whole Body	**Lean Body Mass (LBM)**[a]		**Ventilation at Rest**		**Exercise Effect**[c]
					Males
	55 kg	39 kg	4 L/min	3.4 L/min	$Q_{alv} = Q^*_{alv} + 0.28W_{exer}$
	(density ~1 kg/L)		(Q^*_{alv})		Females
					$Q_{alv} = Q^*_{alv} + 0.16W_{exer}$
			Cardiac Output at Rest		**Exercise Effect**[c]
			6.0 L/min	5.2 L/min	$Q_{art} = 1.25Q_{alv}$
Well-Perfused (WP) Group (organs)	**Percentage of LBM**		**Percentages of Q^*_{art}** [b]		**Metabolism**[d]
	20%		70+7% (no gender or exercise effect)		$\frac{-V_{max} C_{tis}}{K_m(Vliv/V_{wp}) + C_{tis}}$
Poorly Perfused (PP) Group (muscles and skin)	**Percentage of LBM**		**Percentage of Q^*_{art}** [b]		**Exercise Effect**[c]
	56%	41%	20+3%	17+6%	$Q_{pp} = Q_{art} - Q_{wp}$

Notes:(values taken from International Life Sciences Institute. 1994)

a.LBM = lean body mass, which is body weight minus fat.

Percent body fat = $(1.2{*}BW/Ht^2 - 10.8{*}(2\text{-Gender}) + 0.23{*}Age - 5.4){*}0.01$, where Gender = 1 male, and 2 female.

b.Blood flow to fat is distributed to WP and PP compartments to maintain mass balance, and percentages of LBM are adjusted to obtain the same perfusions. These changes increase the perfusion by 10–30%.

c.Represents the increased blood flow and ventilation associated with exercise (W_{exer}, expressed as watts) as developed by Astrand (1983). This is approximately correct up to moderate exercise levels (~100 watts).

d.Metabolism is represented by the Michaelis-Menten equation. Assuming that the WP tissue concentration will have the same relationships as though the metabolism was only in the liver concentration, K_m is adjusted for the relative volume of the two $K_m{*}V_{liv}/V_{wp}$, which maintains the nonlinear relationship with WP tissues.

or other capacity-limited processes that limit the system responses; or (2) when the response or effect from tissue concentration is nonlinear, such as acute effects on short time scales. We discuss the second situation in detail in Chapters 10 and 13. Our goal here is to apply the Goldilocks principle: to find a level of complexity "that is just right." How can we tell if our model has the right level of complexity? We need to answer some questions before we can decide:

Nature of exposures: How intense? Variations in composition? Multiple agents?
Temporal factors: Duration? How variable? Are there short-term peak exposures?
Route of entry: Respiratory? Skin? Ingestion? Multiple routes?
Active agent: Direct acting? Is it metabolically activated or detoxified? Is it relatively lipid soluble?
Target tissue: Is it part of a route of entry (respiratory tract, skin, or GI tract)? Is it an internal organ, a well-perfused tissue?

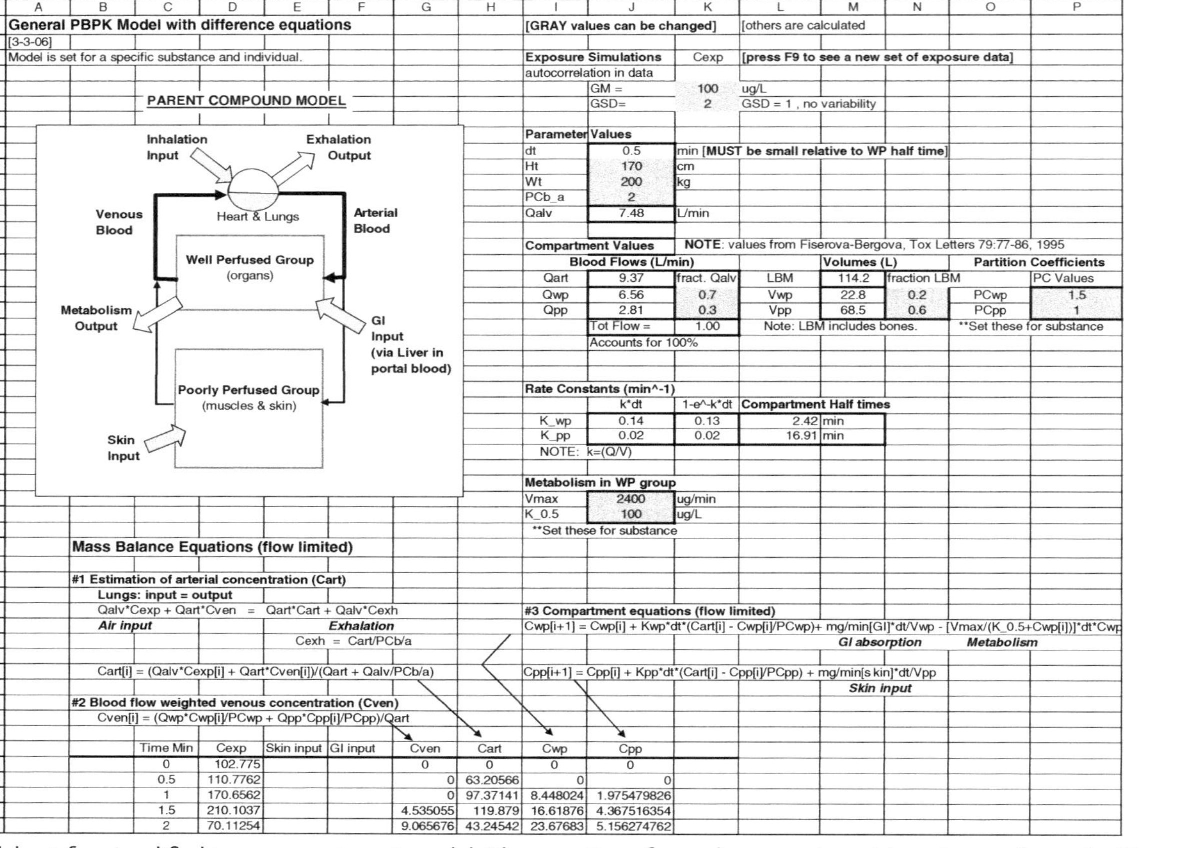

Time Min	Cexp	Skin input	GI input	Cven	Cart	Cwp	Cpp
0	102.775			0	0	0	0
0.5	110.7762			0	63.20566	0	0
1	170.6562			0	97.37141	8.448024	1.975479826
1.5	210.1037			4.535055	119.879	16.61876	4.367516354
2	70.11254			9.065676	43.24542	23.67683	5.156274762

Figure 4.8 Spreadsheet for simplified two-compartment model. The equations for each compartment have inputs from air, GI tract, and skin absorption (only air has input). Parameters were obtained from Fiserova-Bergerova's 1995 paper on modeling.

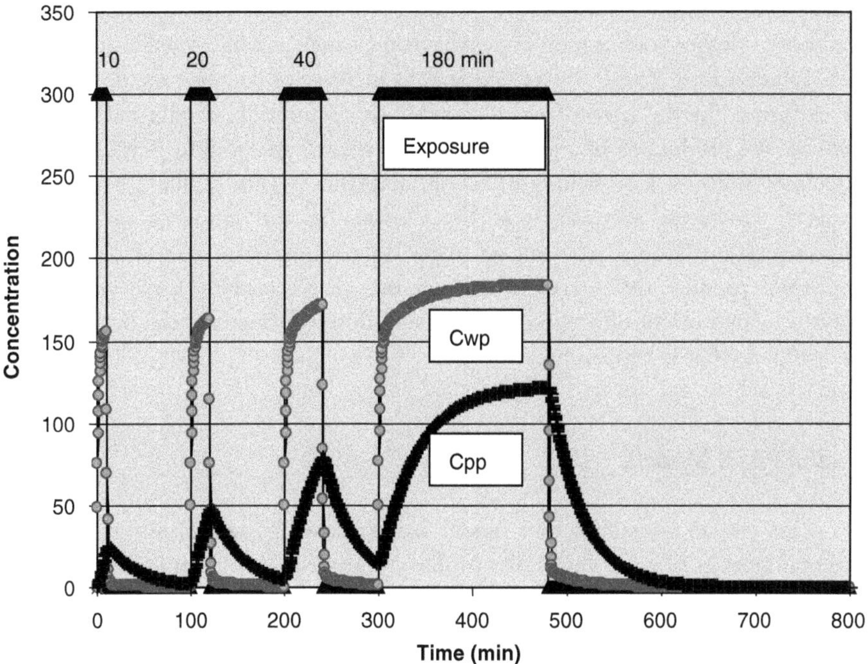

Figure 4.9 Time course of concentrations in a standard male in well-perfused (C_{wp}) and poorly perfused (C_{pp}) compartments given a simulated series of inhalation exposures with different durations. The model shown in Figures 4.7 and 4.8, the parameters in Table 4.4, and partition coefficients $PC_{b/a} = 2$, $PC_{wp/b} = 1.5$, and $PC_{pp/b} = 1.0$, were used for the calculations.

There are combinations of answers to these questions that will suggest what type of PBTK modeling is needed. The four most common situations are:

Airborne: inhalation of a direct acting agent with the respiratory tract as the target tissue.
Skin contact: absorption of a lipid-soluble, direct-acting agent with the skin as target tissue.
Ingestion: absorption of a lipid-soluble or metal direct-acting agent with internal organ target.
Airborne or skin contact: intake of a lipid-soluble, enzyme-activated agent with internal organ target.

For low-level exposures, all of these will be linear systems, and the internal dose will be proportional to exposure. However, because biological processes often have limited capacity, as the concentration approaches the limit the process saturates and stops increasing with increasing concentrations. Enzymatic metabolism is a good example. Other examples are protein binding (highly polar agents), saturable transport, or other nonlinear processes that can saturate and have nonlinear exposure-dose relationships. These effects can work to increase or decrease the dose per unit of exposure. In the case of toxic metabolites, saturation of one pathway may increase the production of another, more or less toxic metabolite. Anticipating the effects of shifts in a branching metabolic pathway is very difficult, even for the experts. However, we can look for changes in the amounts of metabolic biomarker(s) produced per unit of tissue dose, or we can observe that intense exposures produce differences in risk per unit of tissue dose. Clearly, the more ways we can look for effects on the exposure-dose relationship, the better we can understand what is happening.

Useful PBTK Models

To assist you in your choice of model and to develop your intuition about the internal kinetics of toxic materials, we have prepared a simple spreadsheet model that will cover most of the common situations noted earlier. The one-compartment model can be used for direct external contact with an agent in which the target tissues are the eyes, mucus membranes in the nose, mouth, and throat, and skin for a wide range of time scales—seconds to hours. The two-compartment model can be used for materials with all common routes of entry and internal organs as target tissues for time scales ranging from minutes to 24 hours. The three-compartment model can be used for highly lipid-soluble materials for which long-term storage is important. For bone accumulation, the fat compartment can be replaced with the bone compartment. Note that when we are interested in exposures and disease processes that span days to years, the short-term variations (less than ~8 hr) often will be irrelevant. However, effects on very short time scales, such as minutes of peak exposures to carcinogens, can produce risks on very long time scales, which can also be associated with nonlinear risk relationships. We will develop the use of these models more when we explore the specific examples in Chapters 10-15.

When Kinetics Are Not Proportional: Nonlinear Models

In this section we discuss a much more complex issue, which may be beyond the expertise of readers who are at the entry level. At the beginning of this chapter we noted that our goal is to estimate dose. If the target tissue concentration is directly proportional to the exposure (i.e., a linear relationship), then we do not need a PBTK model to form a useful dose metric—we can use a metric based on the disease process model (presented in Chapter 6) and calculate it from the exposure time profile. However, if tissue levels are not directly proportional to exposure, then we need a model to determine how the nonlinearity impacts the time course to make useful dose metrics. One of the most common causes of nonlinear relationships is a system capacity limit, such as for metabolic activation or detoxification. Enzymes are important for activating many compounds into toxic forms and detoxifying active forms. As noted earlier, enzymes have capacity limits, and when those limits are approached the processes become nonlinear. If the metabolizing tissue concentration is less than approximately 20% of the enzyme's Michaelis-Menten constant, K_m, then the process will be approximately linear, so knowledge of enzyme K_m values can guide us in deciding when things may get nonlinear.

There is a growing body of literature on common enzymes that metabolize toxic materials; hence a literature search for metabolic data on the toxic material you are investigating can often give an animal or sometimes a human K_m value. A survey of typical K_m values shows that they range from about 10 to 800 μM ($K_m \times$ molecular weight of the agent gives the result in μg/L).

Nonlinearities in personal dose arise primarily from three different PBTK processes: metabolism, binding, and reactions with tissue components. When one metabolic pathway saturates, another may become important and produce a more or less toxic by-product. Protein binding is important for a variety of processes, many involving transport, this binding tends to have a similar capacity limitation as metabolism and is also a cause of nonlinear relationships, but this is not usually a problem at low agent concentrations. The same is true for tissue reactions. Many toxicants are reactive, and they will react with a variety of biomolecules, including DNA and enzymes. Fortunately, there are many scavengers, such as glutathione, available to remove these reactive materials. As the concentration of the reactive agent increases, the scavengers or the primary molecular target may be depleted, and the agent will shift to other targets, such as critical proteins and nucleotides, perhaps causing damage or changes in function. Passive, diffusion-based excretion is linear, but active excretion and resorption processes have capacity limits. Macrophage transport of particles is limited, too, because alveolar macrophages can only ingest a limited amount of material, which may be reduced if the material is toxic, such as crystalline silica. The clear message from this brief review is that at low levels, processes are usually linear, so dose metrics based on exposure will often be adequate. However, it is wise to discuss with a toxicologist the likely toxicokinetic behavior of an agent you are studying and to anticipate the need for nonlinear modeling if you find that the unit effect per dose changes at higher dose levels.

Population variations in metabolism, transport, and other nonlinear processes produce differences in dose per unit of exposure. Interindividual process differences can increase dose heterogeneity by increasing differences in the amount of active agent formed from the parent compound or the amount of agent that is detoxified. The most common sources are physiologic and genetic differences. Physiological differences in tissue dose per unit body weight among people are usually small, even among different genders and racial groups, because the blood perfusion of the organs is critical and generally proportional to lean body mass. This can be seen by examining Table 4.3.

Table 4.3 Reference data on organs and tissues for males and females with algorithms to personalize estimates

Tissues	Physiologic Processes	Volume		Air and Blood Flow		Perfusion(Q/V)	
		Male	Female	Male	Female	Male	Female
Body Weight (BW^a)		70 kg	58				
Body Height (Ht)		170 cm	160 cm				
Lean Body Mass (without fat)		55 kg	39 kg				
Total Blood Flow (cardiac output, Q_{art})	Blood distribution			6.05	5.25		
Minute Ventilation	Total air in/out			6 L/min	5.2 L/min		
Aveolar Ventilation	Gas exchange			4 L/min	3.4 L/min		
Well-Perfused Group		**Percentage of LBM** 20%(density ~1 kg/L)		**Percentages of Q_{art}** 70%		**Fraction flushed per minute** **~2.6**	**~0.6**
Brain		2.7%		12%	12%	**~5 times/min~1**	
Kidneys	Urine excretion	0.6%		19%	17%		
Liver	Metabolism (Blood from GI)	3.3%					
Poorly Perfused Group		**Percentage of LBM** (density ~1 kg/L)		**Percentage of Q_{art}**		**Fraction flushed per minute**	
Muscles		50.9%	36.9%	16% (changes with activity)	12%	**0.036**	**0.044**
Skin		4.7%	3.9%	4% (changes with temp.)	5%	**0.12**	**0.17**
Very Poorly Perfused Group		**Percentages of BW** (density ~1 kg/L)		**Percentage of Q_{art}**		**Fraction flushed per minute**	
Fat	Energy storage	23%	38%	10%	13%	**0.026**	**0.035**

Notes:(values taken from International Life Sciences Institute, 1994.)
a.Excluding bone density, which is ~2 g/mL.
b.LBM = lean body mass, which is body weight minus fat.
Percent body fat = $(1.2*BW/Ht^2 - 10.8*(2\text{-Gender}) + 0.23*Age - 5.4)*0.01$, where Gender = 1 male, and 2 female.
c.Materials absorbed through the GI tract will enter the liver through the portal vein.

Another important source of dose heterogeneity is physical activity. When there are differences among subjects in physical activity, these can have a major effect on uptake of contaminants from the environment for both inhalation and skin absorption because of the increases in ventilation and blood flow. Alveolar ventilation and cardiac output are proportional to activity, so a doubling of activity also doubles ventilation and cardiac output. Obviously there are some physiologic limits, but at low and modest activity levels these differences can lead to important differences in tissue concentration time course and dose. Questionnaires and other evaluation tools have been developed to assess physical activity, both intensity and duration, that can be used to survey a population and determine their activity patterns (Ainsworth, Jacobs et al. 1993; Steele and Mummery 2003; Reis, Dubose et al. 2005). This approach was used to detect asthma and other pulmonary effects in children exposed to air pollution in Los Angeles, where the risk was compared for children who do sports versus those who are inactive (Peters, Avol et al. 1999).

When the source of dose heterogeneity is polymorphic genotypes for metabolic enzymes, we can sometimes incorporate genetic factors in our PBTK models if we know each subject's genotype and the activity of the enzyme for each type. Then the rate of formation or removal of a toxic material can be included in the model. Biomarkers offer a window on these processes. If there is a change in the relative amount of biomarker associated with high exposures versus low, then we may suspect that some process, enzymatic or binding, is saturating. For example, Rappaport and coworkers found that the formation of a metabolite of benzene showed nonlinearity, which they attributed to metabolism (Rappaport, Waidyanatha et al. 2005).

Dose heterogeneity can also come from differences in exposure time course across individuals. There is an important interaction between the variability in exposure and possible range of variations in tissue concentration. Some variations in exposure will not be apparent in the tissue concentration because temporal changes are limited by the perfusion rate (ratio of the tissue volume to the blood flow): for example, the well-perfused (WP) tissue, (11L)/(4.2 L/min) = 2.6 min^{-1}, which indicates that the half-time for flushing the WP tissue compartment would be 0.26 min, that is, the time needed to reach half of the new steady-state concentration. However, when we have two or more compartments connected to each other, as in Figure 4.7, then they affect each other, as shown in Figure 4.9. Initially the WP concentration increases very rapidly, but as the WP starts to plateau (to reach steady state) the poorly perfused (PP) starts to affect and slow its rise. Figure 4.9 shows the characteristic "shark's fin" time curve for PP tissues changing the compartmental concentration from one steady state to another. The WP tissues show a much more rapid change in concentration than the PP tissues, as expected from their lower perfusion. You can see that the WP tissues can nearly reach the steady-state plateau in 10 min, but the PP tissues have not quite reached steady state in 180 min. Therefore ~10 min peak exposures will produce higher WP tissue concentrations than in the PP tissues. On the other hand, peak concentrations that have duration less than the half time will have less effect on the tissue concentration, unless they are very large. A toxicokinetic model in combination with toxicologic data on adverse effects can be used to define peak exposures that are "sufficiently large," that is, how high for what duration is needed to have an effect; it is not just a function of the height of the peak. A time course of short-duration high peaks may cause fewer effects than more moderate peaks of longer duration. We will revisit this issue of peak exposures again when we discuss discrete responses in Chapter 14.

4.5. SUMMARY

In this chapter we have investigated the physiologic link between external exposures and internal tissue concentrations for an exposed individual. It is possible to construct simplified descriptive toxicokinetic models formalizing a reasonable inference about key kinetic processes for many environmental chemicals. These can give us insight on the ways materials move through the body and are processed. We assert this because the possible routes of entry, possibilities for metabolism, long-term storage and removal pathways are limited and depend strongly on the biochemical and physiological characteristics of the substance. Given even limited data on a substance, a reasonable first-order PBTK model can be constructed, and when biologic monitoring data exist, the time course the model predicts can be checked against measurements. More important, we have observed that as long as there are linear or proportional relationships between exposure and internal tissue concentrations, then we do not need even a simple PBTK model to estimate a dose metric; we can instead use dose metrics calculated directly from exposures (Rappaport, Kupper et al. 2005). It is only when the relationship between exposure and tissue concentrations is nonlinear that we need to use these models.

Simple toxicokinetic models also allow us to visualize how variation in exposure affects the tissue concentrations over time. They also permit us to adjust our dose metrics to compensate for differences in body size, gender, age, and some genetic factors. Depending on the specific characteristics of the target tissue, the temporal exposure variation may or may not be transmitted to the tissue concentration. As has been said repeatedly, tracking the internal concentration variations in response to the temporal variability of exposure is the key to exploring what is happening inside the body. In the next chapter we explore biomarkers of exposure, which are also windows on internal processes. Although there are complexities in the exposure-biomarker relationships, they reflect the toxicokinetics and can be described with models.

NOTE

1. The basic ideas here are taken from human physiology. For further information the reader should consult Guyton's *Medical Physiology* or similar texts.

5 Biomarkers as Indicators of Exposure

Research is to see what everybody else has seen, and to think what nobody else has thought.

Albert Szent-Györgi (1893–1986) U. S. biochemist

Biomarkers are internal substances, such as levels of liver enzymes, or methemoglobin, or genotypes, or lead in blood that can be readily measured and used to indicate the presence of disease, or its risk, susceptibility, or exposure. In this book, we are primarily interested in the biomarkers of exposure and the processes and factors that affect them. Even those biomarkers that are clearly related to exposure are not direct estimates of exposure because they are also affected by an individual's personal toxicokinetics, as we saw in the previous chapter. They are a powerful window on the internal temporal processes that affect tissue dose, such as blood transport, metabolism, and excretion. Therefore, they can be invaluable for understanding internal processes and calibrating our models of those processes. When biomarkers of exposures are taken together with biomarkers of effects and susceptibility, we can have a very useful way to explore the internal interactions and processes leading to disease. Each type of biomarker represents a different part of the exposure-dose-response-disease continuum.

The use of biomarkers is not new. Biological monitoring has been used in occupational exposure surveillance programs since at least the 1940s to assess worker exposures to common hazards, such as lead and arsenic. However, beginning in about the 1970s, there has been concern that biological monitoring is not preventive, because the detection of values exceeding the permissible level indicates that the individual has already been overexposed. On the other hand, it as been argued that overexposure could be the result of poor personal hygiene, which is extremely difficult to measure by environmental sampling. For example, if a worker does not wash his hands before eating or smoking, he can have increased ingestion or inhalation exposures. A partial solution to this problem has been to set an action level, which is half the permissible value. This approach implicitly assumes that the biomarker is not changing rapidly. When a worker's blood or urine level reaches the action level, then action must be taken to control the exposure by installing engineering controls and other effective means. This must include improvements in hygiene conditions, such as providing a place to wash and a separate lunchroom. However, the action-level approach is successful only if the exposures are generally well controlled and workers are monitored frequently enough to detect when an individual's internal levels are increasing so that there is time for an intervention before they reach harmful levels. As a result, biological monitoring surveillance should not be the primary exposure control strategy. Despite its limitations for exposure control, biological monitoring data have been very useful for studies of health risks, which is our primary topic of interest here.

5.1. TYPES OF BIOMARKERS

A biomarker is an internal substance in a readily accessible biological media, such as blood or urine. There are four broad types of biomarkers: exposure, early response, susceptibility (such as metabolic enzyme genotypes), and disease. Because some of these overlap, and because some biomarkers may be used for more than one type of application, we need to be clear about the applications. The utility of a biomarker depends on its selectivity and the exposure situation.

Biomarkers of Exposure

These are internal concentrations of environmental substances or their metabolites, or reaction products within the body, which are used to estimate external exposures during some time interval. Readily accessible materials are used to make these measurements, including urine, blood, saliva, hair, fingernails or toenails, exhaled breath, and occasionally sweat or feces. We have discussed the differences among the routes of entry in Chapter 4. Figure 5.1 shows how these relate to our most common sources of biomarkers. A nonselective biomarker of exposure may be very useful, if there is nothing else present. However, in settings with mixed exposures, interpreting the concentration of a nonspecific biomarker may be difficult. There are three important subcategories with differences in specificity:

1. *Biomarkers of exposure may be a single, specific environmental contaminant,* such as lead in blood, styrene in exhaled breath, or an arsenic metabolite in urine.

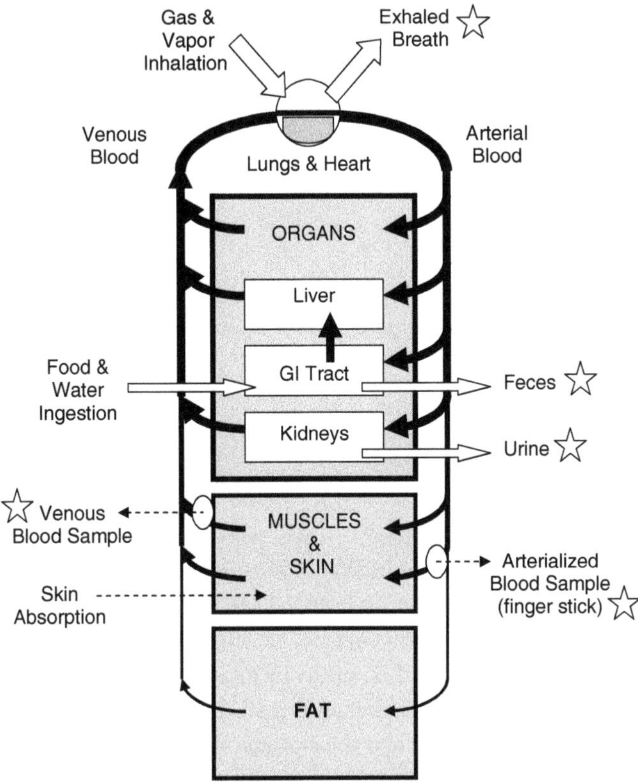

Figure 5.1 Physiologically based toxicokinetic (PBTK) model of exposure routes of entry (bold letters) and common biomarker sampling points (stars).

2. *Biomarkers of exposure to a complex mixture of contaminants may be a single indicator substance.* For some mixtures, such as cigarette smoke, there is a specific source and one component of the mixture that is unique and easy to measure, such as nicotine or cotiniene, a metabolite of nicotine, in saliva. This type of biomarker can have a wide range in specificity for the mixture; it may be highly specific for an emission source, such as nicotine for cigarette smoke, or relatively nonspecific, such as 1-hydroxy-pyrene in urine associated with PAHs, which are found to some degree in all combustion emissions.
3. *Biomarkers of effects also may be used to indicate exposure.* A good example is depression in red blood cell (RBC) cholinesterase by organophosphate (OP) pesticides, or genetic up-regulation of a protein. Again, depending on the setting, the biomarker may be useful or may be prone to misclassification and difficult to interpret.

Biomarkers of Response

These are measures of early changes in endogenous biochemicals and markers of early responses to environmental chemicals that are possibly on the causal pathway to pathological effects. The development of new methodologies to broadly scan for many concurrent effects in the "biome," such as using proteomics and metabolomics, offers considerable power to better define responses to toxic exposures by using arrays of biomarkers.

Biomarkers of Disease

These indicate the presence of clinical disease, which is defined by changes in endogenous biochemicals, genes, gene products, or clinical parameters indicative of pathological processes.

Biomarkers of Susceptibility

These indicate the presence of genetic, metabolic, or physiologic variants that are associated with differences in the risk of a disease (Thier et al. 2003). Some biomarkers indicate the susceptibility of an individual to a particular exposure, such as the genotype of a critical metabolic enzyme, in which one isoenzyme is less effective at detoxifying a contaminant, and its presence indicates an increased risk of a toxic effect. These may be useful in combination with exposure measures for studies of gene-environment interaction. When the amount of a metabolite produced per unit of exposure varies, it may be an indication of differences in enzyme activity.

Epidemiologic Applications

The distinction among these categories is not sharp. A given biomarker may be used for more than one type of application. For example, depressed serum cholinesterase, noted earlier, is used to indicate both intensity of a past exposure to organophosphate pesticides and an adverse response that indicates risk of poisoning. Although it is used for both, serum cholinesterase is a better index of response and disease than exposure, because many individual toxicokinetic factors and the history of past exposure also affect the serum levels. Similarly, DNA adducts may represent exposure by being higher in those who are exposed, or they may represent the internal dose of an activated agent, but they also may represent early effects, such as DNA damage that may be along the pathway to cancer. DNA adducts have been used in all of these applications. The categorization of biomarkers is based less on what they measure and more on the application. This is especially true for biomarkers of response and disease. Two or more biomarkers may be used together, such as biomarkers of exposure and susceptibility, as a way to understand the interaction. In this chapter, we emphasize the application of biomarkers to internal dose assessment, especially as reflected in the time course of changes in biomarkers in response to exposure.

In some of the older scientific literature, you will see that epidemiologic investigators used biomarkers, such as blood and urine lead, as indicators of lead "body burden." They hypothesized that cumulative adverse health effects were a function of the total body burden—the higher the body burden, the higher the risk. Even though lead is highly cumulative in the bones, the blood and urine represent circulating lead, which fluctuates with daily intake and is insensitive to long-term deposits. Later, with the development of an X-ray fluorescence technique to measure bone levels, they obtained a much better index of total amount of lead in the body. However, even bone lead declines slowly over time when there is no exposure. This same dose approach was also used for noncumulative substances whose effects were cumulative; they hypothesized that the effects would be proportional to the total material taken into the body. These initial applications occurred at a time when there was little information on the toxicokinetics or mechanism of the effects. We now know that this is too simplistic, although it was a useful approximation in some situations, such as with smelter workers, in whom a surveillance program measured urinary arsenic over long periods of time. Average urinary arsenic showed a strong relationship with lung cancer risk (Pinto, Henderson, and Enterline 1978).

There is a perception among some investigators that exposure biomarkers can allow them to avoid the need for difficult and expensive evaluations of external exposure, especially in situations in which multiple routes of exposure exist. Unfortunately, this perception is largely incorrect because of differences in toxicokinetics across individuals, and the route-of-entry and biomarker variations caused by exposure variability are important for understanding the risk and for intervention strategies. Biomarkers are good substitutes for exposure measurements only under some limited circumstances, because the exposure-biomarker relationship is defined by complex temporal biological processes that link the biomarker with environment. As we saw in Chapter 4, toxicokinetic processes can reduce some of the exposure variation transmitted into internal levels, but they can also add biases because of systematic differences across individuals in personal kinetics and metabolic enzymes, which will result in systematically higher or lower internal doses of active agents and thereby also affect the risk. Consequently, empirical relationships between exposures and biomarkers tend to be variable between individuals and across time. We explore this more fully in section 5.4 in this chapter.

5.2. CHARACTERISTICS OF BIOMARKERS

A variety of important issues need to be considered when applying exposure biomarkers in health studies.

Time Scale of the Biomarker

The time scale of variations in the biomarker should be approximately the same as the variations in the outcome of interest. This is most evident at the extremes. If the outcome varies on the scale of hours, you should not use a biomarker that varies little from day to day, or vice versa. However, it is less clear when the time scales are closer together. Exhaled breath is an excellent indicator of arterial blood and vessel rich organ concentrations. Its time scale depends on the blood solubility of the substance. Breath levels can change within minutes for substances with modest or low blood solubility that are rapidly cleared by exhalation. This may be useful if we are interested in rapid-onset neurological symptoms, but we need to make carefully timed outcome measurements to track the variations in breath levels. For substances with high blood solubility, such as ethyl alcohol, the rate of change depends on the metabolic rate, and breath concentrations do not change rapidly, so fewer measurements are needed to estimate the relevant level of alcohol for neurological symptoms of intoxication. We need to consider the time scale of the biomarker, exposure, and outcome to get the best match up of the exposure-dose and dose-response relationships.

Relationship to the Causal Agent

One critical factor in the use and evaluation of exposure biomarkers is how they are related to the hypothesized agent of health effects and its causal pathway. If the biomarker is not directly associated with the causal agent, then, not surprisingly, variations in its concentration may not be predictive of the risk. This is especially true for exposures to mixtures. For example, total arsenic in urine for some workers may not be predictive of lung cancer risk because they may have a diet rich in seafood, especially crab (Kales, Huyck, and Goldman 2006). Crab and some other seafood contain high levels of an organic arsenic compound that is not digested in the gut but is absorbed and excreted largely unchanged in the urine. When urine is digested with strong oxidizing acid for analysis of the total arsenic, the bound arsenic is freed and measured as part of the total in the analysis. Thus the high total urinary arsenic after a crab dinner has no relationship to cancer risk from exposure to inorganic arsenic. Analysis of the chemical species of arsenic in the urine can solve this problem. In some cases, the relationship with the causal agent is implicitly assumed, and the biomarker is used to determine whether there is a risk from the exposure—a negative finding is difficult to interpret if the causal relationship is uncertain.

Route of Entry

Inhalation, skin absorption, and absorption from the GI tract are the primary routes of entry into the body, as noted in Chapter 4. Each has its own temporal characteristics and features that control uptake and distribution. Figure 5.1 shows a general physiologically based toxicokinetic (PBTK) model with three routes of entry, which is useful for thinking about the kinetics for the common sampling media. When we are choosing a biomarker for a new study or interpreting biomarker data, we need to be aware of the temporal differences associated with different routes of entry and the properties of the substance to be measured as the biomarker. Just by inspection of the diagram in Figure 5.1, we can see important differences in the PBTK relationships for target tissues in the organs, such as brain, lungs, liver, and kidneys.

Exhaled-breath biomarkers indicate materials in the arterial blood coming out of the lungs, which go directly to the target organs. Exhaled quantities depend on the solubility of the agent in the blood;highly soluble substances may not appear in exhaled breath in high enough concentrations to be useful as biomarkers. The levels in the highly perfused organs can change rapidly (~ 2–4 minutes) when air levels change, such as immediately following the beginning or the end of exposure. Thus the choice of sampling time and its relationship to sharp changes is critical for interpretation of the breath results; real-time sampling is most effective for linkage with exhaled breath samples.

Venous blood samples collected from the cubital vein in the arm are direct indicators of the muscle compartment, but they can be related to the organ levels, depending on the variability of exposure and the approximation of steady state. Large changes in inhalation exposure lasting 10–15 min will affect venous levels, but shorter variations will be smoothed by integration in the muscle compartment in the arm. The serum levels of water-soluble materials can indicate the general contents of interstitial water.

The slowest routes of entry are skin absorption and ingestion. Materials entering by skin absorption must pass through the epidermal barrier of dead cells, keratin, and hydrophobic lipid materials (as shown in Figure 4.5), which is a relatively slow process for most substances, although for a few it can be fast, such as the glycol ethers. Uptake by the skin depends on the thickness, integrity, and lipid characteristics of the skin. Once through the skin, the substance enters the venous blood. Ingestion uptake is relatively slow and depends on the physical form entering the gut and its solubility in GI fluids. From the GI tract, absorbed materials first go to the liver, where they may be activated or detoxified, and then into the venous blood to pass through the lungs and finally into the tissues.

Clearly, the degree to which materials are metabolized, are diluted with venous blood, or may be exhaled on the first pass will vary by route of entry and the characteristics of the substance. These differences by route of entry may enhance or reduce the amount of biomarker formed or removed, which also can affect the risk of adverse effects.

Sample Type/Media Differences

Figure 5.1 identifies all of the common media used for collection of biomarkers. As noted previously, each of these media represents a different time scale of variations.

Commonly a biomarker's time scale is identified as its half-time of removal. The range of half-times is very wide; exhaled-breath levels can change in a few minutes, whereas bone lead may not change much in decades. However, the half-time of removal, usually the fastest phase, can be somewhat misleading, because many biomarkers have multiphase washout profiles when exposure stops. Some may have slow clearance from some tissues, such as the fat. Other biomarkers may represent a specific time period in the past, such as toenail arsenic, which represents the time that inorganic arsenic was bound to nail protein, that is, when the nail was laid down, which may be 3 to 9 months (depending on the toe size) plus a lag for the nail to grow out to the point at which it can be removed for analysis (Kile et al. 2005). Urinary excretion is a function of serum protein binding, blood filtration, kidney resorption, and urine density. As a result, urinary output of a biomarker is often more variable and less precise than blood measurements, but it has the advantage of ease of collection.

Metabolic Products as Biomarkers

Biomarkers of exposure that are metabolic products, such as many of those appearing in urine, have a number of additional sources of variability beyond those seen for substances that do not undergo metabolic transformations. First, the metabolic kinetics depend on the specific isoenzyme produced by the individual based on his genotype and gene transcription, which define the amount of enzyme protein available to facilitate the reaction. Second, many P450 enzymes may be induced or inhibited by dietary components, drugs, smoking, or alcohol usage. Third, there may be several competitive pathways that will all metabolize the agent, but only one produces the biomarker. Consequently, the fraction of the agent metabolized through the biomarker pathway can vary among individuals and within an individual across time. All of these factors will affect the amount of biomarker produced per unit of agent in the blood. When volunteers are exposed under identical carefully controlled conditions in the lab, we find that the amounts and timing of biomarker formation can vary substantially. In that case, a metabolic biomarker of exposure may not be an improvement over a carefully measured external exposure. However, if the metabolite is a suspected agent of the health effects or is on the causal pathway, then those variations may make the metabolite more relevant to risk and make the biomarker the one of choice.

Reaction Product Biomarkers

Reactive chemicals in the environment or those formed by metabolism can react with blood proteins, such as hemoglobin or albumin, or they may react with DNA to form covalently bound adducts (Ehrenberg, Osterman-Golkar, and Tornqvist 1986).

Good examples are hemoglobin adducts by epoxides, albumin adducts by isocyanates, or DNA adducts by PAHs. All possible adducts are not stable, but some important ones are. For example, many DNA adducts are spontaneously removed by depurination or by repair enzymes. Terminal valine adducts on hemoglobin are important biomarkers for ethylene oxide and other alkylating agents.

Our own powerful oxidative metabolism can produce epoxides and other highly reactive substances "naturally" without exposure. Consequently, we also can produce some biomarkers endogenously, such as ethylene oxide adducts on hemoglobin, which can limit the utility of that biomarker for low exposures. Fortunately, the quantity of endogenous biomarkers formed tends to be small. These may not be a problem if exposures are high enough to substantially exceed the baseline for endogenous production or dietary input of the agent. Urinary excretion of the same metabolites that are produced by the metabolically activated agents will introduce noise in the marker levels. For example, benzene metabolism produces phenol, but it is also produced by a common medication, which limits the utility of phenol as a biomarker of benzene exposure. As a result, even with nearly identical exposures, individuals may excrete different amounts of a biomarker.

Markers of Adverse Effects as Indicators of Exposure

As noted earlier, some adverse effects may be used to indicate exposure. One of the best examples of this is use of blood cholinesterase as a measure of exposure to organophosphate pesticides (Wessels, Barr, and Mendola 2003). Another good example is DNA adducts used as an exposure surrogate. In general, the problem is twofold with these markers. First, adverse effects tend to be very nonspecific. For example, a depression in cholinesterase levels can be caused by a variety of disease conditions. Second, the presence of elevated levels of biomarker may provide no warning of an effect, but occur concurrent with the effect.

As we have seen with other aspects of the exposure-risk relationship, we emphasize the temporal relationship between each of the components of the exposure-effects relationship. Biomarker concentrations are a strong function of toxicokinetics of the agent and the formation and removal of the biomarker. We need to keep clear the purpose of the biomarker we are proposing to use. If it is intended to estimate external exposure to an environmental contaminant, a biomarker may do a poorer job than direct measurements, because the biomarker is sensitive to personal physiologic and metabolic factors that will reduce or change the correlation between exposure and biomarker levels when compared across individuals. However, when the goal of the biomarker is to estimate each subject's internal dose from exposure, then the biomarker may do a much better job than direct exposure measurements. In an epidemiologic application, the closer the biomarker is to the internal dose for the hypothesized agent, the stronger the test for a causal hypothesis. However, for this to be appropriate, the biomarker must integrate across a time interval that is relevant for the adverse effects, or multiple samples must be collected. When the agent is unknown, or the relationship to the biomarker is uncertain, then the findings for the biomarker must be interpreted cautiously. This is especially true when no relationship is observed, because there are many possible reasons that there may be no relationship, only one of which being that there is no causal linkage. A positive relationship is a necessary but not sufficient support for a causal relationship, particularly when the mechanism is uncertain. However, when we have a strong test of the hypothesis—that is, the test is specific and clearly developed from a detailed mechanistic hypothesis—then either a positive outcome suggesting a relationship or a negative outcome rejecting the proposed hypothesis are more informative.

5.3. MODELING TEMPORAL VARIATION IN BIOMARKERS

Our interest in using a biomarker usually focuses on one of three aspects: (1) its relationship to exposure, (2) its relationship to internal dose, or (3) its relationship to adverse health outcomes or actual disease. Sometimes we are interested in all three when we try to answer questions about exposure and risk. For example, if hemoglobin adducts in the blood of a woman with ethylene oxide exposure are measured and found to be 4 pmol/g globin, what does that tell us about her exposure or her disease risk? Do the biomarker data tell us more than does the knowledge that she was exposed to 3.5 ppm of ethylene oxide for 4 hr in a surgical preparation area on the day we collected the blood sample? As noted in Chapter 4, internal levels of chemicals vary over time, which limits our ability to interpret the meaning of a single exposure or biomarker measurement and to relate it to exposures. We can use the difference equation models discussed in Chapter 4 to describe the complex temporal behavior of a biomarker's concentration within the body: how it is affected by route of entry, its lipid solubility, its formation and/or removal rates, its storage in slow turnover compartments, such as bone or fat, and its secondary effects on metabolism.

In some cases, several biological measures of exposure may be available, but they represent dramatically different time scales. Modeling can help us determine a biologically meaningful quantitative relationship between the biomarker and exposures. Lead is a good example (Hu et al. 2007). Blood and urine lead are both good measures of recent exposure, but they represent different aspects of internal lead stores. A large percentage of lead initially entering the bloodstream is bound to the proteins of the RBC; only a small fraction remains in the serum. Lead in urine is derived from only the fraction in the serum that can pass through the glomeruli in the kidney, avoid resorption, and pass into the urine. Both blood and urinary lead levels vary with changes in short-term exposure over a few days. Lead is also removed from circulation by deposition in the bones. Trabecular bone (such as the tibia, or shin bone) binds lead tightly, with an approximate 10-year half-life; cortical bone (such as the patella, or kneecap) stores are less tightly bound, with an approximate 2-year half-life. During exposure, blood and urine values will reflect roughly current exposures, but bone lead will reflect the long-term accumulation and will often be poorly correlated with the current blood levels because the latter have had little effect on the current bone lead levels and vice versa.

With steady, approximately constant exposure levels, the biomarker values are easy to interpret, but when there have been temporal variations with periods of intermittent high and low exposures, then an exposure history is needed to fully understand the meaning of the biomarker. For example, during a period of no lead exposure, blood and urine values will decline until they are defined by dietary input and the lead coming out of the cortical bone. With a much longer period of no exposure, the values will decline further until they are defined by diet and the lead coming out of the trabecular bone. Probably before that happens, the tiny daily intake of trace amounts of lead in normal diets will be enough to balance the losses from the bones. The message is that even substances with very long half-times are not permanently stored. It was proposed that body fat levels of dioxin in Vietnam War veterans could estimate exposures received during the war—1 or 2 years of exposure to herbicides sprayed to defoliate the jungles approximately 35 years previously. However, the serum levels measured in 2001 are uninterruptible both because the exposures were unknown and because the half-time of dioxin removal was variable across individuals. As a result, an individual with a relatively high serum level in 2001 could have had a high exposure in Vietnam and normal clearance half-time, or a low exposure and slow clearance, whereas someone with a less than detectable serum level in 2001 may have cleared everything from a moderate or low exposure, or had a high exposure with a relatively rapid clearance. When clearance processes have operated over a long period of time, 4 or more half-times, individual differences in kinetics become important.

Estimating Exposure from Biomarker Data

If we know the quantitative relationship between a biomarker level and exposures, given a measurement of the biomarker, we can estimate the exposure that produced it.

Empirical Modeling for Biomarkers

We noted earlier that there are two types of toxicokinetic models, empirical and physiologic models. When there are human temporal data on clearance, we may be able to fit an empirical model and use it to estimate internal levels. We can sometimes use a one- or two-compartment empirical model to represent the key limiting steps affecting a biomarker's internal concentration, such as metabolism and excretion. For example, in some cases, such as for ethylene glycol methyl ether, which is equally soluble in all tissues, a chemical's concentration in the body can be modeled as a simple balance between input and removal. However, we may also need another model for the uptake and transport of the agent to the location where the biomarker is formed, such as the liver.

In the late 1980s, Rappaport, Droz, and others were concerned about the transmission of exposure variation to the "body burden" of exposed individuals, with the assumption that the risk of effects was proportional to the total body burden (Droz 1993). The estimated body burden levels were expected to represent the levels seen in biological samples, such as blood and urine. Simulation studies were used then and more recently to explore the behavior of internal concentrations with an empirical, one-compartment model of the body (Tardif et al. 2002). Observed clearance half-times for a range of toxic substances defined the removal of the agent from the body compartment. They observed that the variation of biomarkers with long averaging times reflected little of exposure variation happening over shorter times. Conversely, virtually all of the exposure variation for longer averaging times than the half-time of the biomarker was transmitted. Although this knowledge is very helpful, it has some important limitations. First, for many substances, metals and lipid-soluble materials especially, the body does not behave like a one-compartment model; two or three may be needed to make reasonable estimates of blood or urine or tissue levels. In that case, the idea of body burden is too vague. Second, everyone does not have the same half-time for each substance. Although this model may make reasonable estimates of clearance, the estimated internal concentrations may be poor. A model helpful for one purpose, estimating clearance, may not work well for another purpose, estimating internal dose.

An example may make this clearer. Methyl t-butyl ether (MTBE) has been used as a gasoline additive to increase its octane rating and improve combustion performance. However, it has also been found to be a source of water pollution because of leaks from storage tanks. Both the parent compound MTBE and one of its major metabolites, t-butyl alcohol (TBA), have been proposed as biomarkers of exposure via drinking contaminated water (Johanson, Nihlen, and Lof 1995; Buckley et al. 1997; Nihlen, Lof, and Johanson 1998; Prah et al. 2004). If we measured the parent and a metabolite biomarker at a series of times during and after a laboratory exposure to MTBE as shown for blood (MTBE) and urine (TBA) in Figure 5.2, we will observe different patterns associated with the exposure because of the toxicokinetics. We could fit those data with a simple model such as shown in Figure 5.3. Then we could use the model to predict internal concentrations of the parent compound or the metabolite with different exposure patterns, as shown in Figures 5.4–5.6. Figure 5.4 shows the relationship between the biomarker concentrations and past exposures, which clearly depend not only on the exposure intensity and its duration but also on when we measure the biomarker relative to when

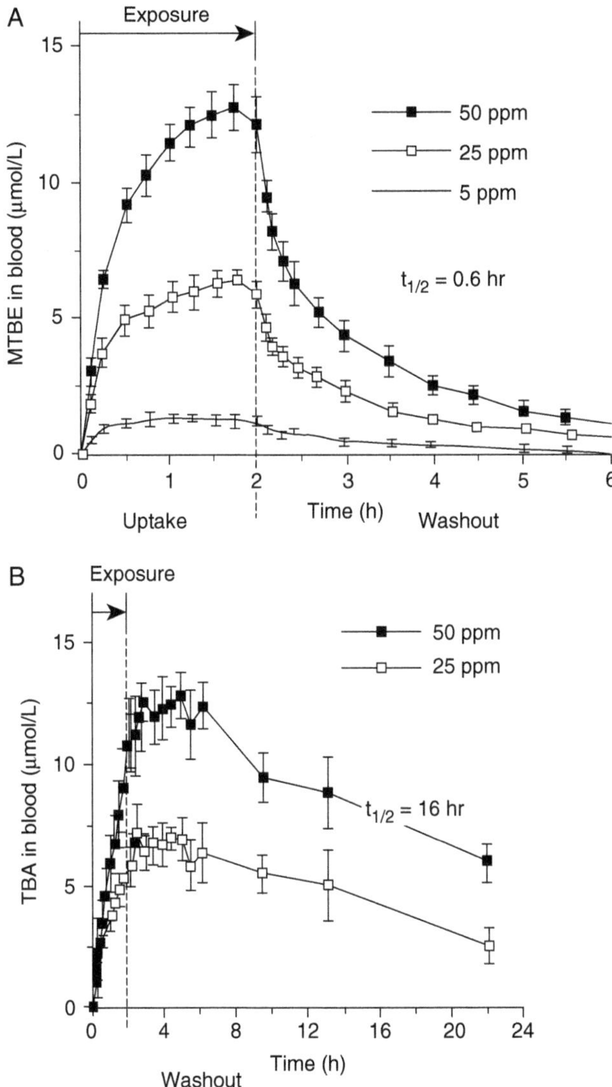

Figure 5.2 Experimental time course for 10 male, healthy volunteers exposed to MTBE by inhalation.
(Figures reprinted with permission from Johanson et al., 1995.)

the exposure occurred. Consequently, if we have only one observation and no other data, we must make many assumptions to define its meaning. If we know nothing about the exposure, then the single value may tell us only that the person was exposed at some time in the past, because there are many possible time courses that can give us the same value at that time point. Thus we must know something about the exposure time course to interpret a biomarker value.

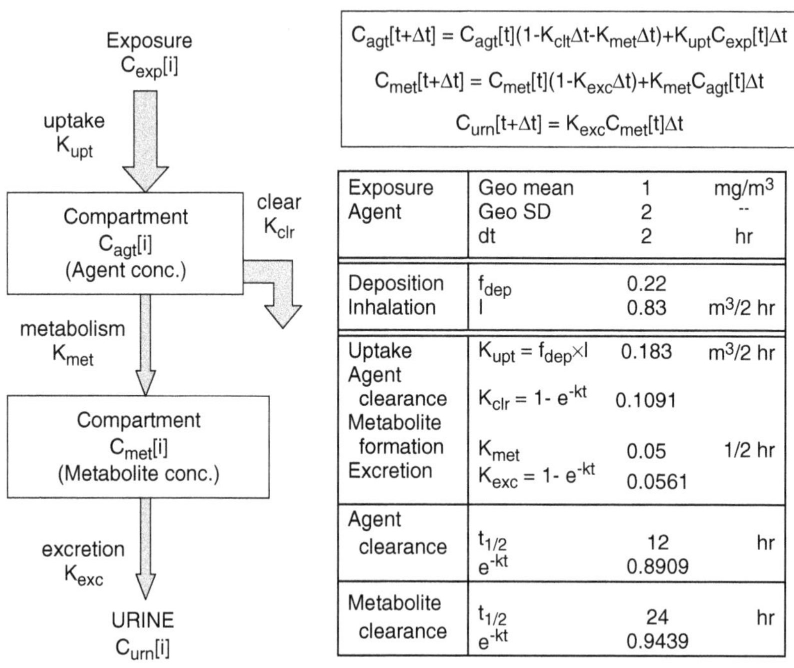

Figure 5.3 Two-compartment empirical model for a simulation of a hypothetical agent (parent) and its metabolite, with a table showing the model parameters.

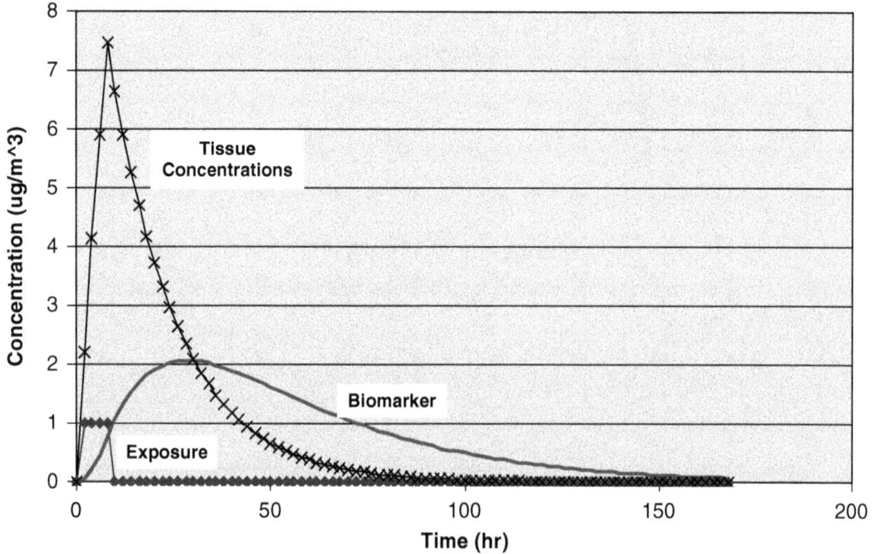

Figure 5.4 Time course for a two-compartment empirical model of exposure to the parent and a urinary biomarker metabolite, which is formed in the central compartment of the model described in Figure 5.3.

Temporal Relationships

The most common assumption, often implicit, used with biomarkers is that the exposure has been approximately constant and the biomarker is in a steady-state balance with the intake. We have made a simple simulation to show you the relationship between exposure to the parent compound and a metabolite used as a urinary biomarker. The time course for the simulation shown in Figure 5.5 is a hypothetical 5-day constant exposure to the parent compound, in which each day is identical. First, as expected from our earlier examination of exposure and dose, we can see that the tissue concentration builds up over time. Second, the tissue concentration after several days of exposure is the sum of residuals remaining from each of the prior days' exposures. You can see that the parent has peak tissue concentrations on Thursday and Friday that are nearly the same, showing there is a balance between input and output, but it is not a smooth curve. The tissue concentration varies strongly because it has a 12-hour half-time, and during a 24-hour period it fluctuates through a range of values. The metabolite biomarker with its 24-hour clearance half-time is still increasing after a week of exposure. After another week of exposure the metabolite will also stabilize, but because of the weekend the metabolite's concentration will fall off, but it will not decline to baseline. If we wished to use a metabolite as a biomarker of exposure, we would want to collect our samples at the end of the week, not the beginning.

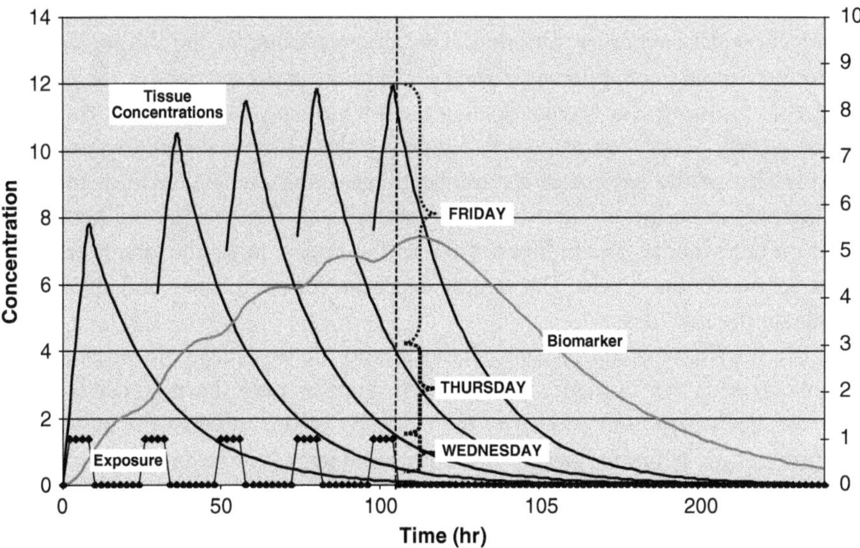

Figure 5.5 Time course of the tissue concentration for the parent and the relative daily contributions to the total urinary biomarker level observed on Friday after 5 equal days of exposure. Each day's exposure adds to the residuals from the previous days. (Model is described in Figure 5.3.)

Figure 5.5 also shows the relative contribution of each day's exposure to the parent compound in the tissues. Assuming that each day's contribution operates independently of the other days, the contribution from each day will decline exponentially or geometrically, with the fraction remaining after a period of time, *T*, equal to $\exp-\{(\ln 2 / t_{1/2})T\}$. The vertical dashed line marks the end of the work shift on Friday, and the distance between the points where the agent's tissue concentration lines cut the dashed line indicate the amount from each day. Given equal daily exposures, Friday makes the largest contribution. Thursday is next largest, and so forth. The total tissue concentration is a weighted sum of the contribution from each past day's exposure. Because the metabolite is cleared more slowly, with its half-time of 24 hours, it builds up over a longer period of time, and the sharp tissue concentration fluctuations are mostly smoothed out. The metabolite continues to increase for several hours after exposure has stopped. It is easy to see that the total metabolite has the same pattern of residual contributions from past daily exposures that occurred for the parent compound, but they would be lagged further in time. What happens to this pattern if the exposures have typical day-to-day variability?

Temporal Effects of Exposure Variability

Figure 5.6 shows a simulation of normal exposure variation and its effect on the agent and metabolite biomarker levels. The 2-hour average exposures in the figure have a geometric mean of 1.0 mg/m^3 and a geometric standard deviation of 2.0, which is modest exposure variation. The large variation in the 2-hour values is smoothed out and not apparent in the agent's tissue concentrations or the metabolite levels. However, the average level of the 8-hour exposure does affect the height of the weekly peaks. You can see in Figure 5.6 that individual high-exposure days sharply change the pattern of accumulated agent and metabolite from that seen for a steady exposure in Figure 5.5. For example, in Figure 5.6A the high exposures on the Monday, and in Figure 5.6B on the Friday, make the largest contributions to metabolite levels. The largest exposure for each week and its residues dominate the next days.

The relative contribution of each day's exposure to a single measurement of a biomarker is shown in Figure 5.6. At each point in time the residuals from the previous day's exposures add up to the total observed, and when the half-time of exposure is 12 hours or longer, there is considerable carryover. Figure 5.6 also shows that the biomarker from a single high exposure may not be evident on the day it is received, but it can dominate the biomarker several days later.

Rappaport and others have explored the mathematical behavior of average biomarker levels given various variable exposure inputs (Rappaport and Spear 1988). They found that over long time periods the biomarker levels were closely related to the average exposure by statistical relationships. The current findings do

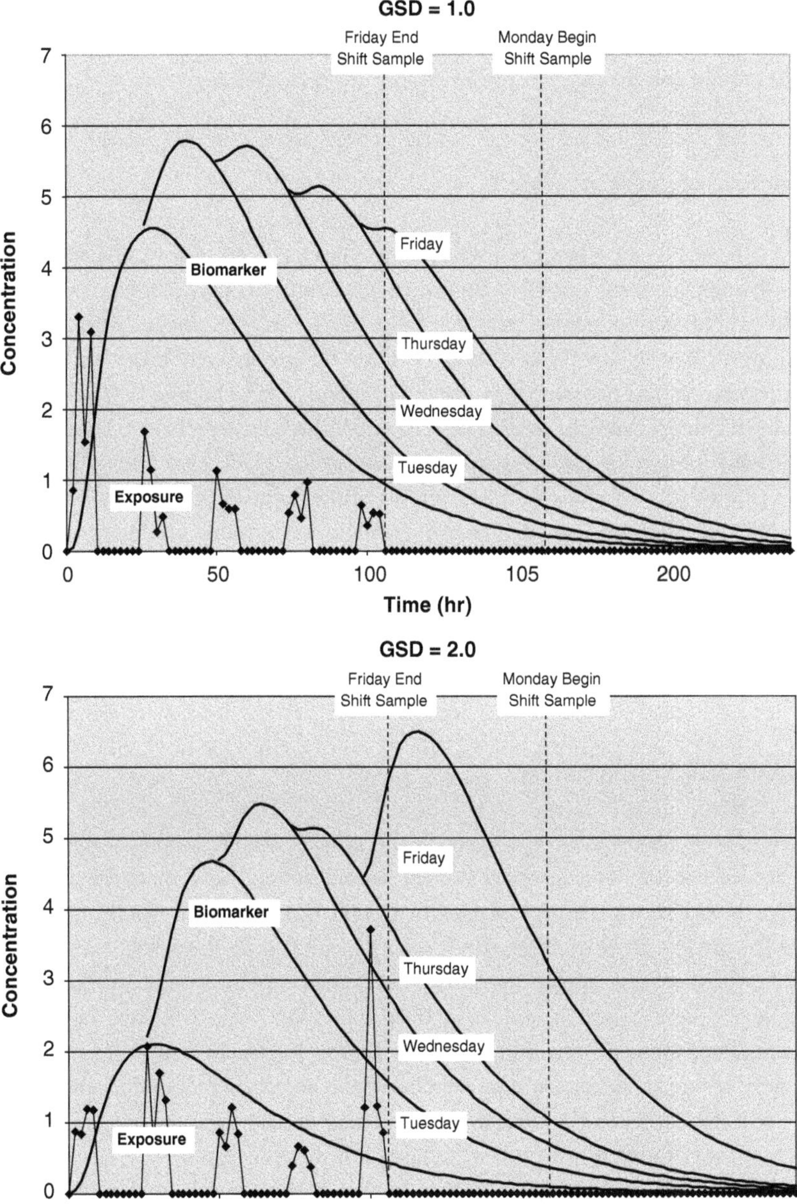

Figure 5.6 Two examples of relative contributions of workday exposures to a urinary biomarker collected on either a Friday afternoon or a Monday morning. The Friday exposures contribute little to the Friday afternoon sample but may dominate the Monday sample. Model responses to 5 days per week of 8 hr of 2-hr random exposures using GM = 1 and GSD =2.

not conflict with their conclusions. But they do demonstrate that over observation intervals close to the half-time (8-hr data vs. 12- or 24-hr half-times) we must be careful about interpreting biomarker-exposure relationships.

Limitations of Empirical Models

These are most accurate for populations similar to the subjects providing the data—young, healthy, white male college students, in many cases. Even so, the model will not provide good personal estimates for every subject. More important, the empirical model estimates may be poor for groups not included in the test population, such as nonwhites, women, the elderly, and children. Empirical models may have limited ability to predict dose for individuals with preexisting diseases. Also, individuals who have variant genotypes that affect P450, glutathione-S-transferase (GST), or other metabolic enzymes may be poorly predicted. Finally, in the situations in which there are nonlinear metabolic or binding processes, the simple empirical models will do a poor job of prediction. The result is that we do not know how well we have estimated the concentrations for individuals who are not members of the originally measured group; the estimates may be good or terrible, but we do not know.

PBTK Model for Biomarkers

A number of researchers have developed complex PBTK models to describe the kinetics of specific substances that form biomarkers. These range from metals—lead in blood and arsenic in urine—to dioxin in serum and urinary metabolites from solvents, such as trichloroethylene and n-hexane. In each case, the model was specific for the agent and route of entry and the biomarker media. Many of these models are very complex and have been validated with extensive lab and field studies of exposed subjects. However, we do not need such a complex model if we are only trying to generally represent how the agent or metabolite changes over time without trying to estimate the actual tissue concentrations, but we do require a minimum complexity to capture the key limiting processes.

PBTK models have the advantage of being more generalizable, but that comes with a price: many more parameters to estimate. We can use the PBTK model with two compartments (Figure 4.7), and we will have the same relationships that we saw in Figures 5.5 and 5.6. All of the processes of uptake, transport, metabolism, storage, and excretion that we explored for tissue dose in Chapter 4 are very important for biomarkers and their relationships with external exposures. The critical question for an exposure biomarker concentration is, What can we infer about the exposures that caused it, or about the critical tissue dose causing adverse effects? There are several points we must consider (Tan, Liao, and Clewell 2007).

1. The parameters for the standard man or woman can be used in the model, but that introduces some error relative to specific individuals. If we can measure a few key parameters for individuals, such as age, gender, height, and weight, then a PBTK model can be individualized and we can reduce error relative to using general parameters, such as the model shown in Figure 4.7 and in Table 4.3 in the previous chapter.

2. Additional important obtainable parameters, octanol/water partition coefficient for the agent and blood and tissue partition coefficient (PC) values, are available for some common substances, or they can be estimated with algorithms (Payne and Kenny 2002).

3. Physiologic algorithms are available for estimating cardiac output, total and alveolar ventilation under various exercise levels, and tissue volumes and blood flows during exercise based on an individual's height, weight, gender, and age (Clewell et al. 2002; Price et al. 2003; Sarangapani et al. 2003).

4. Personal physiologic measurements can greatly reduce error in PBTK models, such as pulmonary ventilation during light exercise, $PC_{b/a}$ for inhalation exposures, and fraction of body fat for lipid soluble materials.

5. Time-course measurements of blood concentrations of parent and metabolites and concentrations of urinary metabolites can be fitted to estimate parameters and provide highly personalized models.

These parameter estimates can reduce error for most people. However, individuals at the ends of the normal distribution, such as those who are very thin or heavy or heavily muscled body builders, will be poorly estimated by the algorithms. There are also ethnic differences in metabolic enzymes associated with genetic background that may be important for estimating metabolite removal or formation rates.

The most important epidemiologic application for PBTK models is cases in which one or more internal processes are nonlinear. Most commonly this occurs when concentrations are "high," which depends on the agent and the individual's enzymes, or transport proteins, or other capacity-limited processes. These can be detected by looking for changes in the exposure relationship with the biomarker, as noted previously.

5.4. BIOMARKER CALIBRATION AND VALIDATION

Cross-Sectional Measurements of Biomarkers

In the past, it has been common to collect validation data by selecting a population with known exposures, observing exposure on one day, and then collecting a biomarker sample at the end of the exposure for each member of the population—a small cross-sectional study. The utility of this approach depends on the half-time of the biomarker. It works best for half-times of 1 to 4 hours. Even at its best, this approach produces very noisy relationships, because there are many sources of uncontrolled variation. In occupational studies, when the exposure timing is poorly matched to the biomarker, sometimes little or no relationship is apparent. The reason for this can be seen in Figure 5.6, in which each day's parent or metabolite concentration is strongly influenced by the prior days' exposures, especially the occasional high values in the upper tail of the exposure distribution. Additionally, the variable internal processing of toxicants across individuals, especially the metabolic processes, will result in large variability in both the exposure and the biomarker relative to each other with cross-sectional measurements.

This problem can be solved with an appropriate prospective sampling strategy that considers both the within-subject temporal exposure variation and the kinetics of the biomarker. Tan and coworkers have investigated the problem of estimating exposures from biomarker data using PBPK models, "reverse dosimetry," for chloroform and found that a simple regression calibration gave poor results (Tan et al. 2006; Tan et al. 2007). Much better predictions of population exposures associated with a given distribution of biomarker data were possible when PBPK modeling, route of exposure data, and Monte Carlo simulations were used.

Prospective Calibration with Repeated Measurements

As noted earlier, the weighting of temporal exposure by a biomarker is not even over time but weighted more toward recent exposures by biological processes. A general relationship between exposure, *E*, and a biomarker, *B*, is shown:

$$B_j = \sum_{i}^{n} [\mathrm{w}_i \cdot k_{form} E_{ij}]$$

Equation 5.1 General relationship for a biomarker with exposure

The weighting for the ith time interval prior to the current time is w_i, $k_{form}E_{ij}$ is the amount of biomarker formed from the *j*th individual's exposure, and B_j is the total biomarker that is measured. This relationship assumes that the internal quantity of biomarker produced is directly proportional to the exposure and that each day's biomarker production is independent of prior days' exposure or biomarker, so that each day's contributions add on to the residues remaining from previous days, as shown in Figure 5.6. This is true for biomarkers in general. One source of problems in the relationship is background levels from any source besides the one of interest, which can reduce our ability to characterize the relationship and/or define effects from the source of interest. Therefore, when we are doing a calibration study, we want to make sure that the exposures are well into the detectable range by conducting a pilot study.

One significant advantage of biomarkers is that we can test hypotheses about a possible causal pathway for an effect, if there is evidence that the biomarker is closely related to the putative agent. Importantly, if the concentration of the biomarker is closely related to the agent, and we suspect that there are nonlinear processes affecting the agent, the level of the biomarker may also be subject to these effects. For example, if an epoxide metabolite is the agent of DNA damage causing a cancer risk and the formation process for the epoxide is saturating with high exposures for some subjects, then using a biomarker of the metabolite may be a better dose metric than something calculated from the exposures because the latter may be overestimating the agent levels.

5.5. POPULATION CHARACTERISTICS OF BIOMARKERS

Cross-sectional measurements of biomarkers gathered from an exposed group reflect the variation in exposures across individual's and in the toxicokinetic parameters across individuals. As a result, it will overestimate the exposure variation. However, if we have repeated measures of exposure and biomarker levels, we can form time courses and examine the biomarker as a response to exposure and internal processes, which will provide us with a powerful way of personalizing the dose metrics in a health study. We discuss this more fully in some examples in later chapters.

6 Disease Process Models

All substances are poisons: there is none which is not a poison.
The right dose differentiates a poison from a remedy.

Paracelsus, 1493–1541

Paracelsus updated:
The dose of the mixture makes the poison,
but differently for different individuals and
differently at different times during growth and development.

Peter Montague, 2002

The hypotheses we accept ought to explain phenomena which we have observed.
But they ought to do more than this: our hypotheses ought to *foretell* phenomena which have not yet been observed.

William Whewell (1794–1866),
English mathematician, philosopher

6.1. OVERVIEW

The dose-response relationship is a fundamental paradigm of toxicology. But in observational epidemiology (unlike in toxicologic experiments), we rarely know with any precision the amount of an agent taken in over time, much less the tissue dose, and we rarely can quantify the time course of physiologic responses, but often must content ourselves with a late sign of morbidity such as clinical diagnosis or death. Our goal in this chapter is to show how we can use basic descriptive information about the time course of observed biological responses and reasonable estimates of the time profile of tissue concentrations to develop summary dose metrics that can be used to study quantitative exposure-response relationships.

Three broad groups of pathophysiologic processes link the tissue burden and damage: (1) direct damage, such as cytotoxicity or mutation; (2) cell signaling and subsequent cellular responses; and (3) tissue repair processes. Although there is an almost infinite number of combinations of specific processes within these three categories (and a great deal that is unknown), we believe that epidemiologic studies can benefit from a dramatic simplification: four basic disease process (DP) models (shown in Table 1.1) that can define the general mechanistic links between dose and response. As noted in Chapter 1, these four result from the combination of two fundamental types of tissue response—proportional and discrete—with two fundamental types of recovery or repair: reversible and irreversible. Many exposure-disease associations which we wish to study in populations can be fit into one of these types. Each of these four DP models has distinct characteristics, and the classification of an exposure-disease association is guided by relatively simple information, particularly about the time course of exposure and resulting disease during and after exposure. Pratt and Taylor (1990) made the same basic argument when they noted that the temporal processes of dose and response reveal essential information about the underlying pathophysiologic mechanisms. In this chapter, we introduce the four DP models, leaving more detailed development of each for later.

6.2. THE NATURE OF ADVERSE EFFECTS

Basic Damage-Repair Model

A very general description of cellular and tissue toxicity was introduced in Chapter 1 (Figure 1.3).[1] This model is based on the sequence of events that typically takes place when a tissue is damaged by a toxic material (Gregus and Klaassen 2001). These events are:

1. Cells are damaged or killed by the agent in the tissue.
2. Cell signaling by injured cells produces inflammatory/immune responses and apoptosis initiators.
3. Depending on the amount and duration of damage, there are different responses. With moderate damage of short duration, apoptosis of injured cells and cleanup of debris by recruited macrophages and other white blood cells reduce inflammatory signals. However, if damage is severe and/or

prolonged, necrosis of damaged cells releases cell contents, leading to an amplified inflammatory response.
4. Signaling by injured and adjacent healthy cells stimulates cell proliferation.
5. New cells form and move into the damaged area, and the tissue matrix is repaired.
6. Repair signals are damped by feedback, and repair stops.

Although this is presented as a sequence, the steps overlap, especially with continuing exposure. Unless toxicant levels are high and the damage extensive, the repair processes are usually not directly modulated by the damage processes (Mehendale 2005). Because of tissue inhomogeneities of cell state and toxic agents, the damage and repair processes also may not be occurring in the same locations in a tissue, especially when they produce focal lesions. All damage and repair processes are controlled by cell signaling, which initiates and stimulates them, and by feedback signals, which damp and stop them. The processes do not occur at the same rates: cell damage happens at varying rates, but can be fast—cell signaling is generally very fast, whereas tissue repair is relatively slow. The damage and repair processes can also be modified (amplified or suppressed) by other processes or signaling. For example, a strong inflammatory response can directly injure cells and lead to a feedback loop, which can prolong and increase the damage response. Misrepair, or inadequate repair, can be an important cause of long-term irreversible effects; fibrogenesis is perhaps the best example.

An Example of a Damage-Repair Process

Adverse effects require time to develop, and the pattern of development is determined by the damage-and-repair aspects of the mechanism. For example, the primary response or effect of a short period inhaling a high concentration of nitrogen dioxide gas is injury and death for cells lining the terminal airways and alveoli of the lungs (Parkes 1983; Crapo et al. 2004). Damaged and killed cells release cellular messengers and leak fluid and serum into the air spaces of terminal airways and alveoli, causing slowly developing pulmonary edema. In response to the release of signaling proteins, there is a recruitment of macrophages and other white blood cells into the damaged areas as part of an inflammatory response. With a relatively high exposure, the subject will have delayed onset (3–30 hr) of lower respiratory symptoms: paroxysmal coughing, wheezing, bloody sputum, chest tightness, and difficulty breathing (Horvath et al. 1978). Pulmonary function tests will show sharply reduced vital capacity and increased airway resistance, and a chest X-ray may show changes consistent with pulmonary edema. Each of these distinct responses may have a somewhat different time course because each has a different mechanism.

If the NO_2 damage is severe, there can be substantial overall pulmonary congestion and dramatically reduced pulmonary function, leading to death. If exposure is less severe, then repair will begin shortly after damage begins, although the former is generally slower than the latter. During long-term, low-level exposures, damage and repair processes will proceed concurrently. Symptoms of lung damage may last several weeks to months before full recovery. In more serious cases, there can also be misrepair and incomplete repair, which does not restore full functionality of the injured tissues. Misrepair of the tissue matrix can produce pulmonary fibrogenesis by forming scar tissue that interferes with normal lung functioning. Repeated short-term, intense exposures also can lead to permanent destruction of terminal airway structure, producing severe effects such as bronchiolitis obliterans.

In this example of NO_2, we have emphasized the time courses of tissue concentration and clinical responses because these reflect the rate-limiting processes and are critical determinants of the health outcomes that will be observed at a particular point in time. A low concentration of NO_2 will probably not be found to be associated with any physiologic response more than a few hours after exposure ceases because repair processes will have already begun, and recovery will be essentially complete (Figure 1.1 can represent the likely pattern of exposure and response in this example). Depending on when and how we look, we may observe a modest irritant respiratory response or nothing at all.

Repair Processes

Given only limited general information on mechanism, plus data on the time course of effects, we can usefully distinguish the type of repair processes by observing whether the adverse effects appear to be *reversible* or *irreversible*, because this distinction will generally determine substantially different temporal exposure-response patterns (Table 1.1). A low concentration of NO_2 will lead to a reversible response. If exposure is "turned on" at a constant intensity for a period of time and then switched off (as in Figure 1.1), the tissue concentration rises until exposure stops, at which time the concentration falls back to zero. The response follows after some delay but also returns to zero once repair is complete. One can see from Figure 1.1 that one would observe a wide range of response intensities at different times postexposure, as the balance of damage and repair shifts.

Rate Limiting Steps

The fact that a pathologic process can go no faster than its slowest steps results in simpler temporal dynamics that can be used to formulate exposure-response models. The cascade of intermediate effects may be short, as it is for a sensory effect such as nasal irritation, or it may be long and intricate, as it is for cancer. But in either case, the observed temporal behavior may follow a relatively simple time pattern. An example of this is the two-stage clonal expansion model of carcinogenesis (presented in detail in Chapter 15), which assumes two rate-limiting steps: initiation and promotion with subsequent clonal expansion of the fully transformed malignant cell (Moolgavkar 1991). It is now clear that cancer involves many more than two steps, but Moolgavkar and coworkers have shown that a mathematical model structured around just two serial steps describes the temporal behavior of cancer incidence to a surprising degree. If this insight is more generalizable, then it raises an obvious question: How can we identify the rate-limiting steps of other disease processes? Putting it another way: If there are only a few rate-limiting steps, what will the resulting temporal patterns look like? Empirically there are many mechanistically unrelated irreversible effects that have chronic cumulative effects. For example, chronic hearing loss from long-duration exposure to high noise levels shows the same temporal pattern as chronic loss of lung function from long-duration exposure to cigarette smoking. This has often been noted by epidemiologists, but few have considered why it might be so. It is our premise that the wide array of possible mechanisms reduces to four basic temporal patterns because of the simplifying effects of rate-limiting steps.

We propose that by examining the type of response (proportional or discrete) and its temporal pattern (repair or cumulative) for a health effect, we can choose one of the four disease process models (Table 1.1) presented in this chapter and developed in more detail in Chapters 10 and 13. Furthermore, the essential parameters of a suitable model can often be estimated from animal studies or clinical observations or by empirical estimation from epidemiologic data. Critical guidance often comes from the observed time course of a response, because the time pattern will reflect the limiting steps. Natural experiments occurring during accidental environmental or occupational exposures can provide useful information on how human responses vary over time and among individuals.

6.3. DEFINING THE TIME COURSE OF EFFECTS

When studying an exposure-disease relationship, we would like to limit our exposure evaluation to the period in which the exposure actually contributes to the observed adverse effect. This period is called the *etiologic time interval (ETI)*. Estimating the beginning and end of this interval is an important step in designing an epidemiologic study. All of the time during which someone is exposed to a toxic agent does not necessarily contribute to risk of the outcome at a later point in time: if there is an effective repair mechanism, for example, then exposures in the distant past may not be relevant to current effects. The ETI does not necessarily continue until the onset of disease; some pathologic processes may lag behind exposure variations, and others, once fully initiated, may continue to develop without additional exposure. For example, once a fully transformed malignant cell has developed, it is assumed that its growth into a clinically significant tumor is determined by factors that are independent of the exposure that caused the necessary initial cell transformations. The relevant period may also include some time after the exposure stops if the agent has not been fully cleared from the tissues and damage and repair processes may continue to modify the type and intensity of response.

Choosing one of the four DP models provides some guidance on the definition of the ETI (Figure 6.1). For the two reversible DP models, the maximum length of the ETI (the period from ETI_0 to ETI_T in Figure 6.1A or 6.1B) can be estimated as approximately 5 half-times of the slowest repair/clearance step. This follows from the general rule that after 5 half-times, only about 3% of an exponentially repaired damage (or exponentially cleared substance) will remain. For the irreversible models, nearly the entire exposure period (lifetime for environmental exposures, "working lifetime" for occupational) defines the maximum ETI (the period from start of exposure to ETI_T in Figure 6.1C). The end of the ETI is generally the moment of incidence of the disease, the onset of symptoms, or the time point at which one measures physiologic function or biomarker of effect. Repeated measures of an adverse effect and its associated exposure may be needed to observe the ebb and flow of a response when exposure changes over time, which may also provide strong data supporting a causal relationship. However, as noted earlier, the ETI may end before this if the development of the outcome lags behind the causal exposure or, stated another way, there is a significant time between the final step in the pathophysiologic process and the diagnosis, onset of symptoms, loss of function, and so forth. The irreversible DP example in Figure 6.1C illustrates this case. For reversible processes such as mild inflammation, the lag may be a matter of seconds or minutes, whereas for irreversible chronic processes such as cancers, it is often thought to be years.

How can we develop a descriptive time course for an exposure-disease association and estimate the ETI? We can start with the kinds of observations in patient files and published case reports, interview data from those affected, and published experimental and epidemiologic studies. The questions we want to answer are:

1. How long after the beginning of exposure do the signs and symptoms begin to appear? Is there a sequence of progressively more adverse effects that appear over time? If so, what is the timing?
2. During a period of relatively stable exposure, do the effects continue to worsen, or do they stabilize at some level or even begin to decline? For discrete responses, does the number of individuals responding increase if exposure intensity increases?
3. When exposure stops, do the effects diminish with time? How long before they begin to diminish noticeably? How long before the intensity of the effect(s) declines by approximately 50% (approximate half-time of repair or recovery)? For discrete responses, do the subjects stop responding, or is there a decrease in the frequency of responses? How long before the number responding declines by 50%?
4. Do the effects eventually disappear, after a period without further exposure? How much of a residual effect remains that does not disappear?

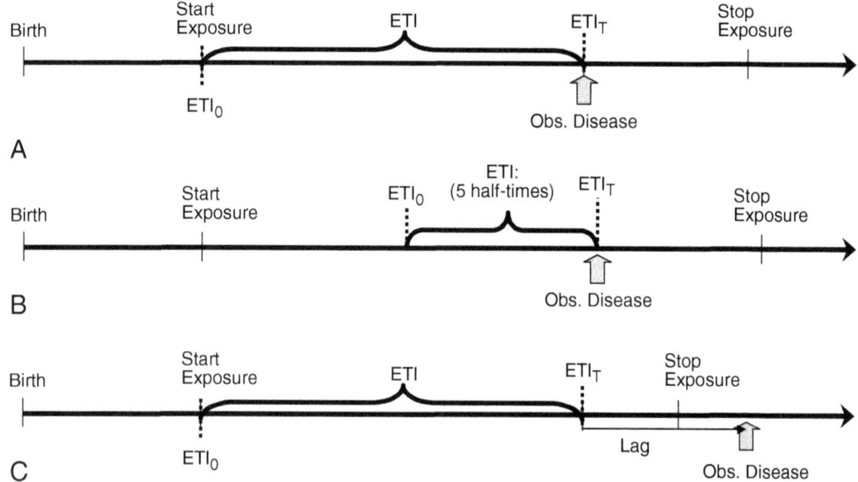

Figure 6.1 Examples of how etiologic time intervals (ETIs) are defined for various exposure-disease processes.
Part A: The generic time course of exposure and disease, indicating key time points and the start and end of a hypothetical etiologic time interval (ETI_0 and ETI_T, respectively).
Part B: The time course of exposure and disease for a chemical causing a reversible disease process with ETI defined by the half-time of repair.
Part C: The time course of exposure and disease for an irreversible disease process with a lag because of disease development.

The answers to these questions—even approximate answers—provide critical raw data to formulate a dynamic model of the key disease steps. Each different physiologic response to an agent (cough, stomach pain, chest X-ray changes, finger numbness, kidney function decrements, for example) may have its own time course, different from others caused by the same agent, especially if the different responses are not on the same causal pathway. For example, modest short-term nitrogen dioxide exposures during silo filling on farms produces a late cough from chest congestion after each exposure, but accumulating irreversible fibrotic changes on chest X-rays are evident only after many such episodes. Someone who suffers from a severe cough after each acute exposure may or may not go on to develop more serious fibrotic changes many years later.

In summary, one of the first steps in understanding the time course for an adverse response is to develop a time line showing: the time between the onset of exposure and the appearance of the first observable effects; the development of the effects and their intensity with continued exposure; and, finally, what happens when exposure stops. Simple descriptive data like these often allow you to provisionally choose one of the four disease process models.

6.4. PROCESS MODELS OF ENVIRONMENTAL DISEASE

The four disease process models provide a rational basis for formulating a biologically based exposure-response investigation of many environmental hazards (Table 1.1), and here we provide a basic summary of each. We provide a simple "iconic" example for each that may help to illustrate the basic mechanism, as well as key assumptions and temporal patterns. We also note examples of epidemiologic studies of exposure-disease associations that fit this type. In these examples, the authors use different terminology and analytic approaches from what we propose, but our intent in pointing to these examples is to show the wide range of environmental health investigations that could benefit from using disease process models.

Disease Process DP-1: Reversible Proportional

Iconic example: ammonia → airway inflammation

Imagine pouring a small amount of ammonia into a shallow dish and allowing a minute for the ammonia to diffuse into the air above the dish. A localized high concentration of ammonia will be produced, perhaps 50–100 ppm. If you were to place your face about a quarter of a meter above the dish, you would, within a few seconds, experience sharp eye and nose irritation and a strong desire to remove your face to a safer location. Once free of the ammonia cloud, your irritation would rapidly subside (Figure 6.2A). This irritation develops and dissipates rapidly, which is characteristic of sensory responses such as odor and irritation. However, if you stayed in the cloud much longer, you could initiate inflammatory responses that are slower to dissipate.

Key Assumptions

1. The rate of damage is proportional to tissue concentration.
2. Damage with short duration exposures is (almost) entirely reversible. Residual damage that may accumulate over time may also occur with longer or more intense exposures, but this process would be modeled differently—probably using DP-2.
3. The rate of repair is proportional to total damage.
4. The damage and repair processes are independent of each other and can run concurrently, in the sense that damage is a response to the tissue concentration and repair is a response to the damage.

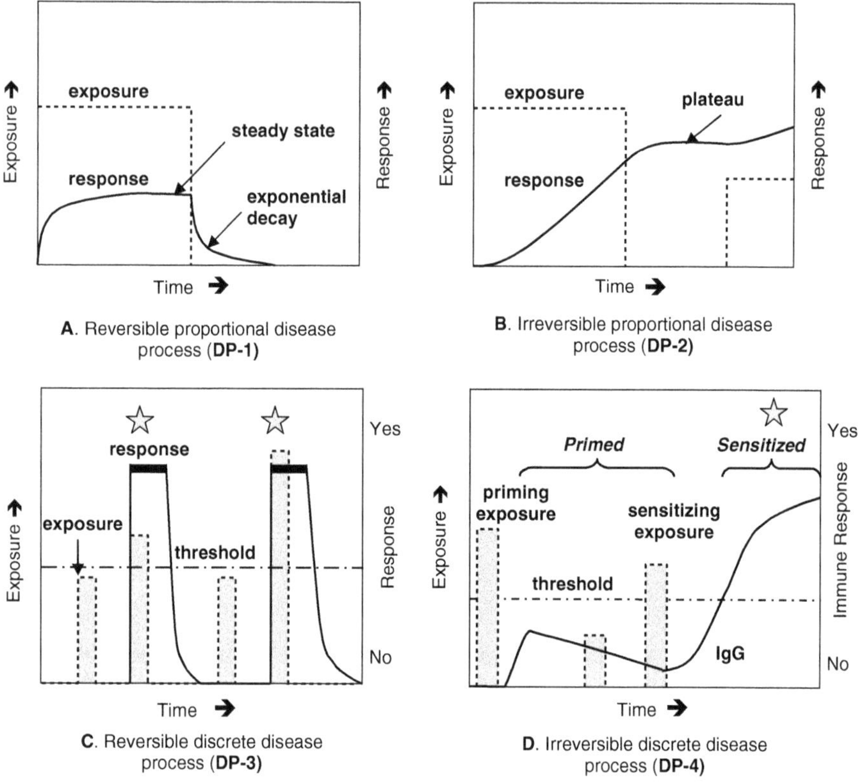

Figure 6.2 Idealized typical time courses for each of the four disease process models (panels A through D). In each, an exposure trend (dotted line) leads to a trend in a continuous measure of response (solid line) or discrete response events (*).

Typical Temporal Patterns

Repair processes are often slower than damage and determine the rate of the entire process. The ETI is often fairly short, as exposures in the distant past may be completely repaired. The start of the ETI can be defined as no more than 5 half-times of the slowest repair or clearance time before the moment of incidence, that is, the time of occurrence of the outcome. The end of the ETI is typically the moment of incidence. If there is a lag between last effective exposure and the response, it is usually very brief.

DP-1 in Practice

Reversible proportional response mechanisms for air pollutants have been fairly widely studied through human experimentation in exposure chambers (McDonnell 1993). Measures of lung function, including spirometry and airway resistance, can be repeatedly and noninvasively measured on volunteers. Nitrogen dioxide and ozone have been studied this way, because low-level exposures are thought to have little or no long-term effects (the response is reversible) and the ETI is typically minutes to hours. The controlled conditions of the exposure chamber make it relatively easy to know what the exposure was in the short period of time immediately before lung function is measured.

Observational studies of reversible proportional exposure-response relationships often suffer from the difficulties of measuring exposure accurately over the short intervals of time that are relevant to the effects. It is fairly common to have to use a longer term average exposure (the daily mean, for example) because no finer temporal discrimination is possible from the exposure measurement methods available. In air pollution studies, lung function measured at a particular time each day is often associated with the day's average exposure. This may or may not be the appropriate ETI for the effect.

It was possible to study short-term exposure and reversible airway responses to formaldehyde in a study by Kriebel and colleagues of graduate students dissecting cadavers in a gross anatomy laboratory (Kriebel et al. 2001). Thirty-eight students were followed over 14 weeks in which they were exposed for about 2.5 hours each week to formaldehyde. Each student's individual weekly exposure was measured, as was the peak expiratory flow (PEF), a measure of lung function that responds rapidly to irritants. PEF was measured at the start and end of each weekly exposure period. Individual formaldehyde measurements, available each day, averaged 1.1 ppm (standard deviation: 0.56 ppm). Formaldehyde exposure in the previous 2.5 hours was found on average to reduce PEF by -1.0% per ppm, although this effect diminished over the first 4 weeks, suggesting at least partial acclimatization.

Additional discussion of using DP-1 in epidemiology is presented in Chapter 10, with examples in Chapter 11.

Disease Process DP-2: Irreversible Proportional

Iconic example: silica → silicosis

Silicosis develops following prolonged respiratory exposure to aerosols of crystalline silica. The evidence from numerous studies demonstrates that it is approximately correct to say that as each silica particle is deposited in the interstitial area of the lungs, it produces a certain quantity of fibrotic tissue after a period of time (a lag), gradually increasing the total amount of fibrosis in the lungs (Figure 6.2B). The fibrosis may be directly detected as opacities on chest X-rays or by its indirect effect on the lungs' ability to hold air (vital capacity) or to exchange oxygen with the blood (diffusing capacity).

Key Assumptions

1. Each unit of the agent causes damage at a rate independent of other units arriving in the tissue before or after.
2. Damage is (almost) entirely irreversible.
3. Often there is a lag between the tissue dose and the irreversible response.

Typical Temporal Patterns

The ETI is long and begins at the start of exposure, as there is assumed to be no repair of the damage done by exposures no matter how distant in time. There is often a lag or latency between exposure and the appearance of symptoms or clinically significant ill effects, so that the end of the ETI occurs before the outcome. A threshold level of exposure (either exposure intensity or duration) below which there is no effect is generally not observed.

DP-2 in Practice

Irreversible proportional disease processes are generally well described by a linear exposure-response model, in which the summary measure of exposure is cumulative exposure (intensity multiplied by duration). The reason is that, if the first two preceding assumptions are correct, the response will simply be proportional to the sum of all the exposure that arrives at the target tissue. It follows that subjects with very different patterns of exposure over time may have the same risk, as long as the products of each subject's average intensity and total duration of exposure are the same. This is approximately true for the risk of chronic lung disease from tobacco: each "pack-year" of smoking (a cumulative exposure measure—the number of packs of cigarettes smoked per day multiplied by the number of years of smoking) causes a fixed amount of loss in pulmonary function. Fibrotic diseases caused by environmental toxins—the pneumoconioses, for example—also seem to follow this pattern.

In one recent example, Kraus and colleagues used cumulative exposure in their study of lung function among workers in the German soft paper industry (Kraus et al. 2004). Among 1,047 workers in nine factories, there was strong evidence of decreasing lung function—especially forced vital capacity (FVC)—with increasing cumulative dust exposure. The confounding effects of age, sex, body mass index, smoking habits, and factory were investigated and controlled using multiple linear regression models. The authors did not report whether or not they investigated the effect of lagging cumulative exposure on the association with pulmonary function.

Use of the irreversible proportional model in epidemiology is discussed further in Chapter 10, with examples in Chapter 12.

Disease Process DP-3: Reversible Discrete

Iconic example: allergen → asthma attack

Asthmatics usually have specific sensitivities to one or more agents—usually allergens. Episodes (attacks) of shortness of breath, wheezing, and coughing are triggered by exposure to an agent to which the individual has become sensitized. The agent is often one that is essentially nontoxic to those who do not have the specific sensitivity. Once an episode has begun, its intensity and duration are generally independent of the amount or timing of exposure (Figure 6.2C). After a recovery period, the individual is usually susceptible once again to a new episode from the same exposure. These dynamics are quite different from those involved in the *development* of asthma—the transition from being nonasthmatic to asthmatic. The latter process is better described by an irreversible discrete model (DP-4).

Key Assumptions

1. The disease process is probabilistic, and, once triggered, the intensity and duration of the response are not proportional to the intensity of the stimuli.
2. For an individual, there is an apparent threshold dose above which the probability of a response rises steeply and below which there is little or no probability of response.
3. The dose threshold varies widely among individuals and may also vary within an individual over time.

Typical Temporal Patterns

The ETI is generally short, as it is for DP-1, the other reversible mechanism. The agents are often irritants or inflammatory chemicals that exert their effects on contact with skin, mucous membranes, or the epithelial linings of the respiratory or digestive tracts. As a result, the ETI is often determined by the kinetics of the agent in the target tissue and is generally shorter than a few half-times of the clearance of the agent from the site of contact—minutes to hours, but not usually longer. The end of the ETI will usually be the moment of onset of the outcome event. For an asthma attack, the ETI may be only a matter of the few minutes immediately preceding the episode. This is why attention is often focused on "peak" exposures as predictors of risk for reversible discrete events.

It is important to distinguish here between the ETI and the length of the observation period; to capture the relationship between peak exposure and a discrete outcome, the investigator will have to study a period considerably longer than the ETI—a day at least, and probably more—to increase the chances of observing a meaningful number of events. Furthermore, unless extensive real-time personal exposure monitoring is available covering this entire period, it will be necessary to develop a model for the relationship between the short-term exposure intensity *distribution* (e.g., the probability of an exposure exceeding some threshold) and the probability of an event. This is discussed further in Chapter 13, with examples in Chapter 14.

Following an event, there may be either a refractory period in which the individual is at decreased risk of another event or, on the contrary, a period of heightened sensitivity. The duration of this period of altered sensitivity depends on the kinetics of the recovery process.

DP-3 in Practice

There is a large literature on asthma attacks and various measures of air pollution in both adults and children. A good example is a study by Yu and colleagues on ambient air pollution and childhood asthma symptoms in Seattle (Yu et al. 2000). One hundred and thirty-three children with mild to moderate asthma were recruited to a study in which they completed daily diaries of symptoms, lung function (peak expiratory flow), and medication use for an average of 58 days. Air pollution monitoring data from central urban monitoring sites were used to estimate each child's daily average exposures to several pollutants. The symptoms-diary data were used to identify days on which a child suffered a significant exacerbation in asthma, and the exposures on these days were compared to those on days on which the child did not have an exacerbation. The dose-response relationship is fundamentally probabilistic; each dose unit has some probability of producing a response, but the response will not be proportional to the dose. The risk of an exacerbation was associated with the ambient carbon monoxide and PM_{10} concentrations. For example, the odds of an asthma exacerbation increased by about 20% for each 10 $\mu g/m^3$ increase in PM_{10}. There were somewhat stronger associations with these pollutants on the day before the exacerbation rather than on the day of the exacerbation, but the difference was not large. Carbon monoxide is unlikely to be causally related to asthma attacks, but it may have been a surrogate for complex particulate and oxidant emissions from traffic. In this example, we would expect that if personal exposure data had been available, and if they had been available for finer divisions of time, an even stronger association with asthma might have been found using short-term exposure data more tightly coupled to the time of symptom onset.

Other exposure-disease processes that might be profitably studied using this model include low back pain and sudden cardiac events such as infarction and certain arrhythmias. You may want to compare your knowledge of these processes against the preceding assumptions and decide whether or not you agree that this model would be appropriate.

Disease Process DP-4: Irreversible Discrete

Iconic examples: air pollution → development of asthma
asbestos → lung cancer

In both these examples, there is a fairly rapid transition from health to disease, and the process is generally believed to be irreversible (Figure 6.2D). In both cases there are probably at least two ordered steps that must occur, and environmental exposures may affect one or more of these steps. It is generally assumed that a long period of exposure precedes disease onset, although it is less clear whether the reason is the need for a certain quantity (dose) of the agent or, instead, a certain amount of time in which essential pathologic processes must occur.

Key Assumptions

1. There is a probabilistic initiating event for the disease process that either does not reverse or reverses very slowly compared with the time scale of the initiation. This event may or may not be caused by environmental toxins, depending on the specific disease mechanism.
2. One or more subsequent stochastic steps must occur to cause disease. After the first step, the probability of subsequent steps depends on the preceding step and possibly on environmental exposures, as well.
3. The magnitude or intensity of the response is not proportional to the dose, but the *cumulative probability* of response generally is proportional to the dose.

Typical Temporal Patterns

In principle, one can imagine ETIs for each of the necessary sequences of steps, but because these are not generally observable, we instead must consider the entire period during which all the steps occur as the ETI. For each discrete event in the disease process, there is a population distribution of the time to event, and not a single fixed time. We are interested in the joint probability that event 1 will happen, followed by event 2, and so forth. If these are low-probability events, the period will generally be long—beginning at the start of exposure (at birth or even conception for widely dispersed environmental toxins). There is often a lag between the last step and the onset of clinical disease, caused by the time needed for the disease to become symptomatic, so that the ETI may end before diagnosis.

DP-4 in practice

The epidemiology of the incidence or onset of new cases of asthma might benefit from application of this DP model. As with the reversible discrete examples, there is often a practical challenge of gathering sufficiently detailed exposure data to permit the use of a biologically based model. Several longitudinal studies of the development of asthma or asthma-related symptoms in children have been published, and these might provide opportunities for applying such a model. For example, the prospective birth cohort study of Gold and colleagues (Gold et al. 1999) followed children from birth with periodic medical examinations and assessments of indoor allergen exposures. Among 226 children under 5 years of age, house dust endotoxin above the median level was found to increase the risk of wheeze by about 50% over 4 years of follow-up (Litonjua et al. 2002). With repeated measurements of exposure and symptoms, it might be possible to use DP-4 to improve estimation of asthma risk in studies like this one.

Perhaps the best developed example of epidemiologic application of DP-4 is in the study of cancer, and this is described further in Chapter 13, with examples in Chapter 15.

6.5. DOSE METRICS

When a patient is told to take one 50 mg pill every 4 hours for 2 days, this is an effective way to describe the "dose" the patient receives. Unfortunately, there are many different ways in which the term *dose* is used, and the resulting confusion cannot be easily cleared up. As noted in Chapter 4, we define the tissue concentration of a toxin at one point in time as the *burden* and the time integral of the concentration over some appropriate interval the *dose* (Checkoway et al. 2004). This formal definition shares with the doctor's prescription in the first sentence the two key components of dose: a quantity and a time interval for a specific material (Klaassen et al. 2001; Rowland and Tozer 1995). As we saw in the previous chapter, even though the exposure may occur at a single point in time, such as ingesting a pill, an internal time profile of concentration, or burden, is produced at the site of action (target tissue), which depends on the relative rates of the pharmacokinetic processes of solubilization, uptake, distribution, metabolism, and excretion. In epidemiologic studies, it is uncommon to have data on the target tissue burdens of the agent of interest for each member of the population. Biomarkers can sometimes provide excellent estimates of burden or dose (see Chapter 5), but even when these types of data are available, they often provide "snapshots" in time or on a subset of the population. Thus we are nearly always left trying to estimate the dose, or something proportional to it, in the etiologic time interval.

The general form of a proportional dose metric is a quantity calculated from the profile of exposure intensity as it varies over time for each individual. The metric can be thought of as a weighted sum of the past exposures in each of the time increments (Δt) in the ETI (Kriebel 1994). The weights are derived from a specific hypothesis about the disease process. There are various ways that such a weighted sum can be constructed, but a useful general form looks like this:

$$DM = \sum_{i}^{N_{ETI}} w_i \left(C_{\exp}[i]\right)$$

Equation 6.1 A simple dose metric for a proportional effect

where the average exposure, $C_{\exp}[i]$, for the ith time interval is weighted by w_i, and these products are summed over all intervals in the ETI. The choice of weights depends on the type of response and repair and their dynamics. Some simple variations include adding a lag, L, and a threshold concentration, C_{Th}, below which the exposure has no effect. Then the dose metric would be:

$$DM = \sum_{i}^{ETI} w_{i-L} \left(C_{\exp}[i-L] - C_{Th}\right)$$

Equation 6.2 A general dose metric for a proportional effect

where DM is calculated only when $(C_{exp}[i\text{-}L]\text{-}C_{th})>0$ and $i\text{-}L>0$. In Chapters 10-12 we examine the two DP models for proportional effects, show a detailed example of each, discuss how to use information about possible mechanisms, and indicate some appropriate weighting functions and dose metrics.

The dose metric, DM, for a discrete effect has a different form because the dose determines the probability of a response, not the intensity of the response, as it does for the proportional responses. The metric is a probability measure calculated as the sum of the weighted probabilities for each dose level, where the dose is the average exposure intensity during the ETI, or $C_{\text{expo}}[_{i\ j}]\Delta t_{\text{ETI}}$. This metric is based on the observation that the probability of a response depends on the joint probability of a dose level times the chance of a response at that level, summed over all dose levels. Dose groups are formed of individuals that have similar dose frequency distributions, very similar geometric means, and GSDs. The DM_j is assigned to each member of the jth dose group as defined:

$$DM_j = N_{obs}\left(\sum_{i=1}^{m} w_i p\{C_{\text{expo}}[ij]\Delta t_{ETI}\}\right)$$

Equation 6.3 A dose metric for probabilistic discrete reversible processes

where j identifies the dose group, m is the number of groups, w_i is the weight for the probability of each dose interval, and N_{obs} is the number of Δt_{ETI} intervals observed. The weighting function defines how much each dose level contributes to the overall risk probability. In many cases, discrete responses appear to have a logistic probability function. We can use the logistic relationship as the weighting function for each dose interval to calculate the dose metric. In this case, the weighting function serves the same role as the weighting used in the proportional response dose metrics. The weighting function is a hypothesis about the shape of the risk relationship, such as a hypothesis of a logistic relationship. We develop this much more in Chapter 13 and demonstrate its application in Chapter 14.

For irreversible discrete responses, such as immunologic sensitization or cancer, we also develop a probabilistic dose metric based on a hypothesized model of the disease process. In many cases, a two-step model with an irreversible final step provides a good fit to the observations. This will be developed in more detail in Chapter 13 and shown in applications in Chapter 15.

6.6. DP MODELS IN THE PREVENTION OF NEW DISEASES

Public health practitioners continue to discover new exposure-disease associations when people become ill for unknown reasons. We believe that it can be useful to explicitly propose a tentative disease process model quite early in the investigation of new disease outbreaks. A recent example is the discovery of a cluster of cases of bronchiolitis obliterans (BO) among workers in microwave popcorn factories in the midwestern United States (Centers for Disease Control [CDC] 2002; Kreiss et al. 2002).

BO is a severe lung disease characterized by destruction of the air exchange region of the lung, leading to irreversible loss of lung function. The disease is rare, and few environmental causes are known. In 2002, eight former workers at a microwave popcorn packaging plant were diagnosed with BO. The disease was concentrated in the area of the plant in which artificial butter flavoring was mixed. Onset of the disease sometimes occurred within only a few months after first employment in the factory. A cross-sectional study of lung function among the plant's current workforce found an association between exposure to artificial butter flavoring aerosols and reduced lung function (Kreiss et al. 2002). The disease has since been identified among workers in other several other factories processing the same flavorings. The National Institute for Occupational Safety and Health (NIOSH) performed experiments in which rats were exposed to vapors from the flavoring mixture and found multifocal, necrotizing bronchitis at concentrations in the range found in the mixing areas of popcorn factories (Hubbs et al. 2002). The pathologic findings in rats were sufficiently similar to BO that the findings were considered relevant to the human disease outbreak. Chemical analysis of the vapors revealed a complex mixture of diacetyl (2,3-butanedione), acetic acid, acetoin (3-hydroxy-2-butanone), butyric acid, acetoin dimers, 2-nonanone, and δ-alkyl lactones. Several lines of evidence suggest that diacetyl is the most important toxic compound in the mixture (Kullman et al. 2005).

Which of the four disease process models is most appropriate for this new disease outbreak? Appealing to parsimony, we first assume that all of the observed effects are manifestations of a single disease process—that is, the acute disease bronchiolitis obliterans, the subclinical loss of lung function in currently exposed workers, and the effects in rats all derive from a single pathologic process.

Is this process reversible or irreversible? It certainly appears largely irreversible, based on the lack of significant recovery among the severely affected workers after removal from exposure. This is supported by the finding of necrotizing destruction of the lungs in the rats, a type of lung damage which is not repairable. The second question, then, is whether the disease process is proportional or discrete. If the only evidence we had was the small number of cases of BO, we might entertain the possibility that this is a discrete process in which a small number of affected individuals (perhaps highly susceptible for some genetic reason) suffer a rapid attack on normal lung architecture. This view is contradicted, however, by the finding of a clear exposure-response relationship for lung function decrements among workers who do not have BO. Thus we would conclude that the most likely view is that this is an irreversible proportional disease process. We would not expect to find a threshold below which no damage occurs and would hypothesize that there may be a relatively smooth continuum of increasing lung damage with increasing cumulative exposure to diacetyl or whatever the active agent may be.

This example shows how one can reason from fairly limited evidence to a formal hypothesis about the disease process. This hypothesis can then be tested in epidemiologic investigations and can also help to guide the design of exposure assessment strategies and exposure-response models. By explicitly posing such a hypothesis, we can also more easily recognize when the data are *not* consistent with our initial assumptions, thus supporting further study and reexamination of the evidence.

6.7. CONCLUSIONS

Disease process models are highly simplified versions of the true complexity of biologic pathways. Whether or not the simplification is worthwhile should be tested with empirical data. If the proposed model does not fit the observed disease time course, then the model may be wrong. This will occur because the hypothesized process model is not appropriate for humans with these environmental exposures, because the model has been constructed or applied incorrectly, or because the data are biased or confounded. But if the data are accurate and our hypothesis about the nature of the exposure-response process is reasonably correct, then the predicted time course of exposure, dose, and responses should fit the observed data.

Disease process models may be thought of as a middle way in efforts to improve the sensitivity of environmental epidemiology—neither the "brute force" empiricism of data-driven methods such as spline curves and data mining nor the theoretically satisfying but rarely practical detail of densely parameterized toxicokinetic and pharmacodynamics models. We believe that simplified process models that capture the key temporal dynamics of exposure and observed responses are practical in many situations and can provide improvements in environmental risk detection and quantification. The examples in Chapters 11–15 develop these ideas further.

NOTE

1. This figure is abstracted largely from information presented in Gregus and Klaassen (2001). This is an excellent general textbook with a suitable level of detail about general mechanisms and specific agents, and it can provide the nontoxicologist with background on biological processes.

SECTION B

Exposure and Disease in Populations

Section B presents the epidemiologic approach to studying exposure and disease in **populations** rather than in individuals. Chapter 7 is a review of the basic epidemiologic methods and key concepts needed to understand how and why disease risk is the fundamental object of epidemiologic investigation. Chapter 8 is an explanation of the many sources of uncertainty in epidemiologic studies and how this uncertainty can be managed. Chapter 9 shows how dosimetry can be used in epidemiology to link exposure assessment to epidemiology through the use of biologically based models.

7 Epidemiologic Evaluation of Environmental Hazards

7.1. INDIVIDUAL VERSUS POPULATION VIEWS OF DISEASE

The first half of the book has described in some detail how toxic hazards enter the human body and trigger disease. These processes occur inside individual human beings, and it is individual human beings who become ill. But, in contrast, many environmental hazards have been detected by studying patterns of disease in populations rather than in individuals, and almost all quantitative exposure-risk relations have been measured through population studies.

In this chapter we review basic epidemiologic concepts and, in particular, try to show how necessary and complex is the shift from the individual to the population when one thinks about the causes and patterns of disease occurrence. This shift is especially important for the readers of this book, because, as we have noted, the exposure assessment investigation is increasingly focused on the individual. If counting cases of disease were as simple as counting money, then there would probably be no need for a separate field of epidemiology. But as we all know from our own personal experience, the development and onset of an illness is a complicated process. And it is fraught with uncertainty. We view disease as a stochastic process. There are two views of why this is so. First, there are those who take a deterministic view of the world, and for them, the stochastic or probabilistic nature of disease derives from the great ignorance that we have about the full set of steps leading from a risk factor to a disease. The second view of disease as a stochastic process holds that there is inherent uncertainty about the outcome of complex systems such as the human body. If the disease process is a long series of branching possibilities associated with the particular makeup of each individual, then it is unlikely that we can ever truly know how an individual developed a disease. And. according to this complex-systems view, no amount of knowledge will ever completely reduce the uncertainty surrounding when a disease will begin and what its ultimate outcome will be for an individual.

For our purposes, it does not matter very much which one of these two explanations holds, because there is so much that is not known and is currently not knowable about the complex biologic, chemical, and physical systems that underlie the development of disease processes. For almost any environmental hazard that we might choose to study, a group of individuals all identically exposed to this hazard will follow different paths or trajectories as their exposure continues. Some of them may ultimately develop disease from this toxic exposure, whereas others will not. The reasons for these differences will remain largely unknown. In other words, predicting an individual's disease status is hard to do. And furthermore, the full causal pathway leading up to an illness in an individual is difficult, if not impossible, to discern. In contrast, patterns of increased and decreased *risk* of disease in populations can be readily detected and quantified, and this is the objective of epidemiology.

Before reviewing the important concepts from epidemiology that the reader of this book needs, it may be useful to look more closely at some of the differences between studying the occurrence of disease in individuals and in populations. First, let us define more clearly the reasons that it is difficult to study disease occurrence in an individual, to identify causes, or to quantify the strength of associations between risk factors and diseases. First, most diseases that are of concern to those studying the modern industrial environment and its health effects have many causes. Cases of cancer, heart and lung diseases, reproductive failures, developmental abnormalities, and many other diseases occur in the absence of any one putative causal factor that may be under study. That is to say, there are background risks in nearly all of these diseases. From this it follows that one will not be able to say for certain that a given case of this particular illness was caused by one particular causal pathway. Second, many of the diseases of interest to epidemiologists are irreversible. That is, the pathways that lead from risk factors to disease generally involve a series of steps or stages whose cumulative effects are difficult to repair. Thus, once an individual becomes ill, he or she never completely reverts to the healthy state. What this means in practice for those who would study these illnesses is that each individual person makes just one transition in his or her life from being free of the illness to being a case of that illness. There are exceptions, of course, and in Chapter 6, we presented models of reversible disease processes and cited several examples.

A third challenge to the study of disease in individuals, the most fundamental, is that time cannot be stopped and rewound. One can never do the ultimate experiment of observing the same life twice, once with some risk factor or exposure present and then a second time in its absence. Without this possibility there always remains some uncertainty about the causal link between a particular risk factor and a particular disease. And yet fundamentally, this link, this *association* as it is called, is precisely what we want to know. We want to know whether the removal of this risk factor or exposure will result in a reduced risk of the disease. Epidemiology provides a partial solution to these difficulties. Instead of studying disease in individuals, epidemiologists compare the pattern of disease occurrence among groups of individuals or populations. This is only an incomplete solution, however, because in shifting our focus from the individual to the population, we have changed the object of study from *disease occurrence* to the *risk or probability of disease occurrence*. This shift means that results of an epidemiologic study, no matter how definitive they may appear, must always be questioned for their relevance to the future health status of any given individual (Rose 2001).

There is an additional fundamental limitation of epidemiology: uncertainty about the comparability of groups whose disease risks are compared. In its simplest form, an epidemiologic study compares the frequency of disease in an exposed group with that in an unexposed group. The disease risk in the unexposed group is assumed to represent the disease risk that the exposed group *would have experienced in the absence of the exposure*. Thus a necessary but questionable substitution has occurred—the (observable) risk in the unexposed group is evaluated as a stand-in for the (unobservable) risk of disease in the study group had they not been exposed. Epidemiologic evidence for exposure-disease associations is called *counterfactual* evidence because of this fundamental limitation (Rothman and Greenland 2008).

7.2. DISEASE IN POPULATIONS

Epidemiology is the study of the occurrence of disease in populations. Let us examine more closely the meaning of "disease in populations." Clearly, it is individual human beings who become sick. Clearly, disease is a pathophysiologic process that occurs inside individual organisms. What, then, is the meaning of a population view of disease? The occurrence of disease in a population is most frequently described by a risk or probability. Risk is a population concept. Individuals either do or do not become sick. Populations, on the other hand, may change their incidence or prevalence of illness, which is a phenomenon related to but fundamentally different from individual illness. In the individual, disease is a state that either exists or does not, whereas in a population disease is represented by a fraction, a probability, or a rate of change. For example, if a population of lifelong cigarette smokers is followed for many years, we know that some of them will develop lung cancer. In fact, current data indicate that approximately 16% of lifelong smokers will develop lung cancer (Peto, Darby et al. 2000). Any given single smoker may or may not develop a disease. Does this number, 16%, represent the risk of disease in a single smoker? Putting it another way, is this figure analogous to the chances of winning (or losing) in a game of cards or other game of chance? To answer that question a judgment must be made. One must decide whether to believe that this figure, this 16%, is relevant to that particular individual smoker, based perhaps on other known characteristics of that individual and the population from which the risk estimate of 16% is derived. It is important to stress that this is a judgment, that to attribute or assign the probability to the individual is never verifiable, because that individual will or will not ultimately develop lung cancer. On an individual basis, the number *16%* has no direct meaning as a measure of the occurrence of disease (Bailar and Bailer 1999).

This distinction between disease as a state and disease as a probability illustrates one important characteristic of disease in populations. More generally, disease in populations can be seen as an *emergent* property. Many complex systems exhibit properties that are not observed in their component parts. For example, atoms have complex behaviors that can generally be explained by certain basic physical laws. However, molecules are made up entirely of atoms but have much more complicated behaviors—behaviors that are difficult to predict based simply on fundamental physical principles and properties particular to atoms. In consequence, the field of chemistry has been developed, consisting of a set of laws or rules which explain and predict the behavior of molecules. Chemistry, then, in its rules and laws can be seen as a set of emergent properties of atoms once they are combined into molecules. Continuing to higher levels of complexity, living organisms are made up entirely of chemicals, but the laws of chemistry prove inadequate to fully explain the characteristics and behavior of organisms. The biologic properties of organisms are, at a still higher level, seen as emergent.

Taking the population view of disease allows epidemiologists to see things which are essentially invisible when one looks at individual cases of disease one at a time. By comparing the frequency of disease in different populations, epidemiologists can detect rather weak causes of disease, causes that we believe act on individuals, by altering physiologic processes. But these weak causal chains may not be observable when one looks only at individuals. Consider, for example, the link between exposure to ultraviolet radiation in sunlight and skin cancer. The incidence rate of skin cancer in the United States increases as one moves south, and the average sun exposure of an average person also increases. The difference in the incidence or risk between the southernmost parts of the United States and the northernmost parts is not large, and it would be very difficult if not impossible to detect this difference by looking only at individual cases of the disease. For example, suppose a dermatologist practiced in both Minnesota, in the far north of the United States, and in Florida, in the south. It is unlikely that she would notice that skin cancer was more common in Florida (unless she has been trained to expect to find this pattern, based on the epidemiologic evidence). The data on skin cancer distribution for the entire population, presented in maps by counties or states, allow one to see rather easily that there is an increasing trend in skin cancer risk with decreasing distance from the equator, and so also with increasing sunlight exposure. When this pattern is combined with other information, such as experimental data about the effects of ultraviolet radiation on the cells of the skin, one arrives at a strong conclusion that the observed association between sunlight and skin cancer risk is likely to be causal and that, as a result, it would be wise to choose to believe and act upon this association by avoiding ultraviolet sunlight.

At the same time that we observe through this example the power of epidemiology to detect risks, the example also illustrates that there are quite substantial limits to what one can learn by this approach. For example, how much sunlight increases the risk by how much? Does it matter whether the sun exposure occurs in very intense bursts? That is, does getting a sunburn increase your risk by a great deal? We cannot know from the simple geographic evidence whether protection from the sun using suntan lotion is effective in reducing skin cancer risk. We do not know much about possible modifying factors, such as skin color and other genetic factors, and we know very little about potential modifying effects of nutrition and other kinds of environmental exposures. More and better epidemiologic studies can help to understand many of these uncertainties, but we are very far from being able to predict how much sun exposure will cause cancer in a given individual.

The skin cancer example illustrates that taking the population view of disease allows us to detect certain kinds of risk factors that are not discernible at the individual level. But this view also imposes certain constraints as well. Two in particular are worth mentioning in the context of this book. First: *there is probably some level of exposure estimation that is "too precise."* That is, there is probably a level of precision in calculating each subject's exposure history that is not worth paying for because it will not increase the precision of the risk estimation, but this level also depends on the size of the study and other factors that define the precision of the risk estimate.

Because risk is a population phenomenon, it is at the aggregated population level that exposure-risk associations are assessed. It follows that some aggregation of individual exposure data must occur as exposure is linked to disease. This is easy to see when categorical methods are used, in which average exposure in groups is compared with the risk or rate of disease in those groups. Modern epidemiology increasingly uses regression models rather than categorical ones, and because the individual observations rather than group means are used, one might be tempted to think that these models are estimating individual characteristics. But, although exposure may have been estimated individually, the measures of association that are the object of study describe population phenomena.

A second constraint of the epidemiologic view is: *there are some aspects of the physiologic processes of disease causation which cannot be studied in populations*. That is, they are invisible or indeterminate from an epidemiologic point of view. Very little has been written on this topic, and additional research is needed in order to clarify which aspects of pathophysiology can and cannot be studied epidemiologically. We return to this issue in Chapter 9.

The failure to recognize the distinction between individual and population views of disease leads to certain kinds of confusion. A factor that is a true cause of disease, by which we mean that it increases the risk of disease, will be detectable in a population only if this factor varies in strength or prevalence within the population being studied. A ubiquitous risk factor cannot be detected. It is also true that variability in the outcome, the disease, is necessary. But this is not as much a problem in epidemiology because most diseases that we study occur to some individuals and not to others. Very rare diseases, ones that almost no one gets, are also hard to study and necessitate very large populations to increase the number of observed cases. At the population level, the apparent strength of a cause is dependent on its prevalence and on the prevalence of other risk factors that are steps along the same causal chain or path to the disease (Rothman and Greenland 2008). A disease that has both environmental and genetic causes will appear as largely exposure related when the genetic factor is common and the exposure is rare, and it will appear as a largely genetic disease when the exposure is common and the genetic factor is rare (Vineis and Kriebel 2006).

How much can be learned about pathophysiologic processes at the population level? As noted, the principles are not yet well identified, but one example illustrates the challenge. Suppose that there is a mechanism by which a toxic chemical causes some ill effect. Suppose that this toxic effect has a threshold, by which we mean that there is a safe level below which this toxic chemical has no harmful effects at all (Figure 7.1). Assume further that each individual in the population has a threshold for his or her response to the toxin. But, like nearly all biologic properties, this threshold varies from individual to individual. Some individuals have very low thresholds above which the toxic exposure begins to have damaging effects, whereas others have much higher thresholds. Now imagine that we study this disease process in an epidemiologic study. What will the threshold look like through the lens of the epidemiologic study? Under certain conditions the threshold may be detectable, but in many cases it will not (Brain, Beck et al. 1988). The reason is that the variability in the threshold from individual to individual will be very difficult to distinguish from a dose- response curve without a sharp transition or threshold (Figure 7.2). Instead, the curve may simply be "shallow" at very low levels of exposure, and it will appear that there is no threshold for the effect.

In general, it appears that temporal aspects of pathologic processes may be easier to detect epidemiologically than are such characteristics as the shapes of dose-response curves. For example, epidemiologists are familiar with the concepts of lags or latency periods between exposure and onset of disease. And epidemiologic studies have had some success in identifying and even quantifying the lengths of these latency periods (Checkoway, Pearce et al. 2004). A more complex but still similar concept is the stage at which a carcinogen is thought to act in a multistage process. Stage information is also likely to be relatively easily measurable at the population level in contrast to aspects of pathologic processes such as the shapes of dose response curves (Thomas 1983).

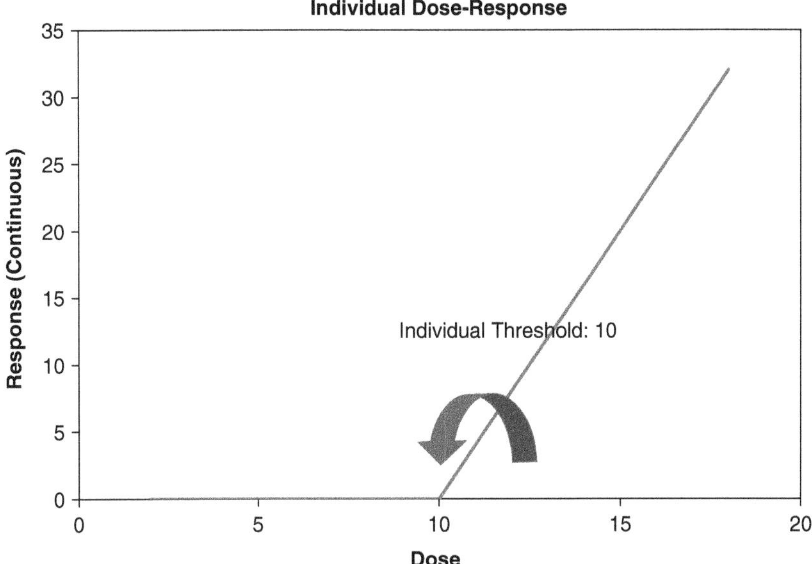

Figure 7.1 The concept of a threshold for a toxic response. For an individual, there is some level of dose (10 dose units in this example) below which there is no response. Above this level, there is a simple linear relation between dose and response.

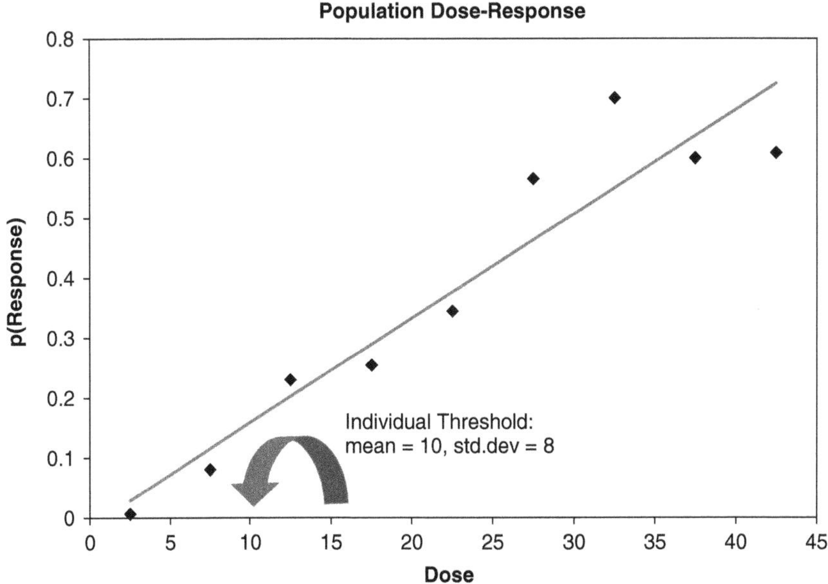

Figure 7.2 A population view of the threshold shown in Figure 7.1. If each individual has a different threshold, drawn from a distribution with a mean of 10 units, the dose-response relation might look like this. The response axis now represents the probability or fraction of the population exhibiting some set level of response.

7.3. MEASURES OF EXPOSURE-DISEASE ASSOCIATION

As noted previously, we would ideally like to compare the disease experience in an "exposed" group with the experience this same group would have had if they had not been exposed. Because we can't know this, we might instead compare the disease experience of our population with that of another, similar one. This comparison is quantified using a statistic called a *measure of association* or an *effect estimator*. The former term is preferred because *effect estimator* seems to imply the existence of an *effect* of exposure, when in fact the study is being conducted to learn whether any such effect actually exists.

John Snow's famous investigation of the causes of cholera in London in 1854 was immeasurably aided by the fact that there were two different private water companies supplying the same districts of London (Snow 1936); (Carvalho, Lima et al. 2004). Snow observed that the rates of death were quite different in households supplied by each of the two companies (Table 7.1), and he drew inferences from this difference about a link between the quality of the water and the risk of cholera. The rate of death in houses supplied by Southwark and Vauxhall was 315 per 10,000 houses, whereas for the houses of the competing Lambeth company, the rate was 37 (whether these numbers are best described as risks or rates is not important to our example). A measure of association can be constructed from these numbers by calculating their ratio: 315/37 = 8.5, or their difference: 315 – 37 = 278. In etiologic research the ratio measures of association are most commonly used. For the comparison of two rates, we calculate the *rate ratio* or, for the comparison of two risks, the *risk ratio*, also called the *relative risk*. All three of these terms can be conveniently abbreviated RR. Risk and rate ratios are calculated differently, but they are interpreted in much the same way. The difference measures—rate difference and risk difference (both abbreviated RD)—have less convenient statistical properties than the ratio measures, and they tend to be more variable from one population to the next than do the ratio measures. That is, the RR comparing an exposed to an unexposed group tends to be fairly consistent when studied in different populations, whereas the RDs tend to vary depending on the population in which the study is conducted.

Table 7.1 John Snow's data on deaths from cholera in London, 1854. Household supplied by two competing water companies are compared

Water Supply Company	Cholera Deaths	Houses	Deaths per 10,000 Houses
Southwark and Vauxhall	1,263	40,046	315
Lambeth	98	26,107	37
Elsewhere in London	1,422	256,423	55

Sources: (Snow 1936; Carvalho 2003)

When comparing populations at many different levels of exposure rather than just two, the measure of association may be a change in, for example, a rate ratio per unit change in the exposure. Regression models lend themselves naturally to this kind of analysis. The overall measure of association across all of the levels of exposure is the slope of a regression line fit to the exposure data for groups with different disease risks. For example, Salvan and colleagues estimated the lifetime doses of the dioxin 2,3,7,8-TCDD for a cohort of U.S. chemical workers that was assembled by the National Institute for Occupational Safety and Health (NIOSH) to evaluate health risks of this chemical (Thomaseth and Salvan 1998); (Salvan, Thomaseth et al. 2001). An elevated risk of lung cancer had previously been observed in this cohort when their mortality was compared with that of the general population (Fingerhut, Halperin et al. 1991). Salvan and colleagues constructed a dose-response curve using individual lifetime dioxin doses and the epidemiologic data on lung cancer mortality (Figure 7.3). In principle, an individual's risk of lung cancer can be extracted from the model, if his or her TCDD dose is known. But it is important to remember that this calculation involves an extrapolation from the population of chemical workers on which the study was based, and it also requires a judgment about the relevance of that population risk for any particular individual. A more complete calculation of the potential future risk of lung cancer in an individual would have to include consideration of many other factors and would

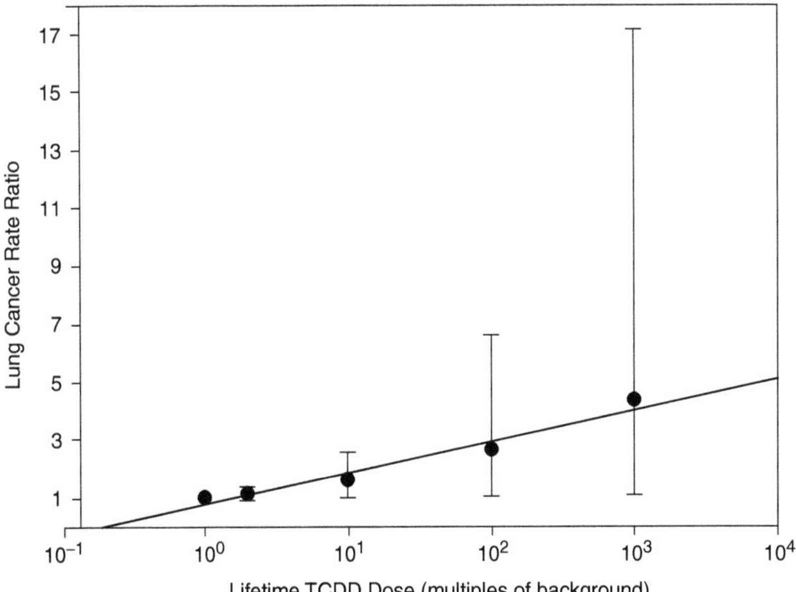

Figure 7.3 Dose-response curve, dioxin (TCDD) lifetime dose, and lung cancer mortality. Dioxin doses are shown as multiples of an average background dose. (Reprinted with permission from Salvan et al. 2001.)

involve substantially more uncertainty than is represented by the error bars in Figure 7.3 (Greenland 1999). The important sources of uncertainty in calculations such as these are taken up in the next chapter.

7.4. THE WEB OF CAUSATION

Diseases are thought of as resulting from a complex web of interacting causes. Take, for example, the problem of preventing adult-onset asthma. We can quickly sketch a box-and-arrow diagram representing some of the potential causes of asthma (Figure 7.4). This web is meant to represent our understanding of factors which contribute to the risk of asthma. That is, the web is envisioned as existing at the population level. In reality, it is probably more accurate to imagine each individual having his or her own web, sharing certain common elements with other victims of the same disease but having a different structure and a widely varying number of different components or nodes. Thus our sketch is idealized or "average." These sketches are also almost always gross oversimplifications, because there are so few diseases that are very well understood. In epidemiology, when we speak of a "cause," we mean that the cases of a disease attributable to a certain cause have webs of causation containing that cause.

Despite all of these limitations, causal web sketches are an important place to start in designing an epidemiologic study. It is generally necessary, though, to pare down the web substantially when the time comes to actually decide exactly which potential risk factors will be measured. Typically, one or a small number of risk factors are isolated, and their statistical associations with the risk of disease are quantified. Other components of the web are either set aside as being unlikely to be important or infeasible to evaluate. And, because these webs are poorly understood,

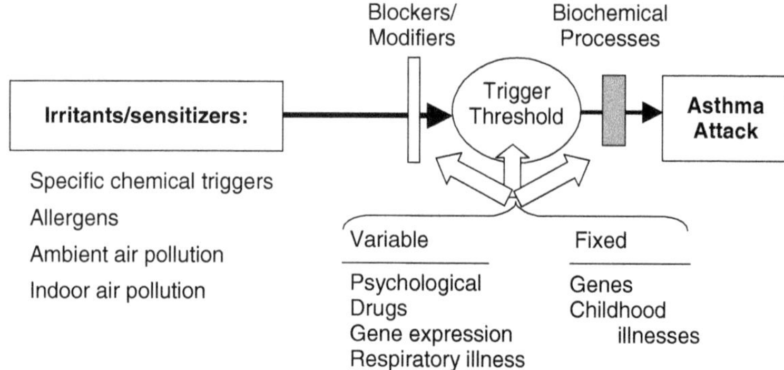

Figure 7.4 A causal web representing some of the causes of adult-onset asthma. The process leading to asthma attacks is very complex, with many separate components. Each of the components may be influenced by a mixed set of fixed and variable factors, determining whether or not a person will have an attack.

there is always the risk that important nodes have been missed or that several factors are codependent and so not separately estimable. In the rare instances in which experiments can be conducted, it is possible through randomization to accomplish a much more effective isolation of a single factor from the web than is possible in observational studies. Observational studies, the mainstay of epidemiology, do not allow randomization in general, and so the main approach to understanding the web has been to use statistical methods to try to isolate a single causal link from the complicating effects of other parts of the web.

There are two principal concerns that arise when simplifying causal webs for epidemiologic studies. These are called *confounding* and *effect modification*, and, although they are very different phenomena, they are often confused.

Confounding

Confounding in epidemiology occurs when the association between an exposure and a disease is distorted by a third factor, the confounder, which is itself a risk factor for the disease and, at the same time, associated with the exposure under study. Suppose, for example, that ambient air pollution and home exposure are both risk factors for asthma, as indicated in Figure 7.4. Suppose further that these two exposures are correlated in the study population. This might mean, for example, that there is a tendency for high air pollution exposures to occur more frequently in those who also experience home exposures. In this scenario, an epidemiologic study evaluating the association between ambient air pollution and asthma risk would obtain a result that was confounded unless proper steps were taken to control this source of bias. To be a confounder, a covariate must meet three criteria. First, it must be associated with the disease, even in those who are do not have the exposure under study. Second, it must be associated with the exposure in the study population as a whole. And third, it must not lie on the causal pathway between exposure and disease.

Suppose that a group of workers is being studied because exposure to a chemical may increase the risk of some disease. Additionally, suppose that there are subgroups within this population, each of which experiences a different level of exposure to the hazard. If the disease being studied is one that is also caused by, for example, cigarette smoking, then the epidemiologist must be concerned about the possibility that smoking may, for reasons of individual behavior or social pressures, be of very different prevalence in the subgroups with different levels of exposure. If this were so, then the effects, if any, of the exposure on the disease would be commingled with the effects of cigarette smoking on that disease. And unless one were aware of this and took steps to control it, the association that would be observed between the pattern of exposure and the pattern of disease would be distorted by cigarette-smoking patterns that are associated with the exposure and the disease.

Table 7.2 Simulated data illustrating the problem of confounding. The relative risks calculated among smokers and nonsmokers (Part A) indicate a modest increase in risk with exposure, whereas the crude table resulting from combining the data from the two smoking groups (Part B) has a considerably higher relative risk

Part A.

	Nonsmokers				Smokers			
	Diseased	Not Diseased	Total	Risk	Diseased	Not Diseased	Total	Risk
Exposed	39	661	700	0.056	262	1039	1301	0.201
Not Exposed	82	1621	1703	0.048	51	248	299	0.171
RR	1.2				1.2			

Part B.

	Combined			
	Diseased	Not Diseased	Total	Risk
Exposed	301	1700	2001	0.150
Not Exposed	133	1869	2002	0.066
RR_{crude}	2.3			

Some simulated data to illustrate this problem are shown in Table 7.2. Assume that these data came from a cohort study in which all subjects were followed for a fixed and relatively short period of time. In this situation, one can calculate the risk or incidence proportion from the data in the 2x2 tables shown. This calculation is performed twice for each table—once for the exposed group and again for the unexposed group. The risk ratios, or relative risks, measuring the strength of the association between exposure (present vs. absent) and disease are the same among smokers and nonsmokers (RR=1.2). Now, when the data for smokers and nonsmokers are combined into a single, "crude" table, we find that the relative risk is considerably larger than in the two groups separately (RR_{crude}=2.3 in Table 7.2b). If we had not planned for the possibility that smoking might confound the exposure-disease association, and so did not collect data on who was and was not a smoker, we would have observed only the crude association, which is biased upward because of confounding by smoking.

Confounding is a bias that distorts our view of the world, and, as such, it must be removed in order to accurately understand associations between risk factors and diseases. In our example, the prevalence of smoking was considerably higher in the exposed group, and so the effect of smoking on disease risk became "mixed" with the exposure risk. Unless we view the data separately for smokers and nonsmokers, we see a biased or confounded picture. Very often in epidemiology we encounter confounding because of the complex behaviors of human beings and human societies. Many aspects of human life are correlated, and so various exposures, behaviors, and risk factors tend to be clustered together when studies are conducted of populations of human beings. It is possible to remove the effects of confounding statistically (either through the simple approach in Table 7.2 of stratifying the data or through more complex methods), and this realization was one of the major impetuses for the growth of modern epidemiology. The statistical control of confounding was seen by epidemiologists as a means to approximate the inferential power of randomized clinical trials using observational data.

There are at least four important ways to control confounding. First, in experimental settings, randomization is by far the most powerful method of controlling confounding. If one is able to randomize the subjects in the population under study into exposure groups, then it is unlikely that there will be associations between that exposure and any other factors. Randomization will, on average, equalize the frequency of all potential confounders across the strata of the exposure. Randomization is the most powerful method of controlling confounding because, unlike all of the other methods, randomization provides a way to control for confounding even by factors that have not yet been identified.

A second method of controlling confounding is simply to restrict the study to the subpopulation with only one level of the potential confounder. For example, if the study in the previous example had been conducted only among the nonsmoking members of that population, then the result would not have been confounded. Note that the study could just as well have been restricted to smokers—again, no confounding by smoking could have occurred. This is often harder to understand because the population would be composed entirely of smokers, and if smoking were a risk factor for the disease under study (as it was in the example), one might think that the effects of smoking would compromise the results of the study. In fact, this is not true: because everyone in the study, regardless of level of exposure, is a smoker, the comparisons in disease risk among those at different levels of exposure would not be distorted by smoking.

A third method of controlling confounding is called *stratification*. This is what was illustrated in Table 7.2. When we analyzed the data from our study stratified on smoking status, the stratum-specific estimates of the exposure risk associations were not confounded. These stratum-specific estimates could then be pooled or combined into an average or overall exposure-risk association, which would take advantage of the full population while, at the same time, being free of any distortion from the effects of smoking. The pooled RR from Table 7.2 was 1.2, simply the average of the RRs for smokers and nonsmokers.

Finally, perhaps the most common way in modern epidemiology to control confounding is to study the exposure-disease association in a regression model in which exposure is one of the independent variables and a second covariate represents the levels of the confounder. A regression model so constructed will simultaneously condition or control all covariate risk associations for the presence of all of the other covariates in the model. So each of the independent variables has an association estimated for it, which is free from confounding by all of the other covariates. Thus control of confounding is accomplished efficiently and simultaneously with the estimation of one or more exposure effects.

Effect Modification

Effect modification is often confused with confounding, but it is a fundamentally different phenomenon. Effect modification, also known as interaction, is not a bias but rather a fundamental aspect of the web of causation and often a reflection of the underlying biology. Unlike confounding, effect modification should not be removed but rather identified and investigated. Effect modification can be detected using either stratification or regression modeling, the same methods most often employed to identify confounding. Stratifying on a variable allows one not only to control confounding by that variable but also to identify the existence of effect modification.

Another simulated data set illustrates effect modification using the same setup as the confounding example (Table 7.3). One can see that there are dramatically different effects of exposure in nonsmokers compared with smokers. If this effect is believed to be causal, then one might conclude that there is some mechanism by which smoking strongly modifies the way that exposure affects disease risk. The crude RR, which in this example happens to be similar to the RR for smokers, would not be used, as it does not reveal the potentially important observation that exposure appears to *lower* risk among nonsmokers or disproportionately *increase* it among smokers. When effect modification is found, the stratum-specific measures of association should be separately reported, not combined or averaged.

If an exposure-disease association is different in one level of the stratification variable than it is in another level, this indicates the existence of effect modification or interaction (the two terms are synonymous) between the exposure and the stratification variable. Regression models are also very useful in identifying effect modification. Typically, an interaction term between the exposure and the potential modifier is added to the regression model, along with variables indicating exposure and the potential modifier (e.g., smoking). If the coefficient that is estimated by the regression model for this interaction term is nonzero, this indicates the existence of effect modification between the two covariates. This regression approach to the identification of effect modification has limitations, however. In reality there are many different ways that two or more factors or covariates may participate in a web of causation. And product terms between two such factors in the same web will be more or less effective at identifying the existence of interaction depending on the exact structure of the web of causation, that is, how the factors—smoking and exposure in this example—biologically interact with each other. The ability to detect effect modification also depends on the prevalences of the two covariates

Table 7.3 Simulated data illustrating effect modification. The relative risks calculated among smokers and nonsmokers (Part A) reveal very different exposure-disease associations, which indicates the presence of effect modification. The crude table (Part B) should not be presented, as it gives the misleading impression that there is a single common effect that applies to both smokers and nonsmokers

Part A.

	Nonsmokers				Smokers			
	Diseased	Not Diseased	Total	Risk	Diseased	Not Diseased	Total	Risk
Exposed	19	513	532	0.036	492	978	1470	0.335
Not Exposed	128	1679	1807	0.071	21	172	193	0.109
RR	0.5				3.1			

Part B.

	Combined			
	Diseased	Not Diseased	Total	Risk
Exposed	511	1491	2002	0.255
Not Exposed	149	1851	2000	0.075
RR_{crude}	3.4			

and on various aspects of the form of the regression model being used. The identification or failure to identify effect modification using a regression model can only be interpreted as a rather weak way to explore the simultaneous contribution of multiple causal factors to a disease. An alternative approach can be developed from the methods being presented in this book. Epidemiologic models that are based on explicit hypotheses about the physiologic processes underlying a disease should be better able to detect and quantify effect modification. This information can be explicitly built into a biologically motivated model, potentially allowing a more sensitive and also more informative way of testing hypotheses about effect modification.

7.5. TREATMENT OF TIME IN EPIDEMIOLOGIC MODELS

To a considerable degree, epidemiologic models are focused on the moment of disease incidence, and in this sense they are largely static models. This can perhaps be seen most clearly in the case control study or in the standard regression method of analyzing cohort data called the *Cox proportional hazards model*. In these two methods of analyzing epidemiologic data, cases of the disease are compared with noncases or referents at the moment of disease incidence. Cases and referents are matched and compared, looking backward from the moment of disease incidence of the cases. If the exposure histories of cases and controls differ on average when compared in this way, then one can say that an association exists between exposure and disease. Such a comparison can, of course, include a great deal of information gathered over long periods of time. However, this information is necessarily summarized in some kind of a statistic or limited number of statistics, and it is these summaries that are compared at the moment of incidence. For example, a common summary measure of exposure in environmental and occupational epidemiology is cumulative exposure. This was described in Chapter 6 and will be taken up again in Chapter 9. Cumulative exposure is the area under the curve of the temporal profile of exposure intensity for each study subject. As noted in Chapter 1, it is easy for two people to have two very different patterns of exposure and yet the same cumulative summary measure. For example, a period of very intense exposure for a short period of time, followed by a long period at very low exposure, could sum to the same cumulative exposure as a moderate level of exposure for a very long time (Figure 7.4). However, these two different patterns of exposure might result in quite different risks, depending on the particular disease process that underlies the development of the disease. In other words, cumulative exposure will be an unbiased summary measure of exposure only under certain disease mechanisms, and not under others.

We return to this point in Chapter 9, but for now we stress that most epidemiologic models summarize exposure information over long periods of time in relatively simple statistics and then compare those statistics at the moment of incidence. This presents a very real limitation to the ability of standard epidemiologic models to untangle the complexities of webs of causation. These webs can, of course, be represented in simple box-and-arrow diagrams, as in Figure 7.4. But when they are presented as such, what is missing from them is the component of time. Biologic processes are fundamentally driven by time, and it is difficult to accurately study them without an explicit representation of time in the model. In this book, we are encouraging the development of more biologically motivated epidemiologic models, and these models, we believe, should explicitly incorporate time in a way that is quite different from the approach used in standard epidemiologic methods.

7.6. DISEASE PROCESS MODELS AND EPIDEMIOLOGIC STUDY DESIGNS

The choice of study design is one of the most important decisions when proposing an epidemiologic study. The major designs—cohort, case control and cross-sectional—are presented in detail in every epidemiology textbook (see, e.g., (Checkoway, Pearce et al. 2004). Important variants that are increasingly used in environmental and occupational epidemiology include the longitudinal or repeated-measures design (Diggle, Heagerty et al. 2002), the case-crossover study (Maclure 1991) and the case-cohort study design (Langholz and Thomas 1990). We do not review these here but briefly discuss how the disease process models may affect the choice of study design.

Using a disease process model in epidemiology will often involve developing a dose metric (see Chapter 6), which can then be used in a standard epidemiologic model in the same way that one would use a summary measure of exposure such as the average or cumulative exposure. By this approach, one can utilize whichever epidemiologic study design seems most appropriate.

We often find ourselves designing a study when it is clear that, for economic or logistical reasons, there will not be sufficient individual exposure data to allow the estimation of a dose metric. Even in these situations, in which exposure may be defined in simple categorical terms or by proxy measures such as the duration of residence in a polluted area, it can be helpful to identify the most likely disease process model because this can help with basic design questions such as the etiologic time interval (ETI; see Chapter 6) or the study design (Table 7.4).

Case control studies (and the closely related case crossover) are most often used when the investigator can take advantage of a distinct case-finding mechanism such as a disease (or mortality) registry or well-defined clinical diagnostic criteria that are applied by health care professionals in the course of caring for the sick patient. In practice the latter is typically represented by an International Classification of Disease (ICD) code that is specific to the outcome of interest. By contrast, case control studies are hard to do when the outcome has a gradual or insidious onset (chronic obstructive pulmonary disease incidence) or when victims do not reliably seek medical care for their condition (low back pain). In these cases, cohort or longitudinal repeated-measures studies are generally preferred.

7.7. SUMMARY

Epidemiology is the study of determinants of disease in populations, and environmental epidemiology generally concerns the identification of health risks from synthetic chemicals or other exposures resulting from the technologies of modern society. This chapter has reviewed the key principles of environmental epidemiology, emphasizing some of the methodologic challenges that arise when quantitative exposure or dose data are used in epidemiologic models. Introductory epidemiology courses often focus almost exclusively on a simplistic "exposed/not exposed" framework, whereas practitioners increasingly combine quantitative exposure assessments with multivariate regression models to yield quantitative exposure-risk associations. There are important advantages to this development, including accurate risk assessment—the estimation of how much risk each additional unit of

Table 7.4 Design options for epidemiologic studies of the four disease processes

Disease Process	Study Design	Example	Exposure Assessment	Outcome Assessment
Reversible proportional	Longitudinal (repeated-measures) cohort	Short-term effects of urban air pollution on air flow obstruction in children	Children carry personal real-time air monitors yielding estimates of short-term intensities of irritants, e.g., NO_2	Spirometry performed daily for several weeks
	Prospective cohort	Cross-shift effects of occupational exposure to an irritant, e.g. ammonia, chlorine, ozone, on change in pulmonary function	Personal dosimetry or exposure reconstruction from fixed monitors yielding short-term exposure metrics	Pre- and postshift spirometry performed on same days as air monitoring
Irreversible proportional	Cohort, retro- or prospective	Effects of an air pollutant on chronic pulmonary obstruction	Lifetime exposure reconstruction of exposure to $PM_{2.5}$	Mortality from COPD
	Case-control	Effects of noise on permanent hearing loss	Lifetime noise exposure reconstruction	Hearing loss recognized in case registry (e.g., workers' compensation)
Reversible discrete	Case-crossover	Effects of prescription drugs on incidence of falls in the elderly	Daily prescription drug use from medical records or daily diary	Falls requiring hospitalization
	Prospective cohort	Exacerbation of COPD from exposure to air pollution	Personal dosimetry or exposure reconstruction from fixed monitors yielding short-term exposure metrics	Acute worsening of COPD, requiring hospitalization and increased medication
Irreversible discrete	Cohort, retro- or prospective	Effects of indoor allergens on development of asthma in children	Periodic in-home measurements of mold, endotoxin, etc. from birth to age 8	Annual clinical exams to diagnose asthma
	Case-control, nested in cohort	Cancer risk from occupational exposure to solvents	Lifetime reconstruction of airborne and trans-dermal exposure	Death certificate identified cases of cancer in occupational cohort
	Case-control, population-based	Childhood leukemia risk from drinking-water contamination	Lifetime reconstruction of waterborne exposure to contaminant	Cancer registry cases of childhood leukemia in town(s) with contaminated water

exposure carries, as well as increased sensitivity for detecting small risks. But there are challenges as well—some arising from the fact that quantitative exposure assessment tends to focus on individuals and their personal exposures, whereas epidemiology remains fundamentally a population-level investigation of risk. The next chapter addresses the principal sources of error and uncertainty in population-level estimates of the magnitude of exposure-risk relationships.

8 Uncertainty in Measuring Risk

> The only way we have of studying the unknown is by pretending that it is like the known. That the unknown is like the known makes science possible; that it is also unlike the known makes science necessary. This conflict is the reason that all theories are eventually proven to be wrong, limited, irrelevant, or inadequate.
>
> *R. Levins (1995)*

Chapter 7 described the principal methods of studying disease in populations, but for the most part, it did not deal with the many sources of uncertainty and error in the estimation of risk in populations. Because so much of what we know about environmental health risks comes from observational studies and not experiments, problems of chance, bias, and confounding can seriously compromise a study's findings. A thorough investigation of potential errors is an essential component of every study. We use the term *uncertainty* as the concept that overarches bias, confounding, chance variation, and all of the other sources of error, or challenges to the validity of a study. There is a growing literature on uncertainty in epidemiology and related fields, including risk assessment (Morgan and Henrion 1992; Levins 1995; Cairns and Smith 1996; Bailar and Bailer 1999; Stayner, Bailer et al. 1999). When drawing a conclusion about a hypothesized association between some exposure and a disease, a researcher should be able to describe how *certain* he or she is about the findings. Rarely is this certainty expressed in a simple, quantitative way, although Bayesian statistical methods are moving in this direction (Carlin and Louis 1996; Greenland 2001).

In this chapter, we review several important sources of uncertainty in epidemiologic studies and summarize approaches to managing these uncertainties. Sometimes we can directly estimate the magnitude of errors that may exist in our quantitative data. At other times, less formal methods are available for qualitatively evaluating the range of possible errors or assessing the sensitivity of our results to the assumptions that have been made during data gathering and statistical modeling. At a minimum, it is important to describe potential errors and our best judgment about their magnitudes, so that the reader of an epidemiologic study will understand the potential limitations in the findings.

There are at least five different sources of uncertainty in measures of association calculated from epidemiologic studies (Table 8.1). First, there are errors in exposure estimation. Both systematic and chance variations can lead to differences between the true exposure and what is measured. Reducing these errors is one of the main goals of the methods presented in this book. Second, the problem of confounding introduces uncertainty because there is always the possibility that either a known confounder has not been adequately controlled or, even more likely, that there are additional confounders whose existence is not known. It is only randomization that enables the researcher to control confounding by unmeasured confounders. In observational studies in which randomization is not possible, therefore, one must always remember that any result may be confounded by unknown factors.

The third source of uncertainty is the possibility of various types of bias, and these are briefly reviewed. A fourth source of uncertainty, often not discussed in epidemiologic texts, is the problem of misspecification of the form of the epidemiologic model and of the models used to estimate doses. Finally, there are both systematic and random errors in the outcome variable. Random or chance variation in the number of cases of disease (also called sampling variability) is the most widely discussed of all of these sources of uncertainty in epidemiology. It can be quantified through the calculation of confidence intervals or p-values.

Table 8.1 Sources of uncertainty in environmental epidemiology studies

Source	Can Uncertainty Be Quantified?	Principal Methods to Evaluate/Control
Errors in exposure measurement	• partially	• validation studies • calibration dosimetry • sensitivity analyses
Confounding	• yes, for known confounders • no, for unknown confounders	• known confounders: stratification, regression • unknown confounders: randomization
Biases	• no	• design study to avoid potential biases
Model misspecification	• partially	• sensitivity analysis
Systematic and random errors in outcome	• yes, for random errors (sampling variability) • partially, for systematic errors	• random errors: confidence intervals • systematic errors: validation studies, sensitivity analyses

8.1. ERRORS IN EXPOSURE ESTIMATION

The mismeasurement of exposure, or of dose metrics estimated from exposure data, is a very serious problem in epidemiology and one that is, to a great degree, the motivation for this entire book. Mismeasurement of exposure is often termed *misclassification* of exposure. This term alludes to the classification or categorization of an exposure, but it is also used when exposure is measured on a continuous scale. There are at least two distinct opportunities for misclassification of exposure in an epidemiologic study. First, the actual measurements of exposure intensity or duration may be incorrect, for one of many reasons. Second, even when the actual exposure measurements are of the highest quality, one may still misclassify exposure as it is used in an epidemiologic study if one chooses an incorrect summary measure of exposure, or dose metric, for the epidemiologic model. Because this second misclassification opportunity is not well appreciated, a brief example may help.

There is a growing body of evidence suggesting that exposure to electromagnetic fields (EMF) may increase the risk of several diseases, including brain cancer and leukemia (Feychting, Ahlbom et al. 1998; Savitz 2001; Savitz 2002; Savitz 2003). A large number of studies have been conducted, some with very sophisticated exposure assessments, based on extensive personal monitoring data for the subjects in large case-control, as well as cohort, studies. This level of detail has been possible because the equipment to measure EMF exposure intensities on different wavelengths and time scales is relatively inexpensive and portable and can gather large quantities of data very rapidly. But the existence of large quantities of EMF data does not eliminate the possibility of exposure misclassification, because it is not clear how the electromagnetic field data should be summarized (Wenzl, Kriebel et al. 1995; Kromhout, Loomis et al. 1997). The problem is that, unless one has a detailed hypothesis of the disease mechanism, it is difficult to choose the summary measure of exposure (Savitz 2003). In the case of electromagnetic fields, the problem is even more complex than for a chemical exposure because there are additional characteristics of fields that may well affect their toxicity. These include different measures of the frequency of the radiation and not just the amplitude or intensity of the field. These choices are difficult to make in the absence of a clear disease hypothesis. It is also difficult to choose appropriate time windows.

Differential Exposure Misclassification

There are two fundamental types of exposure misclassification. The first type is differential or systematic exposure misclassification, and the second is called nondifferential or random exposure misclassification (Pearce, Checkoway et al. 2006). Differential exposure misclassification occurs when exposure data are measured or otherwise misclassified in a way that is different for those who are diseased and those who are not diseased. When this occurs, it can introduce serious bias into effect estimators and, furthermore, the direction of that bias will often be difficult to predict. In a case-control study, differential exposure misclassification can be difficult to avoid if interviews with subjects are the source of exposure information. Patients with serious diseases have often been shown to remember exposures differently than healthy volunteers serving as control subjects. Similarly, interviewers may have difficulty remaining unbiased in their data gathering techniques if cases are seriously ill whereas controls are healthy. This problem is often termed recall bias, or information bias (see later in the chapter), but it can also be thought of as differential exposure misclassification. The best approach to dealing with this source of error is to avoid it through appropriate study design—gathering exposure data through means that do not involve patient interviews, for example.

It is often relatively easy to avoid differential exposure misclassification in cohort studies. The key is to collect all measurements of exposure before disease occurs or with methods that can be conducted blind to disease status. As long as the exposure estimation process is conducted independently of the assessment of disease, then differential exposure misclassification is unlikely to occur. A good example of the separation of the two assessment processes is the way that historical exposure reconstruction is often performed in occupational retrospective cohort studies. One team of investigators, often industrial hygienists, estimates the exposures in each job in a workplace and how these exposures have changed over time (Chapter 3). Any errors in these job exposure estimates will most likely be nondifferential. The job exposure estimates are then assigned to diseased and nondiseased members of the cohort through each subject's job history. As long as these job histories are unbiased, then no differential exposure misclassification will occur. Job histories generally are taken from work records, which are unlikely to be somehow altered by disease status.

Differential exposure misclassification is a serious problem, and one that epidemiologists seek to avoid through appropriate study design. One reason that so much emphasis is placed on avoiding this type of bias in design is that its ultimate effects on measures of association are very hard to evaluate once the data are collected and the study completed. When reviewing a completed study in which differential exposure misclassification appears likely, one worries about the most fundamental aspect of this potential bias—its direction. That is, we often cannot tell whether differential misclassification is more likely to have biased a measure of association towards or away from the null—falsely inflating or underestimating the exposure-disease association.

Nondifferential Exposure Misclassification

The second type of exposure misclassification is termed *nondifferential*. This means that errors in exposure measurement are independent of disease status. The term *random* is sometimes used to describe nondifferential misclassification, but this is not quite accurate, as what distinguishes this type is that there are similar inaccuracies in exposure estimation for both the diseased and the nondiseased members of the cohort. The misclassification may or may not be random with respect to other subject characteristics. Nondifferential exposure misclassification is very common and perhaps universal. There are always errors or inaccuracies in how exposures are measured or categorized. What is important to understand, however, is how serious these errors might be in a particular study and to what degree they may have introduced error into the measures of association.

Nondifferential exposure misclassification will, in general, result in bias of the measure of association toward the null (Dosemeci, Wacholder et al. 1990; Brenner and Loomis 1994; Checkoway, Pearce et al. 2004). That is, if the true association is positive (RR greater than 1.0, for example), then the observed association will lie closer to 1.0 as a result of nondifferential exposure misclassification, and so too will a true RR *below* 1.0 be moved closer to the null by nondifferential exposure misclassification. This is an important principle with implications for the design and conduct of exposure assessment studies. We present two different illustrations of the phenomenon to aid in understanding of how it occurs.

Nondifferential Exposure Misclassification with Categorical Data

Imagine a closed cohort study of 200 individuals, 100 of whom are exposed and 100 of whom are not. The 200 are followed for a fixed period of time, and at the end of that time period, it is determined that 80 of the exposed and 40 of the unexposed have developed the disease (Figure 8.1, part A). One can calculate easily that in these simple data, the relative risk is 2.0. Now, imagine that when we conduct the epidemiologic study of these true data, there is an error in classification of the exposed and the nonexposed. Let us suppose that this error consists of an 80% probability of correctly classifying a person as exposed or unexposed and a 20% probability of incorrect classification. It is not difficult, then, to work out the impact of this error in exposure classification on the resulting relative risk (Figure 8.1, part B). An 80%-correct classification means that, for example, in the A cell of the 2-by-2 table, there will be only 64 exposed diseased individuals instead of the 80 that, in truth, belong to that cell. Similarly, for each of the other four cells, 80% of the original data belong in the table for the correctly classified subjects. To finish the simulation, we must determine the positions in the 2-by-2 table of the 20% of the subjects whose exposure was incorrectly classified. Starting again with the A cell, there are 16 individuals who were, in truth, diseased and exposed but were incorrectly classified as unexposed. These 16 are shifted from the upper left, or A, cell to the lower left, or C, cell. Similarly, the misclassified subjects in the B and D cells are inverted. When the resulting data are summarized, the misclassified relative risk is 1.5 (Figure 8.1, part C). As predicted, this relative risk is closer to the

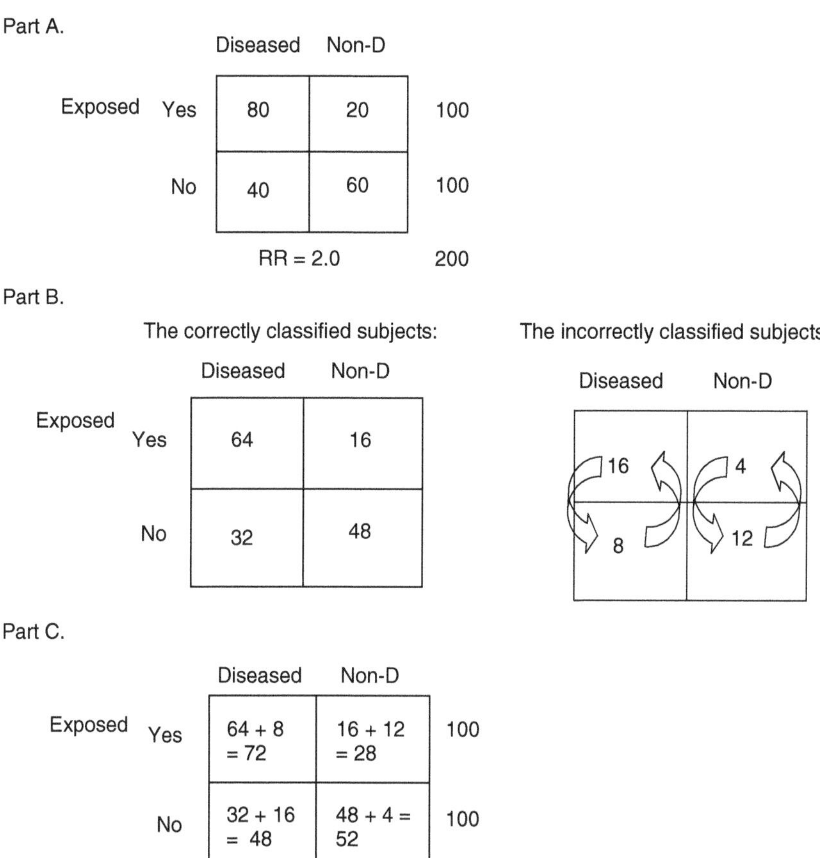

Figure 8.1 Hypothetical data illustrating nondifferential exposure misclassification. Part A: These data represent the "true" exposure and disease status of a closed cohort of 200 persons followed for a fixed period. Part B: Exposure status has been misclassified. For 80% of subjects, exposure classification is correct, whereas for 20%, it is incorrect. Part C: The table of observed data, after exposure misclassification.

null, meaning an underestimation of the magnitude of the association. This pattern will hold for any amount of misclassification of a dichotomous exposure variable.

Several cautions are in order, however. First, one must always remember that this behavior will occur *on average* in data subjected to nondifferential misclassification (Jurek, Greenland et al. 2005). In a small to medium-sized dataset such as the one in Figure 8.1, it is always possible that, by chance, the RR with misclassified exposure data will turn out to lie *further* from the null than the true value. To evaluate how likely this is in any given dataset, one should look at the width of the 95% confidence interval around the RR. This is discussed further later. A second caution that must be mentioned is that if an exposure variable consists of more than two categories (low, medium, high, for example), then nondifferential exposure misclassification can bias measures of association in more complicated ways, with some category risk estimates overestimated and others underestimated. The net effect can, under certain circumstances, bias away from the null (Dosemeci, Wacholder et al. 1990). Furthermore, it is not uncommon for researchers to initially measure exposure on one scale and then later collapse this scale for epidemiologic analyses. If there was nondifferential misclassification on the initial scale, the collapsed scale may end up differentially misclassified if there are more than two categories, leading to bias either toward or away from the null.

Nondifferential Exposure Misclassification with Continuous Data

Now let us consider a slightly more complicated scenario. Imagine that exposure data are continuous rather than dichotomous. For this example, we assume that the outcome data are represented by a continuous measure as well. Let us assume that the outcome is some measure of physiologic function. When both exposure and outcome data are continuous, the measure of association that is typically used is the slope of the regression line between exposure and outcome. What will be the impact of nondifferential exposure misclassification on the slope of a regression model, fit to continuous exposure-response data? To illustrate this, we have simulated a dataset of 100 observations, exposure and outcome, each on an arbitrary scale (Figure 8.2). In the simulation, there is a strong positive association between exposure and outcome (the slope was 0.5, in the arbitrary units of these simulated data). One can see that as exposure increases, response increases also, with only a small amount of variability around the regression line. To simulate the impact of misclassification, random errors were added (or subtracted) to each exposure measurement in Figure 8.2. Each exposure data point was given a certain amount of error, which was randomly chosen to be up to 100% of its true value. That is:

> measured exposure = true exposure ± (error × true exposure), where:
> error is randomly drawn from a uniform distribution with limits (0 to 1).

When the measured exposure is used to regress response in these data, the result is a reduction in the slope or measure of association from 0.5 to 0.24 (Figure 8.3). One sees again the same pattern as in the first example: bias toward the null, or a weakening of the observed strength of association between exposure and response. In summary, exposure misclassification will, on average, lead to bias toward the null in both 2-by-2 tables and ordinary linear regression models, as long as the misclassification is nondifferential with respect to disease status.

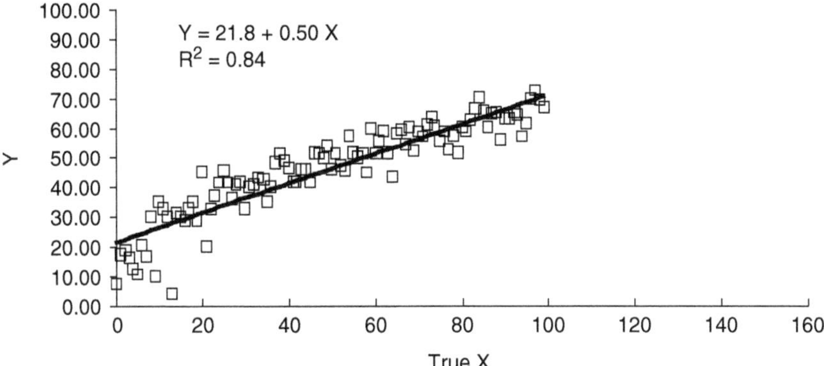

Figure 8.2 Simulated data for illustrating nondifferential exposure misclassification in continuous exposure data. The graph represents the exposure-response relation among 100 subjects, each with a "true" measure of exposure, and a measure of physiologic function, *Y*.

One way to understand the underestimation of an exposure effect that often comes from nondifferential exposure misclassification is to think of it as diluting the exposed group. The error mixes up some subjects who were truly exposed with some who were truly not. If exposure is a risk factor for disease, then the truly exposed will be more likely to be diseased. Thus mistakenly calling some of these subjects "unexposed" moves some diseased subjects into the unexposed category, raising the overall risk in this group and "averaging out" the risks in the truly exposed and truly unexposed groups. The problem will be particularly serious when small risks are being estimated. Most exposure definitions contain some

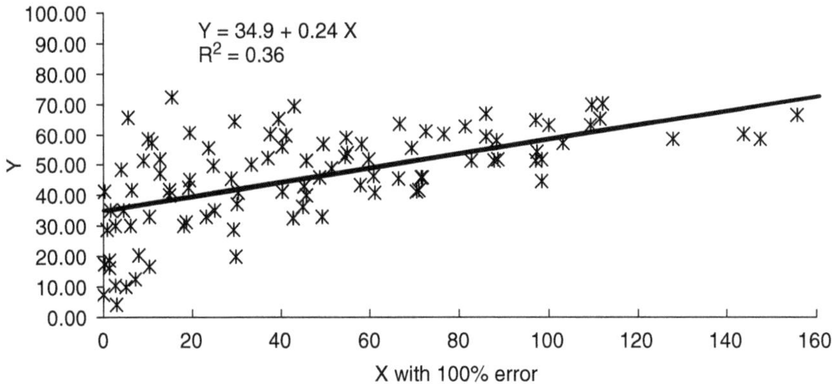

Figure 8.3 Simulated data, as in Figure 8.2, except that the exposure data have been misclassified by adding random errors. The observed slope, or measure of association, is biased toward the null.

error, and if these errors are substantial, they may lead to so much bias toward the null that weak exposure-response associations cannot be seen at all.

Evaluating and Correcting Errors in Exposure Estimation

All standard regression models assume that the independent variables are measured without error. One can see a manifestation of this on many *xy* graphs representing regression models (see, e.g., Figure 7.3). Such graphs often show *vertical* error bars, representing variability in the dependent variable, but rarely present *horizontal* error bars, which would indicate uncertainty in the *x* variable. It should seem obvious from what has been presented in the first half of this book that the implied assumption that exposure measurements are free of error is often violated in epidemiology (Jurek, Maldonado et al. 2006). In fact, it is so universally violated that one may wonder why this assumption is built in to regression—our most widely used statistical model. The answer is that the regression approach was developed in the context of experimental research, in which exposure or "treatment" was fixed by the investigator, who then observed the response and its variability due to sampling or chance.

When exposure data that contain errors, for example nondifferential exposure misclassification, are used in regression models, then the standard statistical measures of uncertainty, such as the confidence interval around the slope, do not include uncertainty that comes from these errors (Spiegelman, McDermott et al. 1997; Jurek, Maldonado et al. 2006). At a minimum, one should keep this in mind in interpreting confidence intervals and *p*-values in regression models. Better methods would include validation studies of the exposure assessment methods, formal adjustment for exposure measurement errors, or sensitivity analyses to understand the impact of plausible degrees of uncertainty in exposure on the study results.

Sometimes validation substudies are conducted through the comparison of two exposure methods: one expensive and highly accurate, the other more economical but less accurate. Chen and colleagues (Chen, Chang et al. 2004) conducted a study of low back injuries in taxi drivers in Taiwan. One exposure of concern was whole-body vibration (WBV). It was not feasible to conduct direct measurements of WBV in more than 1,000 drivers who participated in the study, and so a validation study was conducted for about 250 drivers, using direct WBV measurements in the taxicab. The aim of the validation study was to identify determinants of WBV exposure (the authors called it an "exposure prediction rule") so that the WBV exposures of all study participants could be estimated, based on available information such as the age, make and model of car, size of engine, and so on.

When data from validation studies such as this one of WBV in taxicabs are available, then statistical methods can be used to adjust or correct the exposure-risk associations that are made using the economical but inaccurate exposure measure.

A variety of different techniques are available for combining information from the validation and main studies (Armstrong 1995; Holford and Stack 1995; Armstrong 1998; Chatterjee and Wacholder 2002). Perhaps the simplest to understand is called regression calibration (Spiegelman and Valanis 1998). The method can be summarized as follows. Suppose we use a logistic regression model to quantify the association between a dichotomous outcome *Y* and an exposure X_{approx} that is measured with error, although we will assume that it is not differentially misclassified (see previous discussion). The usual logistic model can be used:

$$\log\text{it}[p(Y=1)] = \beta_0 + \beta_1 X_{approx}$$

Equation 8.1 Logistic model for a dichotomous outcome

where β_0 is a nuisance parameter, and e^{β_1} estimates the odds ratio for a one-unit change in X_{approx} (Kleinbaum and Klein 2002). Now suppose that in a validation substudy, we have measured not only X_{approx}, using the same methods to be used in the full study, but we have also measured X_{true}—or something as close to a "true" measure of *X* as is possible. This true value is often called a "gold standard." Often, a truly gold standard cannot be obtained, but rather something we believe to be better, though not perfect. This is often termed an "alloyed gold standard," and in this case more complicated methods are required (Spiegelman, Schneeweiss et al. 1997). If X_{approx} and X_{true} are not highly correlated, then the odds ratio calculated from Equation 8.1 will be biased downward. This can be corrected in two steps: first, by regressing X_{approx} on X_{true}:

$$X_{true} = \alpha + \gamma X_{approx}$$

Equation 8.2 Calibration regression for X_{approx} versus X_{true}

where α and γ are parameters estimated from the data. In the second step, the regression coefficient from Equation 8.1 is corrected for the measurement error quantified in Equation 8.2:

$$\beta_{corr} = \frac{\beta_1}{\gamma}$$

Equation 8.3 Correction of the regression slope for measurement error

where β_{corr} is an estimate of the strength of the association between *Y* and X_{true}. When there are covariates in the exposure-response model (Equation 8.1), the formulas are more complicated, but the approach is conceptually the same (Rosner, Spiegelman et al. 1990).

Spiegelman and Valanis (1998) presented a good example of an application of these methods, applied to a study of acute health effects of exposure to antineoplastic agents among pharmacy personnel. An exposure and health symptom survey was conducted among 675 pharmacists, covering a 3-month period. Self-reported data on the average weekly number of antineoplastic agents handled and a variety of health symptom information were gathered. A validation study was conducted to assess the accuracy of the 3-month recall of exposure information. A subsample of 56 pharmacists completed 1- to 2-week diaries in which they kept track of their handling of antineoplastic drugs. These data were considered to be close to "true" because the diaries were kept onsite, and filled in continuously.

From 27 symptoms that were initially assessed by questionnaire, Spiegelman and Valanis chose one—fever—for this investigation because it was relatively prevalent (overall prevalence 17%) and plausibly associated with exposure to antineoplastic agents. The crude odds ratio for the association between prevalence of fever and the questionnaire-based number of drugs mixed per week was 1.13 (95% confidence interval: 1.03 to 1.23), comparing the 10th to the 90th percentiles of the exposure distribution. The correlation between the questionnaire exposure estimate (X_{approx}) and the diary data (X_{true}) was 0.70. After adjusting the odds ratio for this error, the corrected estimate, OR_{corr}, was 1.22 (95% CI: 1.04 to 1.43).

The authors noted that there were several limitations to the application of regression calibration in this setting. They were concerned that because fever was fairly common, the odds ratio calculated in the logistic regression model would not be a good estimate of the prevalence ratio (Thompson, Myers et al. 1998). To address this and other limitations of regression calibration, Spiegelman and Valanis also presented a maximum likelihood method of accomplishing the same end, using a model which is probably analytically superior but more difficult to explain (Spiegelman and Valanis 1998).

8.2. CONFOUNDING

Confounding was introduced in the previous chapter, and the standard methods for controlling confounding were reviewed. Here we discuss several aspects of confounding that are of particular concern when quantitative exposure-response relations are being investigated in epidemiologic studies.

To be a confounder, a factor must be associated with the disease, which usually means that it is an independent risk factor for the disease, and the factor must be associated with the exposure. This latter requirement may need further explanation. "Association with exposure" means that there is a different prevalence of the confounder in the exposed and unexposed groups, or among groups with different exposure ranges, if the exposure variable is continuous. If the *confounder* is measured on a continuous scale, then this association with exposure means that the mean value of the confounding variable differs across levels of the exposure. It is also important to remember that a factor which does not vary among members of the study group cannot confound. This fact is the key to confounder control: analyzing the exposure-response association among subgroups who all share the same level of the potential confounder removes any possibility of confounding by that factor.

Confounding is not an "all or nothing" phenomenon; the amount of bias in a measure of association may be small or large depending on the correlations between the confounder and exposure and between the confounder and the response variable. The amount of bias will not be greater in magnitude than the *weaker* of these two associations (Rothman, Greenland and et al. 2008). That is, a weak association between exposure and a confounder cannot explain a strong apparent-exposure effect. This fact has not always been adequately appreciated by reviewers of occupational cancer studies that have not been able to directly evaluate potential confounding by smoking (Kriebel, Zeka et al. 2004).

Large cohort studies constructed using work records and vital statistics information have been an important source of knowledge on occupational cancer risks. But smoking data are rarely available in these studies, because they rely on existing records, and often it is not feasible to contact subjects to obtain smoking histories. If one is using a study of this type to identify a cause of one of the many smoking-related cancers, then confounding by smoking might occur if the prevalence of smoking were different among subgroups with different levels of exposure to the potential occupational carcinogen. In the absence of smoking data, how can the possibility of serious confounding by smoking be evaluated? Several authors have shown that plausible smoking differences among subgroups of a working population will rarely create relative risks for lung cancer greater than about 1.5, and it is even lower for diseases less strongly associated with smoking (Axelson and Steenland 1988; Siemiatycki, Wacholder et al. 1988; Kriebel, Zeka et al. 2004). In other words, weak exposure-response associations might be falsely created by unmeasured confounding by smoking, but even moderately strong associations are not likely to be.

Confounding with Continuous Variables

Confounding is most often illustrated and investigated in the context of categorical exposure data by showing the change in the relative risk when the data are stratified on the confounding factor (see Chapter 7, Table 7.2). However, for the quantitative exposure-response modeling that this book emphasizes, it is important to gain an understanding of how confounding functions when exposure data are continuous (Miettinen 1985; McNamee 2005). Suppose we are studying an exposure-response relation in continuous exposure and response data, much as in the simulation in Figure 8.2 (we would like to credit Professor Olli Miettinen for this graphical understanding of confounding; (Miettinen 1985)). We will use a cartoon of such an exposure-response association in which the data are represented by an oval-shaped "cloud" which indicates the general area in which the data are concentrated (Figure 8.4). The first figure represents data in which there is a strong positive association between exposure and response, represented by the straight dotted line through the middle of the data cloud. To this simple picture, we now add a third factor—gender. Imagine that in both genders there is a similar exposure-response relation, but that for some reason females have a higher background risk than males. This is indicated in part B of Figure 8.4 by two parallel lines, one above the other but with identical slopes (the genders are said to have a common slope). The line representing the females starts at a higher baseline or background response in the absence of exposure, but the increase in response per unit change in exposure (the slope) is the same as among males. The graph in part B also indicates that the exposure distributions in males and females are the same. One can see this by

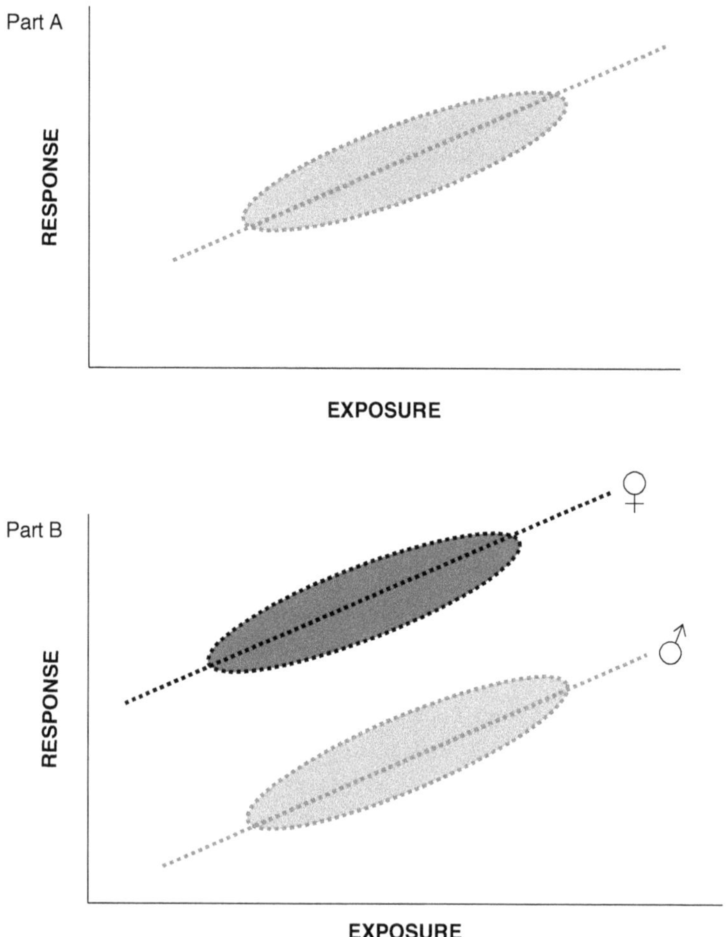

Figure 8.4 Graphic representations of an exposure-response relation in continuous data. The lines represent the exposure-response relation, and the ovals indicate where the data lie. Part A: A simple, linear exposure-response relation. Part B: The same exposure effect is observed in both sexes, but among females, the background risk is higher. The exposure data have similar distributions in males and females. There is no confounding in these data.

noting that the locations on the x-axis of the "clouds" of data for males and females are very similar. If one were to calculate the means and standard deviations of the exposures in the two genders, they would be quite similar. Is gender a confounder of the exposure-response relation in part B of Figure 8.4? No. Gender is associated with response, but not with exposure, and so it cannot be a confounder.

Confounding by gender is illustrated in Figure 8.5, in which the exposure distributions of males and females are different. Notice that the female cloud is

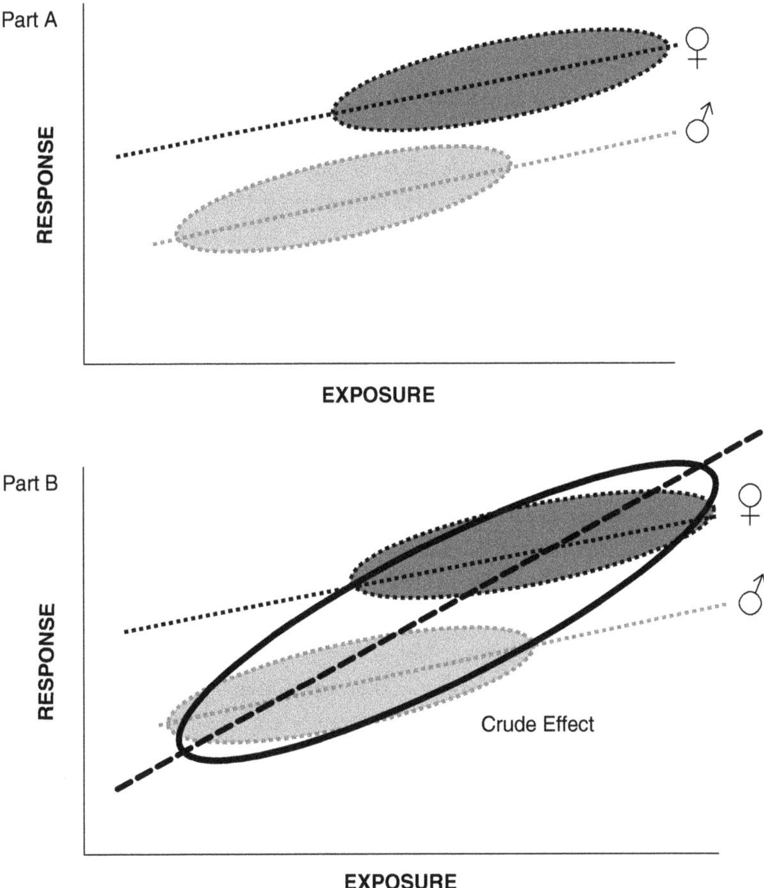

Figure 8.5 Graphic representations of an exposure-response relation in continuous data, showing confounding by sex. Part A: The same exposure effect in males and females as in Part B of Figure 8.4, but now the female exposure distribution is shifted to higher levels. Sex is a confounder, because sex is associated with both risk (females at higher risk at all exposure levels) and exposure (female exposure distribution shifted to higher levels compared to males). Part B: Illustrating the crude, confounded relation between exposure and response. If sex is ignored and a single, crude association is calculated, its slope will be greater than the true common slope for both males and females, when studied separately.

somewhat higher on the exposure scale than is the male cloud—the mean exposure in females would be greater than the mean male exposure. If gender is ignored and a single exposure-response trend is fit in the full dataset, it might look like the black dotted line in part B. The slope of this line is steeper than the true common slope for males and females. To avoid this confounding, one should analyze the data in a way that preserves the view in part A, in which one can see the common slope because the data are analyzed separately for males and females. In practice, this is accomplished through stratification or through a regression model in which gender is an independent variable entered into the model alongside the exposure variable.

Incomplete Control of Confounding

Statistical methods such as stratification or regression modeling can control confounding only by factors for which data have been gathered. If, alternatively, there is an unknown factor that is a confounder, the bias it will cause cannot be controlled in an observational study. Only randomization has the ability to control for the confounding influence of factors that are not measured. This problem is well described in every epidemiology textbook, but in practice, epidemiologists often behave as if unmeasured confounders are unlikely to be a serious problem, when there is no way to really know this. Thus there is uncertainty that derives from what we do not know—and this is by definition impossible to quantify.

Uncertainty deriving from possible *bias* in study design and execution is quite similar in nature—there is rarely any way to objectively know whether a bias exists, nor how strong its effect on the results might be. It is standard practice in reporting the results of an epidemiologic study to say at the end of the article that the results are valid only "if there is no bias and no uncontrolled confounding." But because this is an unverifiable assertion, it is often essentially ignored in the evaluation of the evidence. Unfortunately, this habit may give the nonscientist the impression that the researchers are more confident than they really should be about the study findings.

8.3. BIASES

A measure of association from an epidemiologic study will be biased if there is a consistent difference between the measured association and the true association. Bias is defined formally as the difference between what is measured and the truth, and because the true association is never known, it is essentially impossible to quantify the magnitude of a bias. Rather, biases should be avoided or minimized in the design of a study. Partial assessment of bias is possible, however, when one is able to conduct validation studies—for example, on the exposure assessment as described previously. There are two broad categories of bias in epidemiology, which are reviewed here briefly: first, selection bias, and second, information bias.

Selection Bias

Selection bias is best understood by seeing how it works in the two major study designs, cohort and case-control studies. In a cohort study, selection bias occurs if disease status influences selection into exposure groups. For example, suppose that we are studying an outbreak of childhood leukemia in a town. There is concern that proximity to a toxic waste site may be responsible for an increased risk of the disease. As often happens in these cases, imagine that a certain number of children with leukemia living near the site have already been identified through neighborhood meetings and friendship networks. These are often called the *index cases*. To follow up on this concern, we might conduct a cohort study, identifying all the children in the town and dividing them into "exposed" and "unexposed," according to proximity to the site. A cancer registry or other case-finding system could then be used to identify all the cases of leukemia in the town. Selection bias might occur if we automatically place the index cases in the exposed group instead of using a blind and consistent exposure assignment process for all subjects. Ideally, the assignment of children to the exposed and unexposed groups would be done by someone who was unaware of the health status of the children, so that this knowledge did not influence their classification. In practice, this is often difficult in small studies.

Selection bias operates somewhat differently in case-control studies. In these studies, selection bias occurs if exposure status influences selection into the case or control groups. An example may help to illustrate the problem. Suppose one is conducting a study of cigarette smoking and risk of lung cancer. Who should be chosen as controls? If we were to choose patients with heart disease as controls, we would be introducing a selection bias, and the resulting measure of association between smoking and lung cancer would probably be an underestimate of the true association. Heart disease is more prevalent among smokers than nonsmokers, and, as a result, the contrast of smoking habits between cases of lung cancer and cases of heart disease would not be as large as between lung cancer cases and the general population. To avoid selection bias and produce an accurate estimate of the strength of the association between smoking and lung cancer, the smoking habits of the controls should be representative of the population from which the cases came. This means that controls should either be drawn in some random fashion from the general population or, if they are selected from a hospital population, controls should have diseases that are unrelated to the exposure.

Information Bias

The second type of bias, information bias, is again best illustrated separately in cohort and case-control studies. In cohort studies information bias occurs if information on disease is influenced by exposure status. For example, suppose that the clinicians diagnosing a particular disease have beliefs about the exposures which lead to that disease. They may believe correctly that people with a particular exposure are more likely to get that disease. As a result, they may look more closely for the disease in an exposed population. For example, in a study of asbestos workers, a physician might search more aggressively for cases of lung cancer because of the causal link between asbestos and this cancer. This behavior would be very appropriate clinical practice, but it might create an information bias if clinical data were used to identify cases for an epidemiologic study. If the diagnostic practices are different among exposed and nonexposed subjects, then the observed exposure-disease association may be biased.

Information bias functions somewhat differently in case-control studies. Information bias occurs if the determination of exposure status is influenced by case or control status. This problem was discussed earlier under the heading of differential exposure misclassification, which is a kind of information bias. When participant interviews are used to gather exposure information in case-control studies, then there is often concern that information bias will occur because of the potential for differential recall of personal histories among those who are sick (cases) and those who are healthy (controls). Indeed, minimizing information bias is one of the principal reasons for choosing hospital controls in case-control studies. Hospital controls are, like the cases, ill, and so may be in a similar frame of mind. As long as the control subjects' illnesses are not associated with the exposure of interest, then this approach may avoid both selection bias (see the previous discussion of selection bias from choosing hospital controls) and information bias.

Biases are best avoided, because once data have been gathered in a biased way, it is very difficult to control the bias or to assess quantitatively how large its effects might be. Sometimes one can conduct sensitivity analyses, which can provide useful information on how large the effects of a particular bias might be in a worst case scenario, for example. An example was mentioned in section 8.2, in the context of evaluating the effects of unmeasured confounders. If data on smoking are lacking from a study in which smoking might conceivably confound an observed association, it is possible to conduct sensitivity analyses to evaluate how large the effects of smoking could have been under some assumption about the smoking distribution in the study population (Kriebel, Zeka et al. 2004).

8.4. MODEL FORM MISSPECIFICATION

Even the simplest analysis of epidemiologic data uses a model. The 2-by-2 table is a statistical model, with a set of underlying assumptions that may or may not accurately correspond to the reality that is being summarized. The assumptions behind a typical epidemiologic data analysis are numerous and often difficult or impossible to directly verify (Table 8.2). At a minimum, it is important to be aware of these assumptions, but, when possible, one should also investigate the degree to which errors in these assumptions affect the study findings. In this book we discuss

Table 8.2 Examples of choices made in a typical epidemiologic study and their underlying assumptions.

Category	Typical Choice	Underlying Assumption
Exposure data	Mean of available data	Environmental samples representative & measured without error
Summary measure of exposure	Lifetime cumulative exposure (CE)	CE directly proportional to disease risk
Coding of exposure variable	Continuous variable	Risk rises exponentially with CE over its entire observed range
Confounders	Age, smoking (packyears) measured by questionnaire	Packyears is correct summary measure of tobacco No recall bias Risk linearly related to age No other important confounders exist
Study design	Case control study	Controls representative of study base
Outcome	Lung cancer incidence	Diagnoses are accurate and complete ICD coding is relevant for exposure under study
Latency	Ignore 20 years' exposure prior to disease onset	No effect of exposure after cutoff All prior exposure has same "potency"
Statistical model	Logistic regression	Exposure and confounders have multiplicative joint effects Extreme exposures do not have excessive influence on slope estimate Error distribution is appropriate
Evaluating chance variability	Statistical significance, $p<0.05$	No bias No uncontrolled confounding All of above assumptions are correct

the construction of models to describe the toxicokinetics and pharmacodynamics that govern the biologic pathways between exposure and disease. These additional models may add additional uncertainty to epidemiologic analyses, but they can also be seen as a way to make explicit certain fundamental assumptions about how exposure leads to disease—processes that necessarily underlie even the simplest exposure-response model.

A simple example may help to clarify this point. Suppose that one wishes to investigate the effects of both an air pollutant and tobacco smoke on lung disease risk. The typical approach would be to enter variables coding for each of these covariates into a regression model. If the regression model is a linear one, then the assumption being made is that the air pollutant and tobacco exposures contribute *additively* to change in the dependent variable. If, on the other hand, one uses a logistic or Cox proportional hazards model, then the implicit assumption is that the two risk factors have a *multiplicative* relation to the dependent variable. Although these two types of joint effect are the most frequently discussed in epidemiology, there are in fact many other ways in which two factors may contribute to a dependent variable. Many other possibilities could be represented by disease process models like those described elsewhere in this book. With an explicit hypothesis about the ways that the pollutant and tobacco smoke entered the body, reached the target tissue, and caused cellular damage, one might construct a biologically based model. The nature of the joint effect of the two exposures—whether additive, multiplicative, or something else—would not be a separate assumption but would be a consequence of the particular structure of the disease process model. Although we believe that this approach holds promise, there is as yet little published research about the best ways to construct and apply these kinds of models. It is clear that even simple models can have profoundly different implications for underlying processes. The epidemiologic behavior of disease process models is discussed further in Chapter 9.

Returning now to the world of standard epidemiologic practice, it must be said that here, too, there is little research on the impacts of the choice of a particular statistical model on study results. The choice of, say, a logistic model or a 2-by-2 table is often made for convenience, the availability of software, and other reasons that have little to do with the nature of the underlying exposure-disease process. Using a model that does not adequately capture the behavior of the underlying physiologic processes can introduce an additional source of error into the measure of association or other study results. This topic is very little investigated in epidemiology, but we recommend that investigators begin to study the implications of model selection by varying the model structures that they use in investigating the impacts of these changes on study results (Robins and Greenland 1986; Maldonado and Greenland 1993; Stayner, Smith et al. 1995; Greenland 1996; Maldonado and Greenland 1996).

We recommend that all epidemiologic studies include an investigation of the sensitivity of the results to the choice of model form (Greenland 1996; Greenland 2001). The biologically based models that we advocate in this text are experimental, and there is as yet very little experience with them. It is therefore important that the results from a particular model form be compared with those that come from other approaches to modeling the same data. In the absence of knowledge of the true exposure risk association, the best one can do in this investigation is to report the sensitivity of the final findings to the choices that have been made in the forms of the models.

Investigators from the National Institute for Occupational Safety and Health (NIOSH) have published studies comparing the results of a wide variety of different models fit to epidemiologic data on cohorts exposed to cadmium (Stayner, Smith et al. 1995) and to silica (Rice, Park et al. 2001; Park, Rice et al. 2002). These results demonstrate that the essential or qualitative findings about the risk from cadmium and silica exposures were consistent across a range of different types of epidemiologic models. However, the selection of a particular model could, in some cases, have a large influence on the resulting estimates of risk. This variability from model to model serves as a warning not to rely overly much on any one model form in the absence of strong hypotheses—for instance, based on mechanistic studies—about which one is most likely to be correct. An excellent discussion of the issues involved in choosing among exposure-response models has been published by Steenland and Deddins (Steenland and Deddens 2004).

8.5. ERRORS IN OUTCOME VARIABLES

Both random and systematic errors in the outcome variable must be considered. Random or sampling variability in the number of outcome events is perhaps the most familiar source of uncertainty in epidemiologic studies, and it is investigated with the use of *p*-values and confidence intervals. Despite the emphasis placed on measuring this source of uncertainty and achieving "statistical significance," sampling variability in outcome probably accounts for a minority of the uncertainty in most epidemiologic studies. Systematic errors in outcome variables can be investigated in many of the same ways as systematic errors in exposure variables (Pearce, Checkoway et al. 2006). Systematic errors in outcome variables, although just as important as errors in exposure variables, are not discussed further because our focus is on improving exposure estimation.

Confidence intervals or *p*-values are found in every epidemiologic study, and so it is important to understand what they do and do not measure. In a typical epidemiologic investigation of an exposure-response relationship, one studies a population in which there is some variation in the level of exposure among population members. The quantitative estimate of the change in risk across this range of exposures is typically evaluated using a regression model. The slope of the regression equation estimated from the observed population data is a measure of the strength of the association between exposure and response. We noted in section 8.1 that all standard regression methods assume that there is no error in the exposure variable, because these methods were developed for the analysis of data from experiments in which the independent variables were fixed by the investigator. It is very important, therefore, to always remember that the *p*-values and confidence intervals that one calculates for epidemiologic results do not take into consideration uncertainty in the exposure and other independent variables.

Evaluating Variability in Outcome

The notion of statistical significance and the associated concepts of the *p*-value and confidence interval were developed in the statistics of experiments (Neyman and Pearson 1928). In the experimental setting, it is often possible to randomize observations across treatment, or exposure groups. This approach allows some confidence that confounding has been avoided, including confounding by factors that are completely unknown to the researcher. For this reason, experiments can be designed to formally test hypotheses, and one can reject or fail to reject these hypotheses by quantifying the role that chance variability may have played in the experimental results.

In the absence of randomization, it is much more difficult to view the measurement of exposure-response associations as leading to formal hypothesis testing. Nevertheless, because this view is so widely taught in epidemiology and statistics courses, it provides an important framework for the interpretation of study results, and a brief review may be helpful.

Hypothesis Testing

The essence of the testing framework is the contrast between two competing hypotheses, called the null hypothesis (H_0) and the alternative hypothesis (H_A). In any given investigation, it is generally a fairly simple matter to identify the null hypothesis, whereas there can be many alternative hypotheses. When an exposure-disease relation is being studied, then the null hypothesis will usually specify that there is no association between exposure and disease, or equivalently that the exposed and unexposed do not differ in disease incidence (or prevalence, or some other measure of outcome). Several common alternative hypotheses are: that the exposed show a greater incidence of disease than the unexposed, that the exposed show a lower incidence than the unexposed, or the more cautious hypothesis that the exposed show a *different* incidence than the unexposed.

The role of data in this framework is to allow a test of the null hypothesis. If H_0 is rejected, then this is taken as evidence in support of H_A. Alternatively, if H_0 is not rejected, then it is supported by the data. Technically, one can never "prove" either hypothesis, and one cannot directly test the consistency of the data with H_A—only their inconsistency with H_0. One tests the null hypothesis and evaluates "statistical significance" using a *p*-value. This statistic assesses the likelihood that the data are consistent with H_0 or, more informally, provides an answer to the question: How likely is it that the observed results[1] could have occurred, by chance alone, if the H_0 is correct?

Suppose that we have conducted an epidemiologic study in which we have observed a relative risk of 2.0, comparing some disease risk among exposed and unexposed groups. We can presume that the null hypothesis was that there was no association between exposure and disease. Let us suppose that our alternative hypothesis was the most cautious one, stating simply that exposure affects disease risk, without specifying whether it increases or decreases risk. We could summarize the competing hypotheses like this:

$$H_0: RR = 1.0$$
$$H_A: RR \neq 1.0$$

Suppose that we observed a p-value for this result of 0.02. Because this is less than the traditional 5% threshold, the result can be called statistically significant. But what does *that* mean? The p-value has the following meaning: *there is a 2% chance that the result obtained (or a more extreme result) would have occurred by chance, if H_0 is true*. There are many common misinterpretations of p-values. For example, "there is a 98% chance that H_0 is wrong" and "there is a 98% chance that there is a difference in risk between the exposed and unexposed groups" are both incorrect statements. One of the most difficult concepts to grasp about hypothesis testing is that a p-value has its formal meaning only in a null world. That is, the probabilistic interpretation of the 2% result in this example is correct if and only if there is in truth no difference in risk between exposed and unexposed groups. If the p-value is small, then we can conclude that the data we obtained are unlikely to have occurred in a null world (assuming the study was conducted without bias and that there was no confounding). The p-value does not speak directly to the more interesting question: How likely is it that an alternative hypothesis might be true? For example, how likely is it that exposure *increases* disease risk? The p-value does not tell us this.

The Limitations of Statistical Significance

Before moving on to discuss confidence intervals, which are an alternative way to evaluate the role of chance variation, let us examine more closely what the results in the previous example might actually mean. A study with RR=2.0 and p=0.02 might be the result of any of the following:

There is a true (positive) association between exposure and disease.
There is no association, but by chance, RR=2.0 was observed.
There is no association, but through bias, confounding, model misspecification or some other systematic error, RR=2.0 was observed.

How likely is the first possibility? We lack the data to make a quantitative statement about this. We can say that the second alternative is "unlikely" given the small p-value and assuming no bias or confounding. We cannot quantitatively evaluate the third possibility. Some examples of the assumptions covered by the absence of bias and confounding include: (1) the data must be a random or representative sample from the study base population; (2) the observations must be independent, one from the other; and (3) the measurements of the independent variables must be accurate.

Suppose now that the results were slightly different, so that the p-value was "not statistically significant"; p=0.07, for example. The likely explanations for these findings are:

There is, in truth, no association between exposure and disease.

There is an association, but by chance, a "nonsignificant" *p*-value was obtained.
There is an association, but through bias, confounding, model misspecification or some other systematic error, a "nonsignificant" *p*-value was obtained.

The second possibility is especially likely if the study was small and so did not have much power to detect a doubling of risk comparing exposed and unexposed groups. A key point here is that a "nonsignificant" *p*-value does not distinguish between two very different situations: the absence of an association on the one hand and inadequate evidence of association on the other.

To summarize this brief presentation of statistical significance:

A statistically significant *p*-value does not directly evaluate the likelihood that the observed association is present (true).
A nonstatistically significant *p*-value does not distinguish between the absence of an association and inadequate evidence with which to evaluate the association.
p-values only have their literal interpretations in the absence of bias and confounding, which occurs rarely, if ever, in observational studies.

Because of these limitations, we recommend placing minimal emphasis on *p*-values when interpreting epidemiologic results and avoiding the concept of statistical significance entirely. The task of an epidemiologic study is not to "prove" the existence of an effect. Rather, our interest is in assessing how supportive study results are of one or another of a limited number of alternative states of nature. That is, we are interested in asking how much the data should weigh in our evaluation of a hypothesis rather than in seeing the data as providing a formal test with which we will reject or accept a particular hypothesis.

Confidence Intervals: An Alternative to p*-values*

Confidence intervals provide an alternative method for quantifying the role of chance variability in study outcomes and for avoiding some but not all of the limitations of *p*-values. To understand what a confidence interval means, imagine that a particular study is repeated again and again, in hypothetical repetitions. In this imaginary world, the study is repeated identically, with the only difference from repetition to repetition being the chance variability in the number of outcome events in the exposed and unexposed groups. At the end of each study, the results are analyzed, and a relative risk and confidence interval are calculated. The 95% confidence interval (the only interval that is routinely calculated) around the RR can be interpreted in the following way: in hypothetical repeated samplings, the 95% confidence intervals calculated around the repeated sample risk estimates would include the true risk 95% of the time. Any given confidence interval—for example, the one that we actually do calculate in our real study—may or may not include the true value of the relative risk.

The 95% "coverage," as it is called, assumes that there is no bias and no confounding—an assumption that we have argued is generally difficult to accept. We believe that, given the many uncertainties in observational studies, a looser interpretation of confidence intervals is more appropriate. They should be viewed as representing the "supported range"—the values of the relative risk that are "more likely." If the supported range lies far from the null, then one can conclude that there is fairly strong evidence for the existence of an association or effect. Narrower confidence intervals are interpreted to mean greater certainty about where the true risk estimate may lie. Furthermore, a confidence interval that excludes the null value of the measure of association provides stronger evidence against the null than a confidence interval wide enough to include this value. It is important, however, not to overinterpret the location of the null value as either inside or outside of the confidence interval. Some authors use this distinction as a test of statistical significance (if the null value is included in the 95% confidence interval, then the result is said to be statistically significant), and we believe that such rigid testing is inappropriate in the context of most environmental epidemiology studies. Finally, it is important to remember that a tight confidence interval cannot make up for serious bias or uncontrolled confounding. In short: garbage in, garbage out, no matter how much garbage there is.

8.6. MANAGING UNCERTAINTY

The preceding sections of this chapter have provided an overview of the major sources of uncertainty in epidemiologic studies (Table 8.1). Approaches to evaluating and managing uncertainty vary, from formal statistical methods such as confidence interval estimation to qualitative descriptions of such potentially important topics as the choice of statistical model or the underlying assumptions about the time course of the disease process.

At the end of a long chapter on all the ways that studies can be in error, or uncertain, it is perhaps useful to remember that uncertainty is a positive and essential aspect of scientific inquiry. The uncertainty points the way to new knowledge and further investigation (Levins 1995). There is, at the same time, a strong desire on the part of scientists to be precise, which may come from confusing *uncertainty* of information with *quality* of information—two distinct concepts (Funtowicz and Ravetz 1990). One can have high-quality information about greatly uncertain phenomena.

Much additional research is needed to develop and standardize methods to characterize, express, and communicate uncertainty (Walker, Harremoës et al. 2003; Kriebel 2008). Researchers in each narrow scientific discipline develop professional judgment that they use to assess how strong a particular study finding is and how important the various uncertainties are. But the development of this professional judgment is largely intuitive and not formalized. It is therefore difficult to communicate to outsiders the full complexity of one's assessment of the "weight of evidence" that a study's findings contribute. There is a need, therefore, for research on the characterization and communication of uncertainty in epidemiologic research. Uncertainties that derive from the choice of research methods and mathematical models are especially in need of investigation because so little work has been done in this area.

NOTE

1. Technically: how likely is it that the observed results *or more extreme results* could have occurred by chance under H_0.

9 Dosimetry in Epidemiology

Truth emerges more readily from error than confusion.

Francis Bacon (1561–1626)

This chapter discusses the use of biologically based models of exposure and dose in epidemiologic studies. The previous two chapters presented many of the key concepts in epidemiologic design and analysis for studying environmental and occupational health risks. Now it's time to consider how the biologically based dosimetry from the first section of this book (especially Chapter 6) could be employed in epidemiologic studies. We describe how dose metrics can be used in epidemiologic models and discuss some of the methodologic challenges that arise. Two examples from the published literature illustrate many of the points the chapter raises.

9.1. SUMMARY MEASURES OF EXPOSURE

First we consider the common empirical exposure summary measures widely used in environmental epidemiology. As described in detail in Chapters 2 and 3, exposure is a complex, time-varying quantity, and as we have seen, it must somehow be summarized before it can be put in an epidemiologic model. Given a subject i in a cohort exposed to a toxic chemical, the exposure history of that subject from the beginning of exposure until death, disease diagnosis, or perhaps the assessment of some physiologic function at time T can be summarized as follows:

$$\bar{C}_{expo}[i] = \{C_{expo}[i,1], C_{expo}[i,2], C_{expo}[i,3], \ldots \quad C_{expo}[i,n_i]\}$$

Equation 9.1 Exposure vector for subject i, with n_i exposure intervals

where $C_{expo}[i,j]$ is the exposure level for subject i in time periods $j = 1,2,3 \ldots n_i$, and n_i is the number of time periods of exposure of subject i prior to T_i. The exposure within a time period is assumed to be approximately constant, that is, to have a stable mean and standard deviation (SD).

Most summary measures of exposure used in environmental epidemiology are summarized from this vector of exposures. For example, cumulative exposure (CE) can be written:

$$CE_i = \sum_{j=1}^{n} C_{expo}[i,j] \times t_j$$

Equation 9.2 Cumulative exposure for subject i with n exposure intervals

Cumulative exposure is thus the time integral of exposure intensity. As we have seen, it is widely used in environmental and occupational epidemiology. In Chapter 6, we showed that CE is likely to be proportional to the tissue dose in disease processes involving irreversible proportional damage. Essentially, CE will be proportional to tissue dose in the target organ if one assumes that the fraction of ambient exposure that reaches and is retained in the target tissue is constant over time and among members of the cohort (Smith 1992).

Another common summary measure is simply the duration of exposure. There are many studies in which exposure intensity information is not available or is of poor quality, and so duration may be the only quantitative information on exposure. In occupational epidemiology, exposure duration is often approximated with employment duration, although this will often result in substantial misclassification, especially when only a minority of the workers in an industry work directly with the agent of interest. Duration of residence at a particular address shares similar problems when an ambient exposure is being studied. Although duration clearly ignores the intensity of exposure, it is helpful to clarify the conditions under which exposure duration may approximate target tissue dose. First, compare Equation 9.2 with the duration of exposure. If the exposure intensities all have the same mean, that is, they are approximately constant (expected value of $C_{expo}[i,j] \cong \bar{C}_{expo}[i]$ for all $C_{expo}[i,j]$), then Equation 9.2 can be rewritten:

$$CE_i = \left(\sum_{j=1}^{n_i} t_{ij} \right) \times \bar{C}_{expo}[i]$$

$$CE_i = T_i \times \bar{C}_{expo}[i]$$

Equation 9.3 Calculation of Cumulative Exposure when exposure intensity is assumed constant

If exposure intensities are approximately constant, and if CE is proportional to dose, then duration of exposure will also be proportional to dose (Smith, Hammond et al. 1984). Average exposure intensity may also be proportional to tissue dose in an analogous way if durations of exposure vary little among cohort members.

Choosing a summary measure of exposure can be thought of as choosing weights for each component of the exposure profile, as we discussed in Chapter 6 (see Equation 6.1). One wishes to select weights in such a way that the summary measure is proportional to risk. The disease process models described in section A can be thought of as a biologically motivated method for selecting these exposure weights.

9.2. INCORPORATING PHYSIOLOGIC PROCESSES INTO EPIDEMIOLOGIC MODELING

Epidemiologic models describe the relationship between exposure and disease in one or more populations under study. As we saw in Chapters 7 and 8, their great strength is that they can estimate the risk of disease in a population attributable to an unknown or poorly characterized exposure while controlling for the confounding effects of other factors. Epidemiologic models are population models whose results can be applied to individuals only to the extent that the individuals are similar to the population studied. Similarly, confounding is generally a population concept.

This population view of disease, the cornerstone of epidemiology, has been used with great success to identify most of the established environmental causes of human disease. All of the processes that occur *within* each subject in the study—toxin uptake, distribution, clearance, preclinical responses, and so forth—have in the traditional approach been ignored or imagined as a "black box" with "exposure" as the single input and disease or death as the single output. Throughout this book we have been arguing that increasing the sensitivity of epidemiologic studies will require "opening up the black box" to estimate, for each subject in a study, additional intermediate steps or parameters in the exposure-disease process—for example, the tissue dose, and the effectiveness of repair or detoxification. But how should these *individual-level* parameters and processes be incorporated into a population-level (epidemiologic) model?

Comparison of the New Approach with the Standard Approach

Before explaining the approach that we advocate, it may be useful to contrast it with a standard epidemiologic analysis. In the standard approach to fitting an epidemiologic model, one would first identify all of the "exposure" variables that could be measured—the average or cumulative exposure, the total duration of exposure, the average exposure, perhaps several different measures of "peak" exposure, and so on, perhaps for each of several contaminants. Each of these would typically be included in regression models, alone and possibly in combination, to identify the minimum set of variables with "statistically significant" associations with risk. Various types of diagnostic tests might be run on the final model to look for statistical problems such as outliers, excessively influential data points, or nonlinearities. If these tests found no serious problems, the analytic process would be considered complete, and a "final model" would be accepted.

There are several problems with the standard empirical approach. First, it treats the different exposure-related variables— all of them in some way related to the two fundamental characteristics of *intensity* and *duration*—as essentially independent quantities, at least implicitly all of equal a priori validity, and with little thought given to their biologic relationships to the outcome. One or more of these quantities will be combined with other covariates such as age or smoking status, using linear weights (the β values, slopes, or effect estimators) in an additive model to predict response. As we have tried to show in Chapter 6, the processes by which time and exposure interact with the organism to cause disease are complex and dynamic, and simple additive combinations of terms may not be the best way to describe these processes. One of the likely consequences of this is to introduce considerable nondifferential exposure misclassification by forcing dynamic processes into static additive models.

A second problem with the standard epidemiologic modeling approach is that it often uses an essentially "mechanical" method of choosing which of the available exposure variables to include in the model, and this can produce misleading results. Statistical packages contain algorithms that will select the variables to include in a multivariate model, based on how much each new variable will improve the overall fit of the model. These routines vary in their details but basically come in two varieties: the researcher can tell the computer to do either "forward building," in which variables are added one at a time, or "backward elimination," in which all variables are initially included and then eliminated one at a time until some "best fitting" model is found. Even if these routines are not used, investigators often "manually" build their models in a similar way, essentially using only statistical significance as the criterion for selecting variables for inclusion. Although there is an appealing simplicity and "objectivity" to this approach, it can lead to seriously misleading results (Robins and Greenland 1986; Rothman, Greenland et al. 2008; Checkoway, Pearce et al. 2004). Epidemiologic data are, unfortunately, often limited by unavoidable sources of random error and modest sample sizes. In these situations of "weak" data, the investigator needs to employ clear a priori hypotheses to guide the inferential process; weak data can only help distinguish among a limited number of alternative states of nature, and the investigator must choose those alternatives carefully and explicitly. The disease process models described in this book can be seen as a way to develop these explicit hypotheses.

A third limitation of the standard epidemiologic approach to building exposure-response models is that even when a good-fitting model is developed, it may not yield very useful information about the disease process under study. Suppose, for example, that variables measuring each subject's average exposure and peak exposure (by some specified definition) have been constructed and that these are tested in epidemiologic models. Suppose further that average exposure is found to be the better predictor of response. The poor explanatory power of the model with peak exposure may be due to serious measurement error in assessing peaks, because the definition of what constitutes a biologically meaningful peak has been misspecified or because, in truth, peak exposures do not contribute importantly to the disease. Which of these explanations is more likely? The data probably cannot tell us, but the approach that we advocate may use the available data to better advantage by providing a formal mechanistic hypothesis against which the data can be tested. If we are willing to specify a hypothesis about the disease process and formalize this with a dose metric using a model like those described in Chapter 6, this dose metric can be used in place of the whole set of exposure variables in an epidemiologic model to estimate risk of response. In practice, several different dose metrics might be constructed, each one the result of a different hypothesis about the physiologic processes involved.

9.3. DOSIMETRY FOR EPIDEMIOLOGY: A TWO-STEP PROCESS

The use of disease process models in epidemiology can be broken down into two modeling steps: in the first, a *dosimetric model* is constructed which can represent the hypothesized pathophysiologic processes occurring within each subject in the study. In the second step, an *epidemiologic model* is used to characterize the associations in the population between the disease outcome and the result of the dosimetric model for each subject while controlling for potential confounding factors. The output from the dosimetric model might be, for example, the tissue dose at the time of disease diagnosis or death. This model describes for each subject the critical physiologic and pathologic processes leading up to the single moment in time that is represented by the epidemiologic model. It could also include factors which interact biologically with the exposure-response process (effect modifiers) such as might occur if smoking decreases the pulmonary clearance of a toxin under study from the airways of smokers in the study population. The epidemiologic model would continue to represent one moment in time—often the moment of disease diagnosis or death—because it is at this moment that confounding occurs and at which the relative risk or some other effect estimator can be meaningfully estimated.

It may be useful to contrast the two different types of models in some detail, to clarify their roles and the relationship between them. In the dosimetric model, the subject, or unit of analysis, is each *individual*, whereas in the epidemiologic model, the unit of analysis is the *population*. The *time period* considered in the dosimetric model may be lengthy and is determined by the time course of the pathophysiologic process. The epidemiologic model may utilize cases of disease that occur over a long period (as in a cohort study), but actually the model is concerned with the moment in time when the transition from health to disease occurs. *Independent variables* in the dosimetric model will include one or more exposures, effect modifiers, and host factors that might directly influence the pathophysiologic processes. The final result of a dosimetric model is a single dose metric, summarizing the pathophysiologic processes resulting from an exposure to an individual. This single dose metric is then used as an input (independent variable) in an epidemiologic model. The epidemiologic model includes one or more dose metrics and variables coding for potential confounding factors. The *mathematical forms* of dosimetric models are potentially very diverse—linear or nonlinear models and systems of differential equations, for example. Epidemiologic models are of more limited forms, including linear, logistic, and Poisson regression models and various forms of categorical analysis. The diversity of forms of dosimetric models may be indicative of the very early stage in the evolution of this approach; in contrast, epidemiologic models are better standardized, and a small number of basic methods are now well accepted and widely applied. The *output* of a dosimetric model is a dose metric—an estimate of a target tissue dose, a toxin-receptor complex, or other biologically active agent, based on a hypothesized pathologic process, whereas the epidemiologic model is a statistical model that produces a measure of association quantifying the exposure-response relation, conditional on confounders.

In this two-step approach, the epidemiologic model might not include common time-dependent covariates (time since first exposure, duration of exposure, latency, etc.). These time effects occur within individuals and thus may be more appropriately modeled at the individual level. Of course there *are* population-level time effects, such as secular trends due to long-term changes in mean age of starting smoking, mean age at which employment begins, general nutritional status of the population, and so forth. These would be appropriately included in the population (epidemiologic) model. But more frequently, time-dependent covariates are individual characteristics and should be allowed to modify the tissue dose or other outcome of the dosimetric model.

Throughout this book, we describe and illustrate the use of disease process models in epidemiology as occurring through these two linked steps. But there are examples in which a biologically based model has been constructed using a single step capturing the key temporal dynamics of both the toxicokinetics and pharmacodynamics. The Moolgavkar-Knudson two-stage cancer model is probably the best-known example of such an approach (Moolgavkar 1978; Moolgavkar and Luebeck 2003), and it is presented in Chapter 15.

9.4. USING DOSIMETRIC MODELS

Dosimetric models are not new to epidemiology, although they have yet to gain widespread acceptance, even when quantitative exposure data are available.

Two of the impediments to wider use may simply be a lack of common terminology and a unifying conceptual framework. Examples of published studies that have used dose metrics (even though they generally were not labeled as such) in epidemiology are presented in Table 9.1.

In Chapter 6 we presented the example of ozone and its acute effects on the lungs, and we develop this example further in Chapter 11. Because ozone is a very reactive molecule, it is almost instantaneously transformed when it contacts the epithelial lining of the airways. The "tissue dose" of ozone, then, would be the time integral of the concentration of this reactive oxidant, which attacked fatty acids in epithelial cell membranes in the airway surface lining and produced other oxidants. But this represents only the first and fastest step in the exposure-bronchoconstriction process. Recall that the airway responses to ozone are complex and time varying, including both attenuation and sensitization, depending on the pattern of exposure. These behaviors are probably not driven by this rapid first step in ozone's pathophysiology. As a result, it might not be very useful to construct a dosimetric model to estimate this "tissue dose" of ozone. Instead, we might choose to build a dosimetric model that captures several additional steps toward bronchoconstriction. Suppose that the dosimetric model estimated the dose of inflammatory mediators

Table 9.1 Examples of dosimetric models in epidemiology, classified by fundamental disease process

	Exposure	Disease	Dose Metric
Reversible proportional models			
(McDonnell, Stewart et al. 2007)	Ozone	Airway narrowing	Predicted response
(Krajcarski and Wells 2008)	Forces on spine	Low back pain	Predicted response
Irreversible proportional models			
(Smith 1985)	Silicon carbide	Lung fibrosis	Lung silica dose
(Ballew, Kriebel et al. 1995)	Silicon carbide	Lung fibrosis	Lung silica dose
(Links, Schwartz et al. 2001)	Lead	Cognitive decline	Tibial lead dose
Reversible discrete models			
(Woskie, Eisen et al. 1998)	Sodium borate aerosol	Respiratory symptoms	Effective dose
Irreversible discrete models			
(Richardson 2009)	Benzene	Leukemia	none – 2-stage model fit to exposure-risk data
(Quinn, Smith et al. 2000)	MMVF	Lung cancer	Lung dose of biologically active fibers
(Salvan, Thomaseth et al. 2001)	TCDD	Lung cancer	Serum TCDD dose

released into the intercellular space by inflammatory cells in response to ozone exposure. This dosimetric model is further down the causal pathway and, if designed correctly, may capture more of the temporal dynamics occurring in the disease process. Most important, the model should capture the temporal effects of repair or recovery, which we have seen often are important determinants of the overall response.

This example makes the point that many different dosimetric models might be designed to represent the same toxic process; each one would be a formal statement of a hypothesis about this process, and the model need not estimate a "dose" per se. One investigator might attempt to incorporate more steps along the causal path than another. How much of the exposure-response process should be captured in the dosimetric model and how much left for the epidemiologic model? It is likely that this question will only be answered through experience and may well depend on the disease being studied and the extent of existing knowledge of its mechanisms.

There are a wide variety of different forms for dosimetric models. Most of the models that we use in this book are structured mathematically as either difference or differential equations because these are well suited for describing dynamic processes (see Chapter 6). But sometimes important information characterizing the link between exposure and dose can be used to develop dose metrics by a simpler approach in which exposure intensities are weighted to better reflect biological activity of the exposure. An excellent example is the work of Quinn and colleagues on dose indices for manmade vitreous fibers (MMVF) (Quinn, Smith et al. 2000; Marsh, Youk et al. 2001; Quinn, Smith et al. 2001; Smith, Quinn et al. 2001; Quinn, Smith et al. 2005).

Example: Lung Dose of Man-Made Vitreous Fibers (MMVF)

Quinn and colleagues estimated quantitative lifetime exposures for a large cohort of workers employed in the manufacture of MMVF products as a part of an epidemiologic study, focusing primarily on lung cancer risk in this population. The investigators had available to them a number of quantitative measurements of airborne fiber concentrations collected by the industry, and they also conducted their own air measurement surveys. Previous toxicologic research indicated that fibers having certain dimensions were more biologically active than others (Quinn, Smith et al. 2000). A fiber's dimensions (length and diameter) determine the region of the lung in which it is likely to deposit, as well as important aspects of its cellular toxicity. Because MMVF are produced in a wide range of dimensions, the investigators were concerned that their exposure estimates should consider only the "biologically active" fibers. Drawing on toxicologic research, Quinn hypothesized that the biologically active fibers would meet the following conditions:

1. The fibers must be of the appropriate dimensions to reach the target tissue.
2. The fibers must be sufficiently durable to remain at the target tissues long enough to cause carcinogenic cellular changes.
3. The fibers must be able to cause the cellular changes associated with tumor formation.

Experimental evidence exists to support each of these assumptions, although there remain uncertainties about the details, especially for humans. It is not known, for example, how long a fiber must persist in the target tissue to induce the cellular changes associated with a tumor. Quinn's approach was designed to be quite flexible, so that it can accommodate new experimental findings relevant to each criterion. Consider next each of those criteria.

First Condition: Deposition in Target Tissue

The investigators assumed that the target tissue for fiber-induced lung cancer would be that tissue in which the tumors occur most frequently in fiber-exposed people (Quinn, Smith et al. 2000). In humans, greater than 90% of all tumors are found in the bronchial wall; tumors further into the respiratory tract, in the bronchiolar and alveolar walls, are rare, making up only approximately 2–3% of all lung neoplasms. The airways are conventionally referred to by generation: the trachea is classified as generation 1, the main stem bronchi as generation 2, and so on. About three quarters of all human lung tumors originate in the second-, third-, and fourth-generation bronchi; a small percentage have more peripheral origin in the segmental bronchi (fifth generation), whereas very few are located more distally. The regional pattern of tumors appears to be the same for asbestos-related cancers and others. This regional distribution of tumors may be explained, at least in part, by the pattern of particle deposition: both fibrous and nonfibrous particles are found experimentally to deposit with increased efficiency at the airway bifurcations, where turbulent flow increases impaction and interception of long fibers by airway walls. On the basis of these data, Quinn assumed the first five generations of the respiratory tract were the target tissue for human lung cancer induced by inhaled fibers.

The investigators next had to estimate the fraction of the total airborne fibers in a given exposure situation that is deposited at the target tissue. They used a lung deposition model by Sussman and colleagues based on empirical measurements of the deposition of fibers in the bronchi using casts of the human lung (Sussman, Cohen et al. 1991). This work allowed Quinn to calculate the deposition fractions of fibers in different length-diameter categories, which formed the basis for a fiber deposition matrix (Figure 9.1A) representing the probabilities of fiber deposition in the target tissue, for fibers of different length-diameter dimensions.

A. Fraction depositing in target tissue

Length (μm)	Diameter (μm) .05–.1	.1–.2	.2–.6	.6–1.0	1–3	3–5	5–7
2–5	NA	NA	0.03	0.04	0.04	NA	NA
5–10	NA	0.07	0.11	0.13	0.14	0.17	NA
10–20	NA	NA	0.21	0.24	0.28	0.32	NA
>20	NA	NA	0.55	0.59	0.62	0.72	0.76

B. Probability of persisting for at least 18 days

Length (μm)	Diameter (μm) .05–.1	.1–.2	.2–.6	.6–1.0	1–3	3–5	5–7
2–5	0	0	1	1	1	1	1
5–10	0	0	1	1	1	1	1
10–20	0	0	1	1	1	1	1
>20	0	0	1	1	1	1	1

C. Probability of carcinogenic activity

Length (μm)	Diameter (μm) .05–.1	.1–.2	.2–.6	.6–1.0	1–3	3–5	5–7
2–5	0	0	0	0	0	0	0
5–10	1	1	1	1	1	1	1
10–20	1	1	1	1	1	1	1
>20	1	1	1	1	1	1	1

Figure 9.1 Matrices representing the three conditions for biologic activity of man-made vitreous fibers as a function of their length and diameter. (Reprinted with permission from Quinn et al. 2000.)

Second Condition: Fiber Durability

Whereas asbestos fibers are so durable that they require decades or longer to dissolve, some MMVF will dissolve in the lungs more rapidly. The investigators hypothesized that a fiber would have to remain in the target tissue for some minimum amount of time for it to exert a carcinogenic effect. To determine this minimum time, there were two considerations: (1) the minimum time a fiber must persist to initiate a carcinogenic process, and (2) the fiber properties that determine how long a fiber will persist. The investigators reasoned that to trigger heritable cell damage, a fiber should persist in the cell through at least one mitotic division. The time required for epithelial cell division in the large bronchi is reported to be approximately 18 days, and so this minimum figure was used to define "sufficiently durable." Two aspects of a fiber largely determine its persistence in the lungs—chemical composition and dimensions—and it is relatively straightforward to experimentally determine dissolution rates for different types of fibers. Using such data, Quinn constructed a matrix of 0/1 indicators showing the relation between minimum persistence and length-diameter for fibers made from a glass of known chemical composition (Figure 9.1B).

Third Condition: Tumor Formation

The mechanism of fiber carcinogenicity is not certain. There are several hypothesized mechanisms, based on in vitro and experimental animal studies. Several authors have proposed a mechanical hypothesis of fiber carcinogenesis in which fibers thin enough for cells to attempt to phagocytize and long enough to be incompletely phagocytized have the potential to be biologically active. Following several lines of evidence, Quinn hypothesized that fibers longer than the thickness of the liquid bronchial lining layer (approximately 5 μm) penetrate to bronchial epithelial cells, where they mechanically pierce the cell or are incompletely phagocytized. This fiber may then serve as a "wick" or vehicle for introducing carcinogenic agents into the cell.

The limits of fiber *diameter* consistent with carcinogenic activity are less well defined. Reviewing the available experimental evidence, Quinn hypothesized that the diameter limit for carcinogenicity may be near the dimension that is sufficiently large for a cell to respond to the fiber as a surface instead of a particle (cells tend to adhere to surfaces rather than attempting to phagocytize them). The upper limit for the diameter of biologically active fibers was therefore set as the diameter of the target cell itself, approximately 5 μm for bronchial epithelial cells and macrophages.

The third matrix in Figure 9.1C formalizes these observations on the potential carcinogenic activity of MMVF as a function of length and diameter. The matrix expresses the possibility of carcinogenic activity as simply 0 or 1, depending on length and diameter.

The Quinn dose metric does not include the subsequent processes of DNA damage and repeated cell divisions, each of which has a small probability of causing a cell transformation and ultimately formation of the primary cancer cell. The output of Quinn's work might be used as the input of another model of the carcinogenesis process, such as the one described in Chapter 15.

In summary, Quinn and coworkers developed a clear biologically based rationale for deciding which fibers were likely to contribute to the disease process using what was known about biological effects of fibers. Each of the three biological conditions can be thought of as a set of weights applied to an airborne fiber exposure based on each component fiber's length, diameter, and composition. The weights can be 0, 1, or a fraction. The weights are applied by multiplying the corresponding cells in all three matrices together and then multiplying this product by the numbers of fibers in the length-diameter cells of the fiber size distribution for a particular exposure environment (Figure 9.2). The sum of all the cells in the final matrix can then be converted into an airborne fiber concentration having the same units as the standard exposure metrics: number of fibers per cubic centimeter of air (f/cm^3). However, the new dose metric represents the biologically active fiber concentration. One advantage of this approach to constructing a dose metric is that each of the biologic conditions can be challenged, revised, or experimentally evaluated. Importantly, the approach provides a direct method for integrating findings from laboratory studies into epidemiology in the form of testable mechanistic hypotheses.

Example: A Dosimetric Model for Low Back Pain

This book is mostly concerned with chemical exposures, but an excellent example of applying dosimetric modeling to epidemiology is found in work by Krajcarski and Wells (2008) on risk of low back pain in automobile assembly workers. The exposures were ergonomic and resulted from repetitive tasks performed under

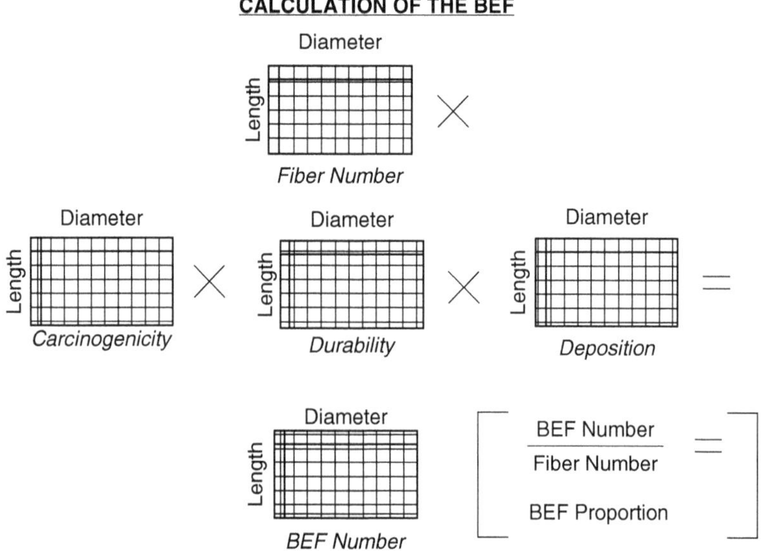

Figure 9.2 Calculation of the Biologically Effective Fibers (BEF) index. For a particular fiber size distribution, the matrices shown in Figure 9.1 are multiplied to yield an overall dose metric for the air sample. (Reprinted with permission from Quinn et al. 2000.)

time pressure on a production line (Wells, Norman et al. 1997; Norman, Wells et al. 1998; Kerr, Frank et al. 2001). The investigators gathered a large amount of information on ergonomic stressors and symptoms among a cohort of assembly line workers engaged in repetitive tasks characterized by a short work cycle. Ergonomic exposures relevant to lower back injury were gathered through a lengthy and detailed analysis of each task performed in each job. The exposures included compression and shear forces at several different points along the spine, as well as peak and static moments (N·m); (Wells, Norman et al. 1997). The tasks were videotaped, and data were extracted from the tapes to fit a 2-dimensional biomechanical model which calculated the estimated forces. In one part of this large study, exposure data were estimated for a sample of 34 cases of lower back pain and 34 controls. Cases were a random sample of the hourly workers who reported new-onset low back pain to plant medical personnel. Controls were randomly chosen from among the hourly workforce in the same factory as the cases. For the investigation summarized here, a single exposure, the L4/L5 extensor moment, was chosen because of previous studies suggesting it is an important predictor of back pain. Two summary measures of the exposure data were calculated for each of the 68 participants: the integrated moment (N·m ·s/shift) across a work shift (a time-integrated measure analogous to cumulative exposure) and the peak moment (the highest estimated L4/L5 extensor moment during the work shift).

These exposure data were used to produce a dose metric using a single-compartment model. The metric had arbitrary units; one can think of it as a kind of "dose," but it does not represent any specific physical quantity. The dose metric was hypothesized to be proportional to the degree of injury in the lower back caused by the forces exerted on the spine during work. The single-compartment model requires a single parameter, called τ, which can be thought of as the recovery time for the repair processes following exertion until the target tissue returns to "normal." The value of this parameter might have been chosen a priori on the basis of experimental studies on the injury and repair processes of the soft tissues of the lower back. The investigators reported that a variety of different physiologic processes associated with muscle fatigue have recovery times of about 5,000 seconds (1.4 hours), including intramuscular pH changes and lactic acid removal. With this recovery time, a dose metric generated from a typical exposure (L4/L5 moment) profile over a work shift suggests a steady accumulation of damage, with little or no recovery over rest breaks (Figure 9.3). By using a much shorter recovery time, 2.0 seconds, the investigators could create a dose metric which represented rapid, presumably reversible effects of short-term exposures. In addition to investigating these a priori values for τ, the investigators also evaluated a wide range of different values. For each different value of τ, a different dose metric was generated. Logistic regression models were then used to estimate the strengths of the associations between each alternative dose metric and risk of lower back pain. These regression models might have used the dose metric as a continuous variable, but instead the data were categorized, and the odds ratio was estimated comparing

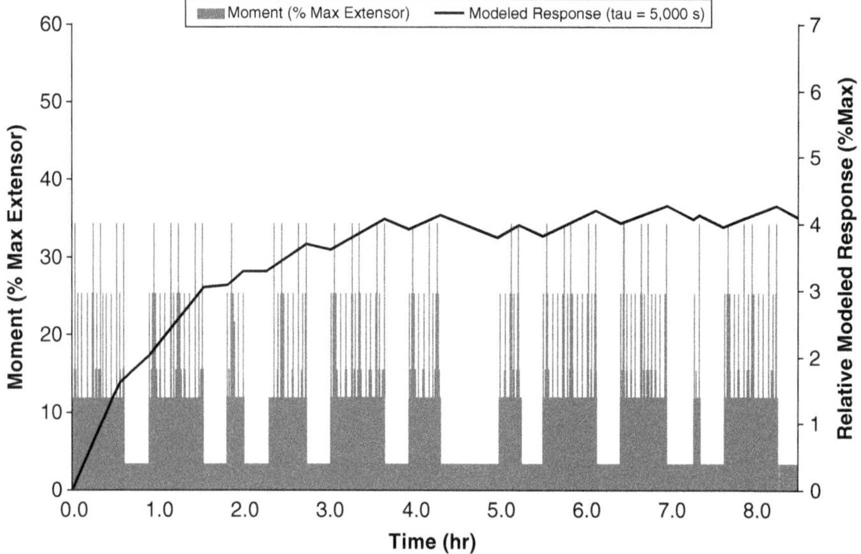

Figure 9.3 Example of exposure data (L4/L5 moments) and dose metric (τ = 5,000s) for one job strut assembly installation. (Reprinted with permission from Krajcarski and Wells 2008.)

those above the 75th to those below the 25th percentiles of the metric's distribution.

When the value of the recovery parameter, τ, was varied over a wide range—from 1 to 100,000 seconds—the strength of the resulting dose metrics' associations with low back pain varied widely (Figure 9.4). When these odds ratios were plotted against the varying values of τ, one sees a smooth trend with two peaks, suggesting that two very different physiologic processes may be at work. The first peak occurred at recovery times of about 2.0 seconds, suggesting that one aspect of the physiologic process linking these ergonomic exposures to back pain has a very rapid response. The other peak—not so clearly defined—occurred at about 1,000 seconds, suggesting that there may be a slower damage/repair process involved as well.

Interindividual Variability in Dosimetry

The utility of a dosimetric model will depend on the quality of data available about study subjects. If we know only that some subjects were "exposed" and others not, then it should be clear that there will be little to gain from a dose modeling exercise. In most of the examples that we present, there are individual quantitative data estimating the personal time patterns of exposures. In these instances, all the model parameters are assumed to represent population averages, and only the model input—the exposure vector (Equation 9.1)—is allowed to vary. But it is

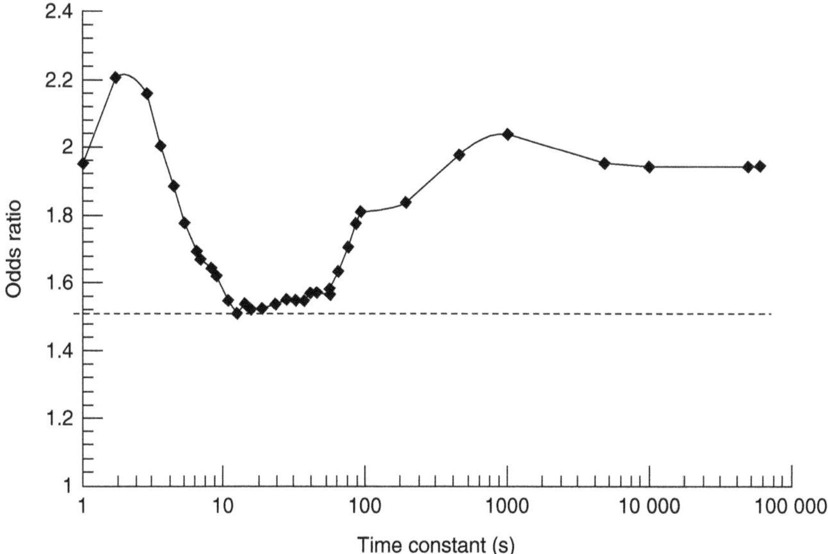

Figure 9.4 Fit of alternative dose metrics to case-control data on low back pain. Odds ratios (vertical axis) are shown for models using dose metrics with varying values of tau, the recovery time (horizontal axis). For comparison, the odds ratio for peak exposure was 1.5 (indicated by horizontal dotted line). (Adapted from Krajcarski and Wells 2008, reprinted from Kriebel et al. 2007.)

important to be aware that the assumption that the model parameters do not vary among individuals is usually made because we have no choice, lacking any data with which to investigate individual variability in such characteristics as a clearance half-time for an aerosol in the lungs or a heavy metal in the blood. In reality, it is likely that all of the physiologic processes described by the dosimetric model *do vary individually*, and our assumption of constancy is a necessary expedient. Where we have been able to look, we have seen variation across individuals, such as in inhalation uptake and metabolic rates for 1,3-butadiene (Smith, Bois et al. 2008). A consequence of assuming population averages for model parameters is probably to introduce nondifferential misclassification into the dose metrics. How serious this misclassification will be will depend on the relative importance of different sources of variability in the population distribution of the dose metric. If exposure variability is, in truth, the principal source of variability among individual doses, then an assumption of constant values for model parameters will not have very serious consequences. But if there is little variability in exposure among subjects but wide differences in a key parameter, such as a clearance half-time or uptake, then the dose metric might be seriously misclassified.

Sensitivity analyses can be conducted to understand how large might be the errors resulting from model assumptions, and we advocate always investigating how sensitive predicted doses are to assumptions about the model parameters.

Example: Serum Dioxin Dose and Lung Cancer Risk

The following example illustrates the use of individual height and weight data to improve dose estimates for an epidemiologic study of dioxin exposure and lung cancer. This example was cited for different reasons in Chapter 7 as well. Dioxin (2,3,7,8-tetrachlorodibenzo-p-dioxin, or TCDD) is a contaminant in several chlorinated industrial chemicals and pesticides, and it is a suspected human carcinogen. A cohort study conducted by Fingerhut and colleagues (1991) found elevated mortality for all cancers combined, as well as for lung cancer, among workers at twelve U.S. chemical plants where TCDD exposure occurred. Salvan and colleagues used serum TCDD data collected at a single point in time on a subset of these workers, combined with occupational history data, to estimate a lifetime TCDD dose (part per trillion-years) for the members of this cohort (Thomaseth and Salvan 1998; Salvan, Thomaseth et al. 2001). They used a single-compartment pharmacokinetic model assuming exponential washout kinetics. TCDD is highly fat soluble, and its clearance from the body is strongly dependent on the amount of fat and the rate of change in fat composition over time. Height and weight data at two points in time were available for a subset of the population, and from these, the relation of body mass index (BMI) and age could be estimated. A published relation between BMI and percent body fat (PBF) was assumed. Serum doses of TCDD were then estimated for the full cohort, using individual data on work history and height and weight (at one point in time). All TCDD-exposed jobs were assumed to have approximately the same exposure intensities, which was a necessary simplification because the only individual data on exposure were the start and stop dates of work in jobs that had been determined to involve exposure to TCDD. This research illustrates how interindividual variability can be incorporated into a dosimetric model; the one-compartment model for TCDD body burden had a clearance parameter which varied within individuals over their life spans and among individuals as a function of individual PBF estimates. The individual serum TCDD dose estimates were then used in a lung cancer mortality analysis. The authors estimated the association between lung cancer mortality rate and TCDD dose. There was a strong linear increase in lung cancer risk with increasing TCDD, from a "background" dioxin level (assumed to be 7 ppt) to doses up to 1,000 times this level (Figure 7.3).

Determining Dosimetric Model Parameters

The parameters of dosimetric models can be thought of as weights applied to the exposure data with a resulting quantity (the dose metric) that may be more closely associated with outcome than the original exposure data. There are many ways to develop these weights, and the approach we advocate bases them on an explicit biologic hypothesis, as explained earlier. Sometimes the values of dosimetric model parameters can be determined from animal experiments or previous epidemiologic studies, and sometimes they can be estimated empirically from the study data themselves. Quinn and colleagues obtained the values of their parameters from external data in the previous example on the biologically active fibers for the study of the health effects of MMVF (Quinn, Smith et al. 2000). They determined, for example, which fiber lengths and diameters were likely to be active based on experimental studies of fiber deposition in a cast of the human lungs. Using experimental data in this way illustrates one potential strength of dosimetric modeling: it provides a formal mechanism by which different kinds of scientific evidence about a potential health hazard can be combined.

Sometimes the values of dosimetric model parameters can be estimated from the epidemiologic data. This may at first sound somewhat circular (if the dose metric is developed for use in an epidemiologic study, then how can you also use those same data to estimate the dosimetric model parameters?). With appropriate cautions in interpretation, you can indeed estimate *both* an epidemiologic parameter such as a relative risk and the dosimetric model parameters with the same data. An example of this is shown in Chapter 15 on applying the two-stage cancer model to a dataset on silica and lung cancer.

9.5. WHICH DOSE METRIC IS BEST?

Epidemiologic studies are time-consuming and costly. Their ultimate purpose is to help prevent disease. We would be remiss, therefore, if we constructed and analyzed only one measure of exposure or dose rather than trying several alternatives. As we have frequently noted, we will often have several different hypotheses that we wish to investigate, and each one may be formalized through a different dose metric, or different epidemiologic model. We will want to compare the fits of these different dose metrics to understand which one(s) are more consistent with the outcome data or which do not fit the data. And we will often want to compare our dose metric(s) to a more conventional summary measure of exposure to assess what has been gained through the dosimetric modeling.

Goodness of Fit

There is no single quantitative measure that can be used to choose the "best" model, or the best measure of dose (Loomis, Salvan et al. 1999). Goodness of fit (GOF) is an important consideration, and there are several different standard measures that are quite useful. The simplest measure of GOF is probably the R^2, or coefficient of determination for linear regression models. This quantity represents the fraction of the variation in the dependent variable that is captured by the model, and it has an appealing simplicity and intuitiveness. Unfortunately, it is also quite sensitive to small changes in influential data points in small to medium-sized datasets, and so modest improvements in R^2 should not generally be given much weight as a reason to prefer one model over another.

In most nonlinear regression models used in epidemiology (logistic, Poisson, or Cox regressions, for example), statistical packages provide several measures of GOF. The -2 log likelihood statistics (-2LL or deviance) is one useful measure (Kleinbaum and Klein 2002), as is a close variant, called Akaike's information criterion (AIC); (Akaike 1973; Forster and Sober 1994). The -2LL is the quantity that is usually maximized in maximum likelihood estimation algorithms. The absolute value of the -2LL for any particular model is not very useful, but differences in -2LL among closely related models can be meaningful measures of relative GOF. In particular, when one model can be said to be "nested" within another, then the difference in their -2LL statistics will have a chi-squared distribution, with degrees of freedom equal to the difference in the number of parameters between the two models. Nesting means that the only difference between a simple and a more complex model is that one or more terms have been dropped from the simple one. The fact that -2LL differences have a known distribution means that one can assign a p-value to this difference, which tests the null hypothesis that the more complex model fits no better than the simpler one. Although we do not advocate formal statistical testing of different dosimetric models, it is nevertheless useful to have the p-value as a guide to what constitutes an "important" difference in GOF.

The alternative models that we want to compare are very often not nested. For example, we may want to ask whether a dose metric fits the epidemiologic data better than a summary measure such as cumulative exposure. This is, in a sense, an "apples and oranges" comparison, and the difference in -2LL will be hard to interpret. The AIC has been shown to be useful in these situations, and this statistic is widely used to compare GOF among models that are not nested. The AIC is essentially a variation of the -2LL that corrects for differences in the numbers of parameters in the different models (Forster and Sober 1994).

Comparison of Measures of Association

Another common method of choosing among alternative summaries of exposure or dose metrics is to compare the sizes of the measures of association (relative risks, for example) that they produce in an epidemiologic model and to choose the measure with the highest effect estimate. This approach is intuitively appealing: it is simple and has a seemingly plausible rationale. If one remembers that nondifferential exposure misclassification will, on average, bias toward the null (see Chapter 8), then it follows that the less misclassified measure of exposure/dose will yield the larger measure of association. Unfortunately, there are problems with this logic (Loomis, Salvan et al. 1999). First, different exposure/dose measures may use different scales of measurement, and so the absolute magnitudes of the measures of association may be very different. For example, how should we compare the relative risks resulting from two models, one using a dichotomous measure of exposure and the other a continuous measure of dose? The absolute magnitudes of the RRs will be based on very different scales of measurement. The second problem is that the model with the largest measure of association will not always be the model that fits the data best (Salvan, Stayner et al. 1995), despite the appealing logic of the misclassification argument. In our experience, it is *usually* true that the better fitting dose metric will also have the highest measure of association (when the scaling problem is not an issue), but Salvan (Salvan, Stayner et al. 1995) has shown that this will not necessarily always be true.

Comparison of *p*-Values

Another common approach that should be avoided is comparing the *p*-values for the exposure/dose variables in competing models and choosing the one with the smallest *p*-value. Recall that *p*-values do not measure the probability that a particular model or variable is true or correct (Chapter 8); they have meaning only in a null world, that is, when the exposure has no effect. But, as with the comparison of measures of association, our experience shows that *usually* the model with the smallest *p*-value for the exposure term also will have the best goodness of fit, all other factors being equal. This tendency probably explains why the comparison of *p*-values continues, despite lacking a sound theoretical basis.

Comparison of Alternative Dose Metrics: Use of R^2

The comparison of different exposure or dose measures is sometimes framed simply as a question of how well two or more measures agree. Clearly, if two measures are nearly perfectly correlated, then we can be confident that their performance in epidemiologic models will be essentially the same, and we will have to use other considerations to decide which one is "better." A logical approach is to calculate the correlation coefficient, r, between alternative exposure/dose measures. This is straightforward, but it raises the difficult question: What value of r is "good"? Values very close to 0 or 1.0 (or -1.0) are easy to interpret, but what does a correlation of 0.6 between two different dose measures mean? It turns out that in this context, it is preferable instead to calculate the *square* of r, sometimes called the coefficient of determination, or R^2.

Lagakos (Lagakos 1988) showed that R^2, which he termed *asymptotic relative efficiency* (ARE), has a useful interpretation when comparing a surrogate and a "true" measure of exposure. If a surrogate measure of exposure differs from the true measure due only to random error, then $(1- R^2)$ represents the loss of power to detect a true exposure effect using the surrogate, compared with using the true measure. To understand the utility of this insight, imagine that one is planning a study and there are two alternative ways of assessing exposure: one expensive but highly accurate, and the other inexpensive but imprecise. Imagine that a pilot study is conducted in which both measures are used. The ARE calculated from the pilot study results could then be used to estimate how much larger the study using the "quick and dirty" method would have to be in order to detect a given effect with the same power as a design using the slow-and-expensive method (Armstrong 1996). If, for instance, the correlation between the two different measures of exposure is 0.7, then the ARE of 0.49 (roughly 50%) can be interpreted in two ways. First, it indicates that if the study is performed using the quick-and-dirty method, it will have only 50% of the power that the more expensive study would have. Alternatively, it says that to have the same power to detect an effect using the quick-and-dirty measure as one would have using the expensive measure, it will be necessary to double the study size.

Unfortunately, we often do not have a "true" measure—a gold standard against which to compare our alternative exposure or dose measures. In this situation, we still advocate using R^2 rather than r as the measure of agreement because one can still reason that if one of the measures being compared is "closer to the truth" than the others, the ARE should still correspond approximately to the loss of power from using a poorer measure compared with the better measure.

When comparing among alternative dose metrics and/or summary measures of exposure, it is important to keep in mind that it is likely that none of them are strictly correct. As just noted, one may be closer to the truth than another, although we will not generally know this. In such a situation, it often happens that several metrics are fairly highly correlated ($R^2 > 50\%$). This can happen for no other reason than that each of the metrics may share a common component; the duration of exposure, for example, is shared by cumulative exposure and by almost any dose metric we will invent for an irreversible disease process. When this happens, it can be tempting to use a regression model to help us decide which is the better metric (using one of the methods critiqued previously, such as comparing p-values). But this is a risky proposition; regression models partition variance, and they will not do this in a reliable way when the alternative covariates (exposure/dose metrics) are well correlated and when all of them are also correlated to the unmeasured "true" metric.

Qualitative Considerations

There are also nonquantitative considerations in choosing among alternative exposure/dose measures. Statistical performance is not the only reason one might choose a particular metric. Other considerations include ease of interpretation and comprehension by other scientists, and by policy makers, as well. Generalizability may also be weighted more or less heavily, depending on the study's broad objectives. For example, if exposure is measured as the air concentration of the toxin, then it will be difficult to use the results to make predictions about risk by other routes of exposure—skin, for example. One might choose a dose metric that would permit estimation of risk from combined exposures through multiple routes of exposure, even if it did not fit the epidemiologic data any better than an airborne exposure metric does.

Our a priori beliefs about the disease process may affect the choice of exposure metrics. Suppose, for example, that in a cohort study of an airborne exposure to a toxin that is assumed to act via a proportional irreversible process (lung damage from cigarette smoke, for example), the average air concentration and the cumulative exposure to the agent both fit the outcome data fairly well, but average exposure actually fits somewhat better. If the improvement in fit comparing cumulative to average exposure is modest, one might still prefer to use cumulative exposure, because we might have a fairly strong belief that the *duration* of exposure and not simply its average intensity is important to risk.

Some dose metrics may be more useful in investigating interactions among different risk factors. Smith (1985) developed an alternative dose metric for studying the effects of silicon carbide aerosols on the lungs. One of the models presented explicitly incorporates the effect of smoking on the clearance of silicon carbide particles from the airways. This method of quantification of the interaction between the two exposures may be preferable to the standard approach of including smoking or a smoking interaction as additional covariates in the epidemiologic model.

Example: Dose Metrics in a Study of Neurobehavioral Effects of Lead

Lead is a well-established neurotoxin. Recent studies have shown neurobehavioral effects from surprisingly low body burdens of this common metal (Wigle, Arbuckle et al. 2007). To detect very low-level effects, it is necessary to have accurate estimates of the brain (or other neurologic targets) dose, which raises the question of what the best dose metric should be. Lead is stored in bone, from where it can move into the blood and then the brain. The most common measure of lead exposure/dose is the blood lead level, but this reflects a complex equilibrium between recent exposure and bone storage of past exposures. Our understanding of the toxicokinetics of lead has improved in recent years, because it is now possible to directly measure bone lead concentrations noninvasively, using X-ray fluorescence (Hu, Rabinowitz et al. 1998).

A study of 535 former organolead manufacturing workers compared alternative dose metrics for their association with measures of neurobehavioral function (Links, Schwartz et al. 2001). The investigators tested the workers, none of whom were currently employed in lead-exposed jobs, on a battery of 19 neurobehavioral tests. They also measured blood lead concentration and tibial lead burden at the time of the examination. In addition to these two directly measured metrics, they calculated two additional metrics: an estimate of tibial dose and the peak tibial lead burden. The dose estimate was the area under the time-tibia burden curve (referred to as area under the curve, or AUC, by Links et al. 2001) for each worker in the study. Using a one-compartment toxicokinetic model (see Chapter 4), the investigators were able to make their dose estimates without benefit of any exposure data nor historic body burden data. They proceeded as follows:

1. They assumed linear kinetics in a one-compartment model with a constant clearance half-time of lead in bone of 27 years. This value was based on previous studies.
2. Because exposure had ceased at retirement, years before the bone lead measurements, it was possible to back-extrapolate from the current tibial lead concentration to a peak tibial lead concentration at the date of termination of exposure using the steady clearance half-time of 27 years.
3. The authors further assumed a relatively constant exposure intensity beginning at the hire date and continuing until the termination date, leading to a linear increase in bone concentration over that period. The slope of the accumulation up to the estimated tibial lead at retirement will be proportional to the average exposure intensity during the work period.

Notice that the only individual data used in the dosimetry were the current tibial lead burden and the hire and termination dates. Both the clearance half-time of lead and the intensity of lead exposure were assumed not to vary among individuals or within an individual over time. This model could not be directly validated, but the relative performance of the four different metrics in predicting neurobehavioral function was investigated. The four metrics were: (1) current blood lead, (2) current tibial lead, (3) peak tibial lead, and (4) tibial dose, or AUC. The workers were tested on the neurobehavioral test battery at two time points, 4 years apart, and the change in performance over 4 years was the outcome variable (a cross-sectional analysis was also performed but is not discussed here). The regression coefficients were standardized so that each represented the change in the test score per 1 standard deviation increase in the lead exposure/dose metric, and all the scores were constructed so that negative values indicated worsening performance.

Overall, the results showed strong effects of lead on neurobehavioral test scores, and especially so when the measures of tibial burden were used. But what concerns us at present is the methodologic issue raised in Links et al.'s (2001) study: How should we evaluate the relative performance of the four metrics, and which one is best? First, we can use the ARE to compare the agreement among the four metrics (Table 9.2). If we assume that the AUC of tibial lead burden over the working lifetime is closest to the true dose, then the table suggests that only

Table 9.2 Asymptotic relative efficiencies (R^2) among alternative dose metrics in Links et al. study of lead and neurobehavioral function in former lead workers

Dose Metric	Current Blood Pb	Current Tibial Pb	Peak Tibial Pb	AUC
Current blood Pb	1.0	0.19	0.07	0.03
Current tibial Pb	-	1.0	0.74	0.49
Peaktibial Pb	-	-	1.0	0.88
AUC	-	-	-	1.0

From Links, Schwartz, and Simon (2001).

the peak tibial lead would be a reasonable proxy measure (ARE = 0.88). Current tibial lead burden and blood concentrations were much more weakly associated with AUC. To illustrate, the ARE of 49% for current tibial lead burden means that one would need twice as large a study in order to have the same power to detect an effect of lead, as would be needed using the AUC measure. This would seem to be a strong argument for the dosimetric modeling work that Links and colleagues (2001) performed.

A related question is: Which of the four dose metrics fits the outcome data best? This is not straightforward, because of the battery of 19 separate tests that were used to assess neurobehavioral function. As a result, there were $4 \times 19 = 76$ different regression models, each of which yielded a regression coefficient (slope) measuring the association between one dose metric and one test. The investigators used two main criteria to conclude that the AUC appeared to be the best fitting metric (Table 9.3). The signs of the regression coefficients were nearly all (95%) in the hypothesized direction for AUC, and the AUC also had the highest percentage of regression coefficients that were "statistically significant" at $p < .05$. The peak tibial lead burden performed nearly as well, whereas current tibial and blood lead

Table 9.3 Comparing the performance of four dose metrics in predicting neurobehavioral function on 19 tests among 535 former organolead workers

Percent of Regression Coefficients:	Current Blood Pb	Current Tibial Pb	Peak Tibial Pb	AUC*
With expected sign**	74%	91%	100%	95%
With $p < 0.05$***	11%****	16%	21%	26%

* Area under the curve—a measure of tibial dose.

** Percent in which measure of association indicated worsening of performance with higher lead exposure.

*** Percent in which p-value from test of null hypothesis that the regression coefficient = 0 was less than 0.05.

**** Two of 19 (11%) p-values were < 0.05, but one of these two indicated a "statistically significant" *improvement* in performance with lead exposure.

concentrations performed less well. These results seem consistent with the findings from the AREs (Table 9.2). We might have preferred that the investigation of the relative fit of the different metrics use goodness-of-fit statistics rather than *p*-values for the reasons noted previously, but we doubt that the results would have been substantially different.

Before leaving this example, let us briefly consider the biologic implications of the evidence that the AUC seemed to be more strongly associated with neurobehavioral performance than current body burdens. If a measure of recent or current exposure (blood lead concentration) were more strongly associated with the outcome measures, this would argue for an acute reversible damage process and a short residence time of lead in the brain. This might suggest that neurobehavioral effects were largely due to recent releases of lead from bones, as the clearance half-time of lead in blood is about 30 days and the cohort members had not been exposed for years prior to the study. The association with AUC, in contrast, reflects the dynamics of uptake, clearance, and progressive damage from the accumulation of irreversible effects, as well as continuing decrements from ongoing releases of lead from bone. Personal estimates of exposure intensity for each participant during his working life would have been useful, but their absence did not prevent some important insights into the biologic basis for the neurobehavioral deficits observed among these former lead workers.

9.6. CONCLUSIONS

At this point, the reader may be asking: Can a more elaborate dosimetric model contribute anything that a simpler and more conventional summary measure of exposure cannot? Can we justify the considerable effort that went into constructing and using a dosimetric model? At least three criteria might be used to answer this.

1. The better model will show a better fit to the epidemiologic data, or more clearly reject the hypothesis of a relationship.
2. The better model will accommodate such secondary characteristics of the exposure-response relationship as interactions with other agents (effect modifiers) and time effects such as latency.
3. The predictions of the better model can be more confidently generalized to other exposure situations.

Thus there is no simple answer to the "Is it worth it?" question, but we believe that the examples throughout this chapter and in the rest of the book meet one or more of these three criteria. Some of the examples suffer from not having been designed to use temporal data to determine how the body is responding to the changes in exposure or dose over time, which is a powerful probe of dynamic relationships.

We believe that the added complexity of many dynamic disease processes can be better addressed by an additional step in the investigation of exposure-response relationships: the dosimetric model. Dosimetric models may be viewed as mathematical expressions of formal hypotheses about the temporal nature of the physiologic and pathologic processes underlying exposure-disease relationships. The great strength of a formal hypothesis is that it may be refuted, thus narrowing the range of likely explanations for the phenomena under study (Platt 1964). There has been much debate in recent years in epidemiology over the proper role of refutation in the field, but few dispute the advantages of more formal hypotheses (Rothman and Greenland 2005). Because disease models represent explicit hypotheses about the processes involved, they provide stronger tests than simply proposing that some exposure-disease relationship exists. If the data are inconsistent with a specific biologic hypothesis, this result advances our understanding by suggesting that we seek alternative explanations for the disease pattern in the data.

A practical limitation that will often frustrate even the most enthusiastic advocate of dosimetric modeling is the small sample size and limited quantitative detail in many epidemiologic datasets. It will generally be feasible to estimate dosimetric model parameters only when the datasets are large, when both exposure and outcome data vary across time, when there is little exposure or disease misclassification, and when the exposure range is substantial. In addition, the number of dosimetric model parameters should be small: eight or nine parameters could probably not be estimated reliably from a few hundred subjects.

SECTION C

Practical Applications of Disease Process Models

Section C presents a series of six chapters covering **practical applications** of disease process models. The first three Chapters, 10 through 12 cover what are called proportional disease processes, and the final three, Chapters 13 through 15, cover discrete disease processes. In these Chapters, the details of building and applying a disease process model for an epidemiologic investigation are presented and illustrated through examples. A final Chapter 16 discusses some implications of the approach for risk assessment and for disease prevention interventions.

10 Modeling Proportional Disease Processes

"I would have seen further,
but giants were standing in the way," noted the Jester.

"If we knew what it was we were doing,
it would not be called research, would it?"
Albert Einstein

10.1. PLAN FOR SECTION C

The essential mechanisms by which environmental exposures lead to symptoms and disease can be classified into four broad categories, as we have shown in earlier chapters (see Table 1.1 for a summary). This section of the book, which includes Chapters 10 through 15, is devoted to a detailed presentation of the four basic disease process models and illustrations of their applications in exposure assessment and epidemiology. There are two basic types of disease process: proportional and discrete types. We will begin by describing the key elements of proportional disease processes, which is the objective of this chapter. Then we illustrate these through examples of the two variants of the proportional model, reversible and irreversible, in Chapters 11 and 12, respectively. The same pattern then is repeated for the category of discrete processes; Chapter 13 provides the key steps to modeling, whereas examples of reversible and irreversible discrete models are presented in Chapters 14 and 15.

When applying the disease process model approach to a new adverse effect or disease, there are five steps we will need to develop and justify. We need to:

1. Identify a tentative type of the process model, such as proportional reversible.
2. Identify the key effect, and choose one or more measures to define the effect.
3. Define the etiologic time interval for the effect.
4. Select the hypothetical agent(s) and a measurable marker for exposure.
5. Define a summary dose metric that represents our mechanism to be tested in the epidemiologic study.

As we develop each of our examples, you will see that we follow this stepwise approach. The more information we have about the adverse effect and the likely agent, the stronger our rationale can be for each of the five steps. This approach can be applied post hoc to a study, but it works better as a study design approach.

There are two types of proportional disease processes, reversible and irreversible. Both are common, and each has distinctive features, but there is a close relationship between them.

10.2. MODELING PROPORTIONALITY OF RESPONSE

Proportional exposure-response relationships are about the easiest to model and the most forgiving of inappropriate model forms. These are the mechanisms which we think of first when imagining an "ideal" exposure-response pattern. Put simply, the more the exposure, the greater the response. This simple pattern actually holds across a very wide range of exposures and disease mechanisms. It is generally what is meant when a relationship is called "linear" (although the word is sometimes used to mean other patterns as well). If both exposure and response are measured on continuous scales, then a linear regression model can be used to quantify the strength of the association, and the slope or change in response per unit of exposure will adequately describe the effect. Averaging of exposure and response, both over time and across individuals, will often not create serious distortions of the association either, although we will discuss limitations to this rule of thumb later.

10.3. REVERSIBLE PROPORTIONAL DISEASE PROCESSES

These disease processes can often be decomposed into two opposing independent physiologic processes, called either "damage-repair" or "response-recovery."

Reversible proportional disease processes are very common. We have all experienced environmental exposures that produced symptoms and other effects in direct proportion to the amount of exposure, which abated after the exposure stopped. The intense eye, nose, and throat irritation caused by the smoky campfire in Chapter 2 is a good example. A more common example is the irritant respiratory effects of ground-level ozone associated with photochemical smog, which can be viewed as a simple reversible proportional process. This example also illustrates how this simplest of models may be modified to capture more complex behavior if the simpler model is found to be inadequate to describe the joint time patterns of exposure and response. For example, some agents produce multiple effects, some of which are not completely reversible.

Example: Ammonia and Eye Irritation

Let us suppose that we wish to model the adverse effects of ammonia exposure on the eyes. Ammonia is highly soluble in the watery fluid in the eyes, nose, and throat, and when it dissolves it makes a basic solution; neutral pH ammonium salt solutions are not toxic. As the air concentration increases, the resulting solution in the eye becomes more and more basic (caustic). When the pH reaches >13, associated with air levels >1,000 ppm, rapid tissue destruction occurs. Because the surface of the eye is directly exposed to the environment and gas diffusion is very rapid, there is a rapid equilibration between the air concentration and the eye's fluid (see Figure 10.1). Thus the eye concentration can be estimated directly from the exposure.

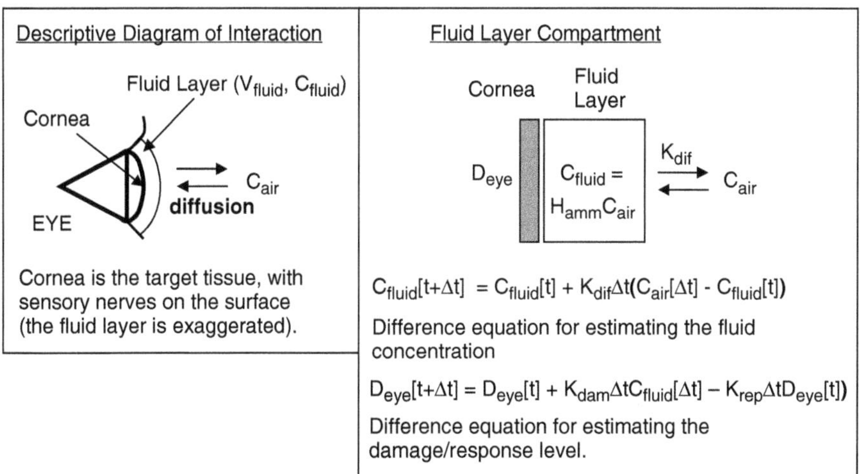

Figure 10.1 Simple one-compartment PBTK model for tissues in contact with ammonia in the external environment. (See the text for the sources of the K_{dif} and H_{amm} values.)

Most of us, at some time or other, have gotten a good strong whiff of ammonia, maybe while cleaning windows, for example. You can therefore imagine what might result from the following "thought experiment" (do not try this at home) on the irritating characteristics of ammonia and the kinetics of reversibility. Imagine pouring a small amount of concentrated 24% ammonium hydroxide solution from the local hardware store into a shallow dish and allowing a minute for the ammonia to diffuse into the air above the dish. A localized high concentration of ammonia will be produced—perhaps 150–300 ppm (associated with an eye pH 11). If you were to place your face about 20 centimeters above the dish, you would, within a few seconds, experience sharp eye and nose irritation and a very strong desire to remove your face to a safer location. Once free of the ammonia cloud, the irritation would rapidly subside. This irritation develops and dissipates rapidly and leaves no lasting effects, as shown in Figure 10.2, which is characteristic of sensory responses such as odor and irritation.

If you were to work regularly in an environment with 50–100 ppm, you might become acclimatized so that you could tolerate the exposure without irritation. If you were trapped in a location with a very high ammonia concentration, such as near a train wreck where a tank car of anhydrous ammonia had ruptured, > 500 ppm, serious eye damage, including corneal cell damage and destruction, edema, and secondary inflammation, could occur rapidly because of the corrosive effects of high pH. In the worst cases, there could be damage to the internal structures of the eyes. With high air concentrations and prolonged exposures, respiratory tract damage is also very likely because ammonia is absorbed by all moist surfaces. Thus the nature of the response depends on the concentration and the duration of exposure, as is the case with many other acute effects.

Response-Recovery Processes

If you were to do this ammonia experiment more systematically, using an apparatus that could control the exposure level and duration, and you had a way to record the intensity of irritation you perceived each second, you could explore how your responses changed with exposure conditions. You would need to allow sufficient time for complete recovery between exposures. After some experimentation, you would find that there was some minimum exposure intensity, a threshold level, needed to provoke a sense of irritation. As you continued testing beyond the threshold, you would probably find three characteristic phases in your responses. First, each time you were exposed above the threshold, there would be a rapid initial increase in irritation, which would stabilize in intensity after a few seconds and remain approximately constant until the exposure concentration was changed. Second, when the exposure level was changed, the intensity of your irritation would quickly rise or fall to a new steady-state level proportional to the exposure. Third, as long as the exposure was not too high, the response would be fully

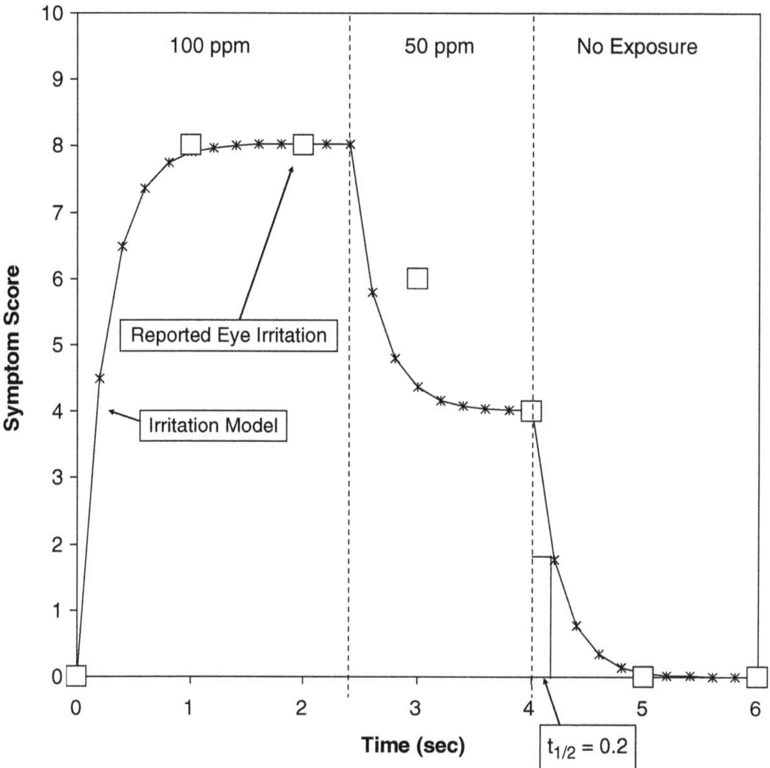

Figure 10.2 Simulated experiment with eye irritation from two levels of airborne ammonia. Subject reported intensity of irritation each second during a 3.5-sec exposure to 100 ppm followed by 50 ppm ammonia and then a short period of no exposure. Based on the data points for 4 and 5 sec, full recovery after exposure stops takes approximately 1 sec (about 5 half-times for 97% recovery). Therefore $t_{1/2}$ is about 0.2 sec, which makes the repair rate constant ($K_{rep} = 0.693/t_{1/2}$) equal to 3.5 sec^{-1}. At steady state, the subject indicated an irritation of 8 when exposed to 100 ppm. This can be used to estimate the damage constant because at steady state there is a proportionality between the repair rate constant, the level of damage, and the provoking concentration ($K_{dam} = K_{rep} \times$ damage/conc.), which is 0.28 symptoms/ppm/sec in this case.

reversible—your level of irritation would rapidly and entirely disappear once the exposure stopped.

The irritant response is virtually instantaneous and directly proportional to the membrane concentration in which the sensors are imbedded, and the (very brief) delays at the beginning and end of exposure represent the kinetics of diffusion of the ammonia into and out of the membrane surface fluid, determining the concentration on the surface of the eye. If you repeatedly tested yourself at moderate ammonia levels, allowing a suitable rest period after each exposure, you would find that the response pattern was repeatable. However, if you tested someone else, you would likely find a different threshold and degree of response per unit of exposure because of interindividual differences in sensor sensitivity. If you retested someone too soon after a test, you would probably find that the person's sensor sensitivity declined as a result of acclimatization. There are several mechanisms of acclimatization (Klaassen 2001). For example, irritant receptors can fatigue, and their responsiveness can diminish because of depletion of signaling chemicals at synapses. Alternatively, acclimatization can occur because of inflammatory tissue reactions that cause swelling of membranes and increased exudation of fluids coating and protecting epithelial membranes. We are uncertain whether acclimatization is protective because it may allow an individual to tolerate a higher exposure, which may cause tissue damage. Whether or not this is true probably will depend on the particular irritant chemical and the degree to which it can cause damage by another mechanism with a longer time scale—a cumulative irreversible response, for example.

Model for Reversible Proportional Disease Processes

A generalized model representing the reversible proportional disease process has a pair of opposing independent processes: *damage*, represented by a parameter k_{dam}, which is the amount of damage caused per unit concentration per minute (or hour); and *repair*, k_{rep}, a fixed fraction of the damage removed per minute (or hour). An example is shown in Figures 10.1 and 10.2. Although we believe that these processes actually exist, they may not have been specifically identified. Nevertheless, we assume that they *could* be identified given suitable clinical and laboratory studies. Because these two processes oppose each other, a steady-state condition can occur when the amount of new damage during a time interval is balanced by an equal amount of repair (the plateaus in Figure 10.2). The steady-state condition is dynamically stable and appears to be unchanging as long as the tissue concentration remains constant. Once steady state is achieved, the prevailing level of damage will be independent of the duration of exposure. This characteristic is a useful way to identify cases in which one is dealing with a reversible process: the level of "damage" will be independent of the duration of exposure after some minimum time period.

A toxic agent often causes more than one type of tissue effect; there may be a spectrum of effects associated with different tissues, dose levels, or time since the onset of exposure. When we are investigating a specific effect, we will choose a damage process that is appropriate for that effect. Different types of effects are associated with different processes and causal pathways for the same agent. For example, some toxic agents produce primarily reversible effects, but they may also produce small amounts of *irreversible* damage by another mechanism. For example, repeated acute exposures to nitrogen dioxide (NO_2) among farmers working in grain silos can lead to residual fibrosis after the subject has completely recovered from the acute effects of the NO_2. The fibrotic process is not reversible and can produce long-term permanent losses of lung function. When two or more different pathophysiologic processes are triggered by the same toxin, these will generally need to be modeled separately, particularly if they are operating on different time scales.

The time required to reach steady state is a critical parameter of damage-repair processes. Time to steady state and time to recover are generally symmetric—during exposure damage will increase at some rate, when exposure stops damage will dissipate at the same rate (Figure 10.2). For direct sensory irritants such as ammonia, about 1 or 2 seconds are typically required for these rising and falling phases of the time course, when the total change is modest. Why should the rising and falling phases be of equal duration? This occurs because the repair process largely determines both phases: time to reach steady state during exposure and the time needed to remove the damage by repair. As explained in Chapter 6, we generally express recovery as the time it takes to repair half of the damage, the half-time, $t_{1/2}$. Recall that this follows from the principle of proportional change, because biological processes tend to act at rates that are proportional to the quantities that are acted on—a large amount of damage requires a lot of repair, and slight damage needs a little repair. This is not always true, of course, but when the rate of a physiologic process is not proportional, it will generally be the result of a specific, identifiable characteristic of a rate-limiting step. For example, when a key enzyme system is overwhelmed, then the rate of the overall process will cease to follow simple proportional change.

We can often obtain data on the approximate clinical time needed to recover from adverse effects relatively easily, and these data can be used to make a rough estimate of the repair half-time (Figure 10.2). Five half-times of repair will remove 97% of the damage present at the start of the time interval (see Chapter 6), and so one can estimate the half-time by dividing by 5 the time required to completely eliminate the effect. Estimation of the half-time by this approach is approximate and may not be suitable if the underlying processes are more complex. But, if a more complex model is needed, then a simple damage-repair model will probably not fit the observed time-course data well, no matter what half-time is used.

The simplest temporal model for damage-repair processes is a linear difference equation, which represents the change in current damage during each short time period as a function of new effects caused by current exposure and the residual damage remaining after repair. New damage is formed at a rate proportional to the tissue concentration of the toxic material. The damage rate is defined by a constant, K_{dam} (cells damaged per mg/L per min), that is characteristic of the toxin—highly toxic materials have large rate constants. The repair process is assumed to be independent of the damage process and proportional to the total amount of damage, that is, a little damage produces a little repair and so on. The repair constant, K_{rep}, is the fraction of damaged cells replaced with new cells per unit time, such as 0.01 (1%) per hour. The general form for the equation for a short time interval, Δt, is shown following, first as a word equation and then in the form of a difference equation (example in Figure 10.1).

Damage(at t plus Δt) = Damage(at t) – Repair(during Δt) + Damage(during Δt)

$$D_{tis}[t+\Delta t] = D_{tis}[t] - K_{rep}\Delta t D_{tis}[t] + K_{dam}\Delta t C_{tis}[\Delta t]$$

This can be rewritten in a less intuitive, but mathematically more parsimonious, way as:

$$D_{tis}[t+\Delta t] = D_{tis}[t](1-K_{rep}\Delta t) + K_{dam}\Delta t C_{tis}[\Delta t]$$

Equation 10.1 General difference equation for the reversible proportional processes

This relationship indicates that the damage at the end of an interval is the sum of two processes: residual damage remaining after repair plus the new damage formed during Δt. When the overall amount of new damage and repair are equal, then the level of damage will remain constant, "at steady state." As exposure conditions are rarely fixed, repeated-measure data on exposure and outcome provide a powerful probe of the system's dynamic behavior and can be used to estimate the relative rates of damage and repair.

Repair of damage is most easily studied after the tissue concentration has dropped to zero postexposure, so no new damage is being formed. The amount of damage present at the start of each interval (Δt) declines to a fraction ($1 - K_{rep}\Delta t$) of its initial value as a result of repair, and it declines by $(1 - K_{rep}\Delta t)^n$ after n intervals of repair. As a result, repair of damage is modeled as a geometric process. Using the repair expression for n intervals, if there was a rate of repair of 0.1 units of damage per minute and Δt was 1 minute, then after 7 minutes (7 intervals), 48% of the initial damage would remain. After 25 minutes, 7.2% would still remain.

Here is a simple but practical example of estimating the rate of repair. Consider the respiratory symptoms of a person who has lived in Los Angeles for 10 years and moves to Santa Fe, New Mexico, to escape the smog and the symptoms that it produced, cough and chest tightness. We will assume that we have the right model. Based on limited interview data we have collected from people who have left Los Angeles for clear skies, we know that it will take about 6 months for the respiratory symptoms to disappear. Many felt that after a month in Santa Fe they felt substantially better. If we assume that the symptoms were only half as bad after 1 month, then the half-time of repair is about 1 month, or $K_{rep} = 0.5$ per month. Based on this rate, then by 6 months only 1.6% of the starting damage would remain, which is consistent with the report that the symptoms were gone (undetectable) after about 6 months. (The approximate rate of repair can often be estimated by this approach using clinical and other observational data.)

Now, what happens to someone in Santa Fe without symptoms who moves to Los Angeles and is exposed to the smog? We will use the approximate repair rate estimated previously, $K_{rep} = 0.5/\text{mo}$. To estimate the damage rate we will also use observational data. It is reported that that during typical smog levels, 0.1 ppm ozone, after 1 month newly exposed people have an average respiratory symptom score of 2 out of 10 possible, so we will set K_{dam} = 2 per 0.1 ppm per month = 20/ppm/mo. Because the exposure is a constant 0.1 ppm, the amount of new damage each month (Δt) is (20/ppm/mo) × (0.1 ppm) × (1 mo) = 2 units. At the end of the first month the symptoms are at level 2. At the end of the second month, new damage produces 2 units, but half of the preexisting damage, 2, is repaired, and the total is 3 units. Similarly for the third month—new damage produces 2 units, but half of the preexisting damage, 3, is repaired, and the total is now 3.5 units. After a few more months, we will approach the steady-state value of 4 symptom units, at which the amount of damage removed, 2, is equal to the amount of new damage formed each month, 2. The other useful thing about this way of formulating the model is that the monthly tissue concentrations do not all need to be the same; they can vary at random with the exposure.

At low levels of total damage, each increment of damage behaves as separate and independent of the others, and it is repaired separately as time progresses. This may be counterintuitive because it seems likely that the damage process should affect the repair process. However, even when we cut our finger next to a previous cut that is healing, we can see that repair (healing) is operating separately on each instance of damage. There are limits on this assumption of independence, and when there is a priori knowledge that the rate of repair is modified by the amount of damage, then this should be incorporated into the model. But in many instances, it will be adequate for epidemiology to assume that the processes do not interact.

The total damage is the sum of the residuals from each past exposure. As we look at the damage from tissue levels in more distant intervals, there has been more time for repair, and more remote intervals contribute less and less relative to their initial levels of damage. This relationship can be summarized in the expression that follows. Thus the level of symptoms we observe after a period of exposure ($N\Delta t$) will be the weighted sum of the effects from all of the time intervals during the time period.

This expression represents our DP model for reversible proportional processes. It is very simple, but it exhibits a time pattern matching the descriptive information we have on the development and dissipation of symptoms, assuming the damage is not too extreme. The difference equation can be formally fit to observational data, with the values of the two unknown parameters K_{dam} and K_{rep} estimated either from existing human data, animal studies, or empirically from the epidemiologic data. We will estimate these parameters for an example in Chapter 11:

$$D_{tis}[N\Delta t] = \sum_{i}^{N} K_{dam}\Delta t C_{tis}[i](1 - K_{rep}\Delta t)^{i-1}$$

Equation 10.2 Reversible proportional disease process model

The expression in Equation 10.2 also suggests that as N becomes large, or the exposure is long term ($N > 5t_{1/2}/\Delta t$ repair), the contributions of the oldest tissue concentrations to the observed effects become negligible because $(1 - K_{rep}\Delta t)^{i-1}$ becomes very small when i becomes large.

The process model also implicitly contains the setting for the etiologic time interval (ETI), which represents the period of time during which the exposure can contribute to the effect. Assuming that the exposure quickly produces tissue concentrations and no extreme historic tissue concentrations (they might produce effects over a much longer time period); then, as a rule of thumb, we can set the ETI to approximately 6 times the repair half-time, which means that less than 1% of any past effect would remain. A lag period might also be needed to shift the ETI to line up the exposures with their effects.

Key assumptions

As noted in Chapter 6, four key assumptions underlie this model:

1. The rate of damage is proportional to tissue concentration.
2. Damage with short duration exposures is (almost) entirely reversible.
3. Repair is proportional to total damage.
4. The damage and repair processes are independent of each other and run concurrently, in the sense that damage is a response to the tissue concentration and repair is a response to the damage.

These assumptions have several implications that must be kept in mind when using the model.

First, the term *damage* in the model refers to only one type of tissue injury, which is defined by how we measure injury. For example, acute NO_2 exposure produces changes in pulmonary function measurements because of airway obstruction and pain on breathing; alternatively, chest radiographs show fluid in the lungs; pulmonary function and chest X-rays measure different aspects of lung injury and do not show the same relationship with dose. When the amount of damage is large, such as that caused by an acute, severe exposure, several types of injury usually occur simultaneously, and feedback mechanisms may limit or enhance the response. For example, effects of some reactive agents are limited by glutathione or other materials. If those become depleted, the proportionality can change. Consequently, as much as possible we need to be clear on what we mean by damage and carefully consider what injury our measurements represent.

Second, apparently reversible effects may leave small amounts of residual permanent damage, such as fibrotic tissue, that may be undetectable until they accumulate sufficiently. However, those effects are caused by a different process and should be modeled separately—probably using DP-2, discussed in Chapter 12.

Third, assuming repair is proportional to total damage effectively means that each unit of damage is repaired independently. It also means that the "damage" being repaired is all the same kind of injury. As noted, some agents produce several kinds of damage, such as cell killing and mutations, which have different repair processes and must be treated separately. When we begin the modeling analysis of damage from an environmental agent, it may be difficult to determine whether the damage is all the same type, but we must be alert to the possibility that it is not.

Fourth, independence of damage and repair processes was discussed in Chapter 6, but it is worth reemphasizing. Simulations clearly show (in Chapter 11) that even a small effect of the agent inhibiting repair produces a nonlinear concave upward time curve where steady state is not reached. This is most likely to be true if there is a massive exposure and extensive damage. A good example comes from studies of liver damage by chlorinated hydrocarbons in rats by Mehendale and coworkers (Mehendale 2005). They showed that there was a level of damage that overwhelmed repair and recovery with fatal effects, but as long as damage remained below that threshold, repair was proportional to damage. In most situations that will be studied by environmental epidemiologists, the exposures will be low to moderate and severity of effects will also be low to moderate, so the assumption of independence will usually be reasonable.

Dose Metrics for Proportional Disease Processes

Simplest Dose Metric: Average Exposure

A dose metric should be an easily calculated parameter that is proportional to the effect over time and among individuals. For reversible proportional processes, this means a metric that will increase when tissue concentration and damage increase and that will decrease when tissue concentration decreases and damage decreases because of repair. We can often make one useful simplifying assumption: if the exposure intensity has been fairly constant for a period of time that is long relative to the clearance half-time, the tissue concentration will be proportional to the exposure intensity. This rule of thumb will not hold if the substance is bioaccumulative (dioxins, PCBs, mercury, for example) or if a rate-limiting metabolic process is becoming saturated (see Chapter 4). We will first consider the dose metric for an individual and then look at the implication of applying this metric to a population study.

In Chapter 6 we learned that with a steady exposure, damage-repair processes will reach a steady-state balance between damage formation and repair. At the simplest, this implies that the level of the response will be proportional to the level of the tissue concentration and the average exposure. Therefore, the simplest dose metric for this type of reversible effect is the average exposure. For example, a cigarette smoker's levels of respiratory symptoms tends to be proportional to the number of packs of cigarettes he or she smokes per day, roughly independent of how long he or she has smoked (U. S. Department of Health and Human Services [DHHS] 1990).

Although this metric is simple and easy to apply, there are some important caveats. As just noted, the exposure should be reasonably stable and not vary in intensity too rapidly if average exposure is to be a meaningful dose metric. Ideally, the tissue burden of the agent will be at steady state during most of the study period. This does not mean that exposure will not vary but that it will vary gradually, with time for tissue concentration to reach steady state soon after each change. Based on our earlier discussion of damage-repair processes, the time to reach steady state is a function of the individual's repair rate—it takes about five half-times of repair. In the case of the smoker, he or she must have been smoking at least 5 months, during which his or her cough has gotten gradually worse until it stabilizes at some level. Other responses may take more or less time to stabilize. If the smoker varies the amount he or she smokes per day, sometimes one pack, sometimes two, or only a half pack, the irritation the smoker experiences will vary, and so may his or her symptoms, which will not be at steady state with exposure. If he or she has not reached steady state, then the proportionality between exposure and response will change over time.

Exposure Variability

When the temporal variability of the exposure is too large to be considered approximately steady within the time scale of the response, then the simple average exposure may not be a good dose metric. By definition, the level of the effect can change by a factor of two during the repair half-time, $t_{1/2}$. We want the exposure variation during consecutive half-times to be "small" so that the effect remains approximately steady. The damage-repair model suggests that a twofold change in exposure during a time interval equal to $t_{1/2}$ will produce a 25% increase or decrease in the response level. If we arbitrarily decide we want the effect to remain steady at ±10%, then we do not want the exposure to change by more than about 20% during a $t_{1/2}$ interval. Thus we can check the exposure variation on the $t_{1/2}$ time scale of the repair, as shown in Figure 10.2. The next Figure, 10.3, shows how the shape of the time course changes as we change the damage and repair constants. You can clearly see how we could adjust them to fit measured time points. The advantage of using a difference equation for the model, instead of a differential

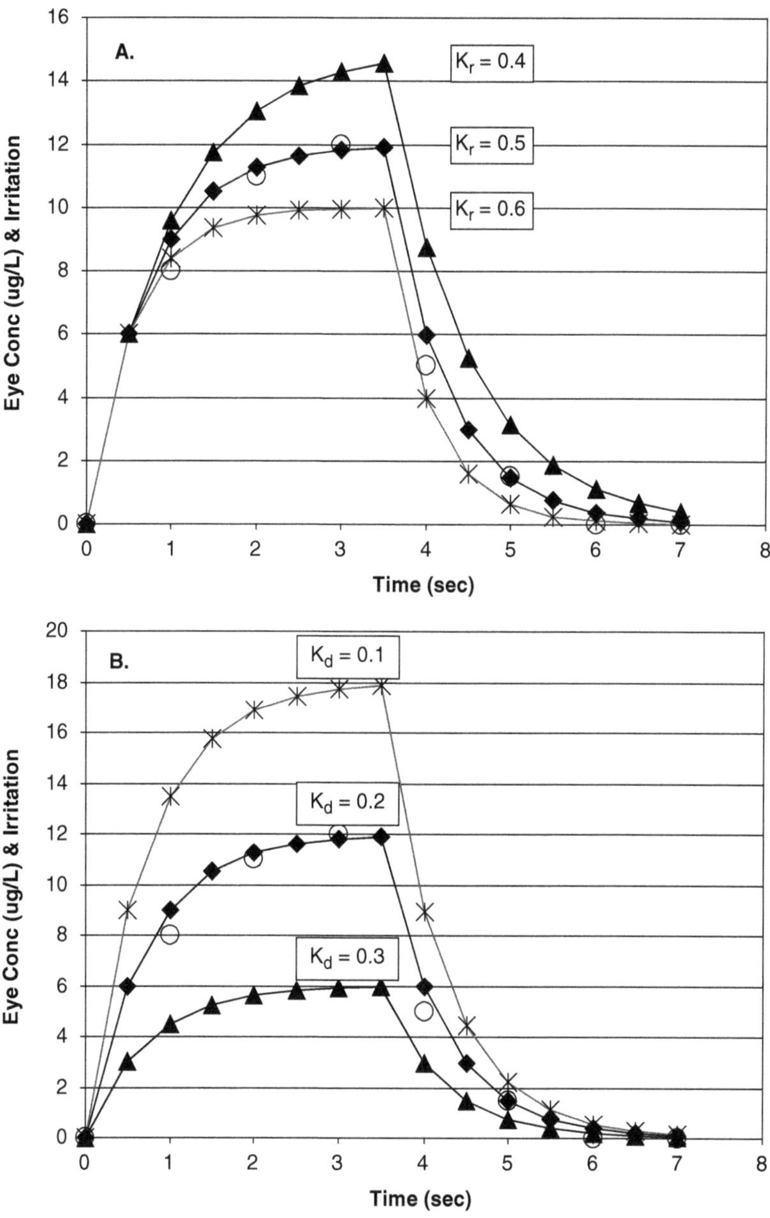

Figure 10.3 Effects of changes in K_{rep} and K_{dam} in the ammonia model for 100 ppm exposure. The rate values can be estimated by fitting the observed data (circles), which can also evaluate the suitability of the model. Part A: Three values of the repair rate constant, K_{rep}, which controls the shape of the curve. Part B: Three values of the damage rate constant, K_{dam}, which controls the height of the curve.

equation, is that we do not need smooth functions to define the time course; we can use discrete steps with any type of statistical distribution for the exposure data.

Exposure variability is a particular challenge when the disease process is rapid because it requires precise temporal measurement of exposure and responses to see the relationship. Ammonia exposure leading to eye irritation is a good example because irritation occurs within seconds of exposure and then dissipates within a few minutes after exposure stops (a recovery half-time of ~20 seconds). Suppose we asked ammonia-exposed individuals about the occurrence of eye irritation over 30-minute intervals during which we also measured average exposure. Suppose further that the ammonia concentration was quite variable from moment to moment. Because of the short half-time of recovery, we would not know exactly which very short exposure "peak" within the 30-minute interval immediately preceded an incident of eye irritation, because we measured only the 30-minute average. And, because the average exposure may be only weakly correlated with the short peak exposures, we might not observe a very strong association between reported irritation and 30-minute-average exposure. There would be misclassification of exposure, likely biasing the association toward the null. The same problem occurs on a longer time scale with air pollution studies of conditions such as cough or wheeze, which may have recovery half-times of days or weeks. If exposure is measured only as the annual average, then there will quite likely be underestimation of the strength of the association between exposure and risk of periods of cough and wheeze due to the annual average being a misclassified estimate of the short-term (weekly, say) average exposure.

A further challenge in population studies is that some subjects (new hires in a workplace study, for example) may not have been exposed long enough to have reached a steady-state response. Inclusion of those subjects may lead to underestimation of the response per unit of exposure. Alternatively, if long-term exposure has declined for some subjects but their response has not yet declined to steady state with their new level of exposure, then their response per unit of exposure will be overestimated. For example, smokers who have recently started smoking or have recently increased the amount that they smoke per day may report fewer symptoms than expected for their current exposure intensity. Conversely, recent quitters will continue to have symptoms reflective of the smoking habits of the recent past, not their current level.

There are at least three strategies for dealing with highly variable exposure situations: (1) perform the study in a different population, one in which there is relatively little temporal variation in exposure; (2) measure the time profile of exposure for each subject with an averaging interval equal to or less than the $t_{1/2}$ of the response with a more precise time resolution of the response; or (3) use a dosimetric model. Strategy 1 may be difficult because exposure situations with stable exposure intensities are often not available outside of the laboratory. Strategy 2 offers much better resolution of the temporal linkage between the exposure and response. However, the approach may be prohibitively costly, and even when it is not, the complex temporal patterns of exposure variation are not easy to use in standard epidemiologic data analyses. Complex statistical models can, in principle, use as much temporal detail as possible, but the analyses can be challenging and time-consuming. A simpler and more direct approach may be to use strategy 3—a dosimetric model.

A Better Dose Metric

We can use the expression in Equation 10.2 to make a dose metric (DM) for reversible proportional disease processes. If the uptake and clearance processes for the toxin are rapid, which is often the case for irritants that exert reversible effects, then K_{dam} in Equation 10.2 can be assumed to be relatively constant over time and among individuals. In these cases, a dose metric can be formulated using only the repair parameter, K_{rep}, in weighting the past exposures (Equation 10.3).

$$DM = \sum_{i}^{ETI} C_{exp}[i](1 - K_{rep}\Delta t)^{i-1}$$

Equation 10.3 Dose metric for reversible proportional disease processes

DM is calculated from the exposure-time profile for each subject, a series of consecutive measurements made during short time intervals Δt minutes or hours in duration. It has only two unknown parameters: ETI and K_{rep}, which both can be estimated with the half-time of repair. Notice that this metric is not static but changes as the exposure changes in the ETI. As a result, it does not require the assumptions that the effect has reached steady state, nor that the exposures have been stable over time; it allows for the development and decline of the effect as the exposure changes. For this DM to work, the exposure profile needs to be measured on a time scale on which Δt is no longer than the $t_{1/2}$ of repair. Additionally, we have to have measured exposure during at least five or six consecutive $t_{[½]}$s before we attempt to relate the DM to the observed effect; this provides enough time for DM to become synchronous with the variations in the effect.

Epidemiologic Applications: Population Variations

In an epidemiologic application, the assumption of an identical repair rate for all subjects is unreasonable. It is more likely that each subject's K_{rep} would vary with his or her personal characteristics, such as age, gender, or genetic background. Given sufficient personal data on exposure and effect, we can estimate each subject's K_{rep}—the one that provides the best fit for the DM. Then we can explore the distribution of K_{rep} values across subjects as a function of the characteristics of the subjects. The fraction of individuals with slow repair is an important characteristic of the population because they will have greater risk from exposure than individuals with fast repair.

Additionally, we can examine the response ratio, which is the ratio of DM to the intensity of the response for individuals with similar K_{rep} values. This is an indication of the sensitivity of the subjects to the exposure. This measure also may vary as a function of the characteristics of the individual, such as family history of asthma or allergies or prior history of exposure. We can also learn about the nature of the effect by noticing when systematic changes in the fit of DM occur with changes in exposure features. If the fit changes with different exposure profiles, then the dose metric provides a means to examine how the biological processes are changing. For example, the response ratio may not be constant across the range of exposure intensities. Epidemiologists have long recognized that short, intense exposures may produce more adverse effects than long-duration low exposures. This can be seen when low values of the DM produce little or no response and intense exposures with large DMs produce large response ratios. Intense exposures may affect the damage-repair process, causing the recovery half-time to increase because of slowed recovery, so the estimated K_{rep} values decrease (less repair per unit of time). One of the most important reasons to use this approach is that we can learn a lot about the nature of the response by looking for heterogeneity of responses in a population relative to the exposures. This is implicit in the biologically based dose metric, even for the reversible proportional model. In Chapter 11, we develop in more detail this interaction between exposure variability and the ability of the biological system to respond.

Discussion of Proportional Disease Processes

Another example of a reversible proportional process is the toxic effects of organophosphate pesticides, such as parathion. These are caused by rapid binding to the cholinesterase enzyme. With the enzyme blocked, acetylcholine accumulates around cholinergic nerves, which produces characteristic respiratory, gastrointestinal, muscular, and central nervous system symptoms. Recovery (without medical treatment) is a considerably slower process than the damage process, because it depends predominantly on the removal of red blood cells with the blocked enzyme and the production of new enzyme proteins. Treatment involves dephosphorylation of the enzyme by administration of pralidoxime (2-PAM; Klaassen 2001). The intensity of the response is approximately proportional to the amount of blocked acetyl cholinesterase enzyme, and the rate of recovery depends on the speed of replacement of the blocked enzyme. Because the untreated recovery is so slow, repeated exposures can rapidly block a dangerously high fraction of cholinesterase, putting an individual at risk of serious symptoms and even death. However, it is still possible for an individual with a very low-level chronic exposure to come to a steady-state level of blocked cholinesterase if the exposure is low enough.

In some cases, the actual mechanism causing the response is obscure and poorly understood, or it is the result of a complex and shifting mixture of different agents, as in the case of respiratory symptoms during cigarette smoking. When the effects show a typical reversible proportional time course, one may still be able to adequately estimate a dose metric without a clear understanding of the specific causal agent. The severity of respiratory symptoms, cough and phlegm, in long-term cigarette smokers is proportional to the packs-per-day they currently smoke, independent of how long they have smoked, as long as they have been smoking for a year or two (U. S. DHHS 1990). This strongly suggests that the processes causing the symptoms are being balanced by repair processes. When smokers stop smoking, their symptoms typically decline, with an apparent half-time of 1 or 2 months. Thus, even when the exposures are very complex, we may be able to construct a useful dose metric if we observe the characteristic temporal pattern for a reversible proportional process.

10.4. IRREVERSIBLE PROPORTIONAL DISEASE PROCESSES

Irreversible effects are those that are permanent; death is a good example. Once the damage is done, it is not repaired or removed. When nerve cells are killed, they do not regenerate. Fibrosis in the lungs is scar tissue; it does not dissolve over time. If cancer cells are not killed by one's natural defenses or chemotherapy, the cancer will grow and spread; the cells will not revert back to normal. Sometimes drastic medical procedures can remove the damage and replace the damaged or failing organ or tissue with a healthy replacement, but that is uncommon. More often, irreversible damage accumulates, and the result is eventually fatal for the subject. In general, irreversible effects are the ones we want to prevent, because that's all we can do.

Many toxins produce irreversible effects, and if they cannot be removed, they will accumulate over time; they are "cumulative" effects. Often this occurs when there is cell killing which is not adequately repaired; scarring is perhaps the simplest example. Fibrogenic effects are observed from environmental chemicals in many parts of the body, and these can often be adequately described by a simple modification to the reversible proportional disease process described in the previous section. Essentially we assume that there is no repair or recovery from the adverse effects; there is only accumulation while the active agent is present in the tissues.

Example: Silica and Lung Fibrosis

Silicosis is an ancient disease of miners. It was recognized by Pliny the Elder (23–79 AD; Hunter 1969). He reported that slaves forced to work in mineral mines encountered very dusty conditions. They developed difficulty breathing after a relatively few years of working in the mines, became incapacitated, and died relatively soon afterward. Isbrand von Diemerbroeck (1609–1674), a professor of medicine in Holland, observed that at autopsy the lungs of stonecutters were like bags of sand, full of scar tissue (fibrosis) and nodular lumps (Hunter 1969). Diemerbroeck and later others found that if the lungs were ashed by burning off the carbonaceous materials, a large amount of inorganic matter remained. In modern times, research has shown that the mass of silica in ashed lungs from miners is roughly proportional to the amount of fibrosis seen on chest X-rays (Churg and Green 1998).

Mechanisms of Damage

Although the precise mechanism of fibrogenesis is somewhat uncertain, some aspects of its complexity are known, as diagrammed in Figure 10.4. Particles of crystalline silica deposited in the alveolar region of the lungs are ingested by pulmonary macrophages, which store them in small compartments (lysosomes). The macrophages aggressively but ineffectively attack these particles with digestive enzymes and reactive oxygen species, a defense mechanism that can be effective for bacteria, but not for sand. The macrophages also attempt to carry the particles out of the lungs via the lymph nodes, which the macrophages may do if they have not ingested too many silica particles. The particles damage the macrophages' internal lysosomal membranes, causing the proteolytic enzymes and active oxygen compounds to leak out, injuring and sometimes killing the macrophages, injuring surrounding tissues, and interfering with the removal of the dust. The dead macrophages release their particles, which can be ingested by other macrophages, causing further release of damaging components. This damage process also initiates a complex cascade of cellular events leading to fibrogenesis and scarring. The scarring eventually encapsulates the silica particles, creating characteristic nodules of fibrotic tissue in the lungs and lymph nodes. There is also an immune system response to the pathogenic process that affects the extent of the response to the silica. Pulmonary fibrosis caused by crystalline silica particles slowly encapsulates the particles, and once all of the free particles are fully encapsulated, the process stops.

Removal of particles by alveolar macrophages is a capacity-limited process. The macrophages can be overloaded by particles, a process that can slow and, if severe, stop the removal of particles. Depending on the toxicity of particles, they may have a large or small effect. As shown in Figures 10.4 and 10.5, we can model this in a way analogous to the Michaelis-Menten kinetics of enzymes (Smith 1985; Yu and Rappaport 1997). The effect is to slow down the removal of particles as the exposure intensity increases, which can also induce a feedback loop. This kind of behavior has been seen to lead to large accumulations of coal dust and crystalline silica in the lungs of miners.

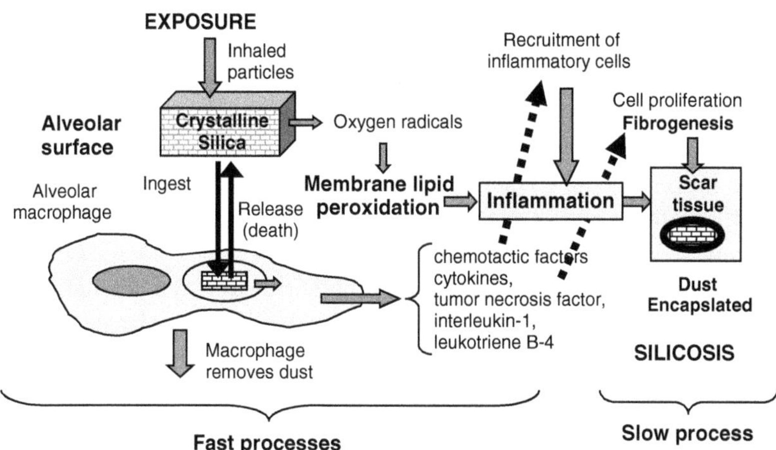

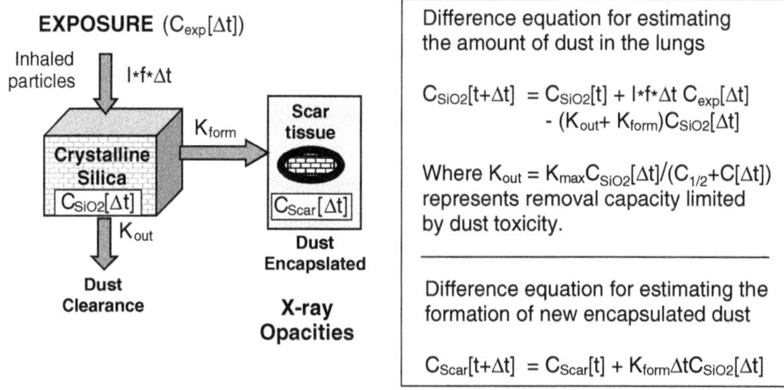

Figure 10.4 Diagram of the descriptive process model and the corresponding difference equation model for fibrogenesis and chest X-ray opacities. The parameters are: Δt = 1 day; I = 10 m^3/day; f = 0.2; K_{max} = 0.2 per day; $C_{1/2}$ = 10 mg; and K_{form} = 0.03 mg fiber/mg dust/day. (Note: $C_{1/2}$ defines the toxicity of the dust.)

At low levels of daily exposure, the macrophages are able to clear most of the particles entering the alveoli. However, some of them escape capture and penetrate into the interstitial area, where they are harder to remove. The slowest aspect of the disease is the encapsulation of particles by fibrogenesis. Particles are encapsulated in the interstitial area, and some are carried to lymph nodes, where they induce fibrogenesis characteristic of the disease. The scar tissue appears as opacities (spots) on chest X-rays, and it stiffens the lungs, causing loss of vital capacity that is observed in miners and others with prolonged exposures to crystalline silica.

Model for Irreversible Proportional Disease Processes

Despite uncertainties in the precise steps and complexity in the pathogenic process, macrophages are clearly target cells for crystalline silica, and it has long been known that the fibrogenic response is proportional to the amount of silica dust reaching the alveolar region of the lungs. We can use this information to formulate a simple process model for fibrogenesis characteristic of the disease, as shown in Figure 10.4B. Irreversible proportional effects can be represented as if each unit of the material reaching the target tissue causes a unit amount of damage that is not repaired. For crystalline silica, each particle of dust deposited in the lungs is encapsulated in a unit of scar tissue. Additionally, there is a latency or lag period between the deposition in the tissue and the development of the scarring, because it takes time to form the new tissue. This simple conceptual process is consistent with the observed temporal behavior of the fibrogenesis aspect of silicosis, and it can be used to represent fibrogenesis even though it does not reflect all of the subtleties of the biological processes that produce the effects—as long as we are willing to assume that they are irreversible. The total tissue damage then will be proportional to the amount of dust deposited in the tissue each day (A_i, μg/day) times the duration (N, days) of exposure, plus the lag, $n_{lag} = L/\Delta t$, needed for the development of the effects. If we measure the total damage at the end of the exposure, we will underestimate the total because the exposure received during the last n_{lag} days will not have caused any effect. We can use a proportionality constant, K_{dam}, to indicate the amount of damage (scar tissue) produced by each dose unit, such as a microgram, that is deposited in the target tissue:

$$Total\ Damage[N + n_{lag}] = \sum_{i=1}^{N} K_{dam} A_i \cdot \Delta i$$

Equation 10.4 Irreversible proportional disease process model

We also add a constant, L (days, months, years), to represent the lag or latency needed for the effect to fully develop. Note that the lag is a biological constant that applies to each dose unit individually; it is not an overall constant, which is not the usual way that lag is conceived in an empirical epidemiologic model. Additionally, we must be clear that this model for fibrogenesis may not be a good model for other aspects of silicosis, such as losses of pulmonary function or immunological responses. In all applications of the model, we must carefully examine the time course of the response we are modeling to be sure it is appropriate.

In this model of irreversible effects, as with other process models, the slowest rate-limiting step defines the quantitative relationship. Because the final irreversible outcome is often the result of a protracted process, there can be a significant lag or latency between the causal exposure and the final effect, such as death from the damage. The latency period is a time interval that applies to each unit of dose.

Thus the consideration of time is complex, because it is both part of the dose calculation and a marker for the progress of the developing effect. As noted earlier, one way to simplify this is to consider that each unit of dose has its own latency period. In essence, as long as there is ongoing exposure, even intermittent exposure, the latency for some of the deposited material has not occurred.

A difference equation can be used to represent the process previously described in a way that is analogous to Equation 10.2 for irreversible proportional processes:

$$D_{tis}[i+n_{lag}+1] = D_{tis}[i+n_{lag}] + K_{fib}\Delta t C_{tis}[i]$$

Equation 10.5 General difference equation for the irreversible proportional processes

where D_{tis} is the amount of damage (fibrosis) in the lungs, C_{tis} is the concentration of dust deposited as free particles in the lung tissue, $L = n_{lag}\Delta t$ is the lag interval between deposition of the particles and the appearance of the fibrotic tissue, and K_{fib} is the amount of damage produced per unit of exposure. During each interval, the tissue concentration produces some small amount of fibrosis that is proportional to the new dust concentration, K_{fib} (the amount of fibrosis generated per unit concentration per Δt), but it does not appear until after the lag interval.

This description suggests a simple additive model in which each interval of exposure adds an increment to the total damage. The total damage caused during a period of exposure, $T = N\Delta t$, is just the sum of all of the damage during the lagged intervals before the observation at T. Because exposure may vary randomly over time, we will use a more general form, $K_{dam}\Delta t\ C_{tis}[i]$ for the damage in each interval. The equation for this sum, assuming exposure begins at i=1 and exposure continuing to the time of observation, is:

$$D_{tis}[T] = \sum_{i}^{N-n_{lag}} K_{dam}\Delta t C_{tis}[i]$$

Equation 10.6 General irreversible proportional disease process model

The sum of all interval concentrations is the same as the number of intervals times the mean, and the interval number times the interval duration, $(N - n_{lag})\Delta t$, is equal to the total duration, T-L, which is approximately T if the duration is long. The ETI is the whole time of exposure if the duration is long (years) and if the lag is assumed to be short (months). A critical assumption underlying this model is that the amount of damage is a fixed proportion to the dose. In the case of lung fibrosis from silica, large amounts of particles may produce more fibrosis or initiate other pathologic processes, in which case this model will not be adequate such as for acute silicosis (Klaassen 2001).

Key Assumptions

1. Each unit of the agent causes the same proportional amount of damage at a rate independent of other units arriving in the tissue before or after.
2. The final damage is irreversible (although it may be preceded by a reversible effect).
3. Often there is a lag between the tissue dose and the development of the irreversible response.

Each of these assumptions has important implications.

As we saw for the damage-repair model, it is necessary that each unit of exposure and its matching tissue concentration (dose unit) have an independent effect for them to add up in a linear sum in the disease model. If the effect of next dose unit is modified, increased or decreased, by the previous unit(s), then the resulting damage may be more or less than expected. One relatively common situation occurs when an intense exposure deposits a high dose that changes the pathological processes – perhaps by overloading a protective process. In that case, a different type of disease process model will be needed. If each unit of dose is not independent, we can test the hypothesis by comparing the amount of fibrosis, or other effect, in subjects with different patterns of exposure intensity over time. This model implicitly assumes that the amount of damage formed by each unit of dose has a fixed ratio to dose, which will produce a simple, linear accumulation of effects. There can be no threshold concentration that must be exceeded to have an effect. As tissue concentration increases, there is an increasing possibility that the response may enter a nonlinear range or the agent may alter earlier effects. This has the effect of changing the magnitude of K_{dam} and necessitating a more complex model. If the proportionality constant varies widely, then periods of equal exposure will not contribute equally to the total effect. This behavior is seen in late-stage effects because it becomes more likely that the later dose units will come in contact with tissue that has already been affected by previous dose units. So the proportionality of late-stage effects may change. It is not uncommon to observe that the highest and longest exposures do not show a proportional increase in risk, perhaps a representation of this phenomenon.

The lag in our model is not the standard epidemiologic conception of latency because the lag is an ongoing part of the damage process for each dose unit. The dose deposited early may have reached its final conclusion, but the effects of more recent dose units are still developing. Thus the latency is not just the final time period required for the effect to develop; the latency is part of the dynamic process of developing irreversible damage. If the nature of the disease process is unknown, then various lags can be tested in the model to determine that sufficient time has elapsed for the process to reach its conclusion. If the lag used in the model is insufficient, then the dose-response relationship will be distorted, and the risk per unit exposure may be underestimated. For example, a study of irreversible effects in retired workers can have different proportionality from a study of workers currently exposed, because the former have had sufficient time to develop the full extent of the effects.

Simple Dose Metrics for Irreversible Proportional Disease Processes

First Approximation: Cumulative Exposure

Exposure response processes fitting this model are those in which each unit of active agent reaching the target tissue for a period of time produces a unit of irreversible damage; for example, each μg/g per day of methyl mercury in the central nervous system (CNS) kills a fixed number of neurons. Then the total amount of nervous system damage will be proportional to the average concentration of methyl mercury maintained in the target cells in the CNS for a period of time. Loss of CNS function may not be proportional to damage because most organs, including the brain, the location of the damage is important, damage in some locations may not detectably affect function, additionally organs have reserve capacity and may be able to adapt to small amounts of damage. The total tissue dose is proportional to the intake rate (R_i, μg/day) times the exposure duration ($N\Delta t$, days), where the increment in time, Δt, is one day and the total duration of exposure is T ($N\Delta t$). The average intake rate ($\overline{R}$) is proportional to the average concentration in environmental media ($\overline{E}$ in water, air, food each day) times the quantity of media taken in (I L/day).

$$Total\ Dose = \sum_{i=1}^{N} R_i \Delta t \quad = \overline{R} \cdot N\Delta t \quad = I \cdot \overline{E} \cdot T$$

Equation 10.7 Total dose for a cumulative effect

If most people take in about the same amount of air, food, or water per day (I), we do not need to estimate the intake of media because it is just a proportionality constant. This leads to the dose metric generalization in Equations 10.3 and 10.8, which is proportional to the intake rate times the duration of exposure:

$$Dose\ Metric = \overline{E} \cdot T \quad \approx \overline{R} \cdot N\Delta t$$

Equation 10.8 Dose metric for irreversible proportional effects

The final form, average exposure intensity times duration, is cumulative exposure, which we discussed in Chapter 5.

For the fibrosis associated with silicosis, this model means that each particle of inhaled dust deposited in the lungs would contribute an increment to the accumulation of fibrotic tissue apparent on chest X-rays. This is only proportionally true because some dust is removed from the lungs by macrophages and other clearance processes.

The dust deposition rate (mg/min) is defined by the concentration of respirable dust (mg/L) inhaled × the minute ventilation (L/min) × the fraction of dust deposited in the alveolar area. The total dust deposited (mg) is the deposition rate (mg/min) × duration of exposure (min). Because a subject's minute ventilation rate usually is not known, nor his or her fraction deposited, we assume that those factors were approximately constant over time among the study cohort members. Based on these considerations, the dose metric would be the concentration of respirable dust (mg/L) inhaled × duration of exposure, expressed as mg/L×years. Again this is, *cumulative exposure* (average exposure intensity times duration), which has been widely used as a dose metric for a variety of chronic irreversible health effects (Smith 1992; Smith, Stewart et al. 1995).

A Better Dose Metric

The relationship in Equation 10.5 can be used to formulate a more general dose metric, as we did for the reversible proportional process model. Although we may not know the value of the proportionality constant, K_{dam}, nor how fast the damage process proceeds, the total damage will be proportional to the duration times the average intensity, as long as the time since beginning exposure has been large. Because our goal for a dose metric is that it be proportional to the magnitude of the effect, an adequate dose metric for irreversible proportional processes such as fibrogenesis is:

$$DM[ETI - L] = \sum_{i}^{ETI-L} w_i C_{exp}[i]$$

Equation 10.9 General Dose metric for irreversible damage processes

where ETI = etiologic time interval, L = lag, and w = a weight representing contribution of C to disease process.

Note that this model is written as if exposure is still continuing. If the exposure has stopped, then we must consider how the time since exposure is related to the lag period. This dose metric is based on exposure, whereas the DP model in Equation 10.6 used tissue concentration. The dose metric will be approximately correct when C_{exp} and C_{tis} are proportional. This will occur in many situations in which the exposure intensity is low or moderate and when transport and metabolic processes are not close to their capacities. If this is not the case, one could construct a simple toxicokinetic model to estimate tissue concentration from exposure (see Chapter 4).

The dose metric in Equation 10.9 will give the same result for a short-duration high-intensity exposure profile and a long-duration low-intensity exposure pattern. This assumes that there is no threshold for the effect—for example, that 1 particle will produce fibrotic tissue as readily as 10,000. In the case of silicosis, however, we know that very intense exposures to silica dust can produce acute silicosis, which is a progressive lethal disease, after an exposure of only 6–18 months. As a result, the dose relationship does not apply for all levels of tissue concentration, and for a population with a wide range of exposures we should verify that the most highly exposed appear to have approximately the same response per unit of exposure as those with lower levels. This is important when a cohort has had early periods of intense exposure followed by long periods of low-level exposures, because the early exposures may initiate pathogenic processes that will not occur with low exposures but can be maintained by continued low level exposure. The history of exposure matters.

Limitations of Cumulative Exposure

Although the cumulative exposure metric is widely useful, it has to be used carefully to be sure it is not violating any of the key assumptions (Smith 1992). One important limitation of this dose metric is it does not directly consider the temporal processes inherent in its relationship to outcomes. Consider each assumption, where it might be violated, and how that might affect the dose-response relationship:

1. The effects of each dose unit are independent. This assumption is best met when the intake rate is not too high, so that the units will rarely overlap. The definition of a "too high" intake rate must be determined empirically. It is possible that laboratory studies may give an indication of a shift in mechanism or interaction, which may be used to define "high" intensity exposures.
2. Each dose unit produces the same amount of damage, that is, K_{dam} is a constant. As noted earlier, when there are large amounts of damage, then new dose units may not produce the same amount or type of damage as earlier ones. High-dose-rate conditions can also affect the value of K_{dam}. Again, this can only be determined empirically. Individuals at the upper end of the distributions of K_{dam} will have more effects and may be more at risk of interactions.
3. The latency is a constant that applies to each dose unit independent of total damage and dose. Effects on latency are the result of modifications of the disease process, which may also be related to the dose rate and the amount of total damage. When either of them is high, then the development of the damage may also be affected and latency changed. Most important, if the subjects are being exposed at the time of ascertainment of the effects, there will not have been time for the full development of the effects from the recent exposures. If a lag is introduced, it will partially compensate for this. However, the development process for each unit will overlap as each dose unit is deposited. Thus the use of a single lag overcompensates for incomplete development of effects of recent dose units. Ideally, each dose unit should be weighted by its time since deposition up to the full lag.

Discussion of Irreversible Proportional Disease Processes

What processes are irreversible? Most commonly, they are formation of scar tissue (fibrogenesis) as a result of tissue damage or cell death, such as by carbon tetrachloride's liver toxicity, or cell killing in the central nervous system through effects of manganese. At low levels of exposure, some chemicals produce effects that are reversible, but at higher exposures they cause cell killing and irreversible damage. An example of this is seen in the kidney effects of lead, which include both reversible renal tubular disorders when exposure is of short duration and irreversible interstitial nephropathy when exposures are chronic. An important class of irreversible neurological effects of lead occurs in children, who have higher sensitivity than adults. Childhood ingestion of lead particles in soil or paint chips can lead to

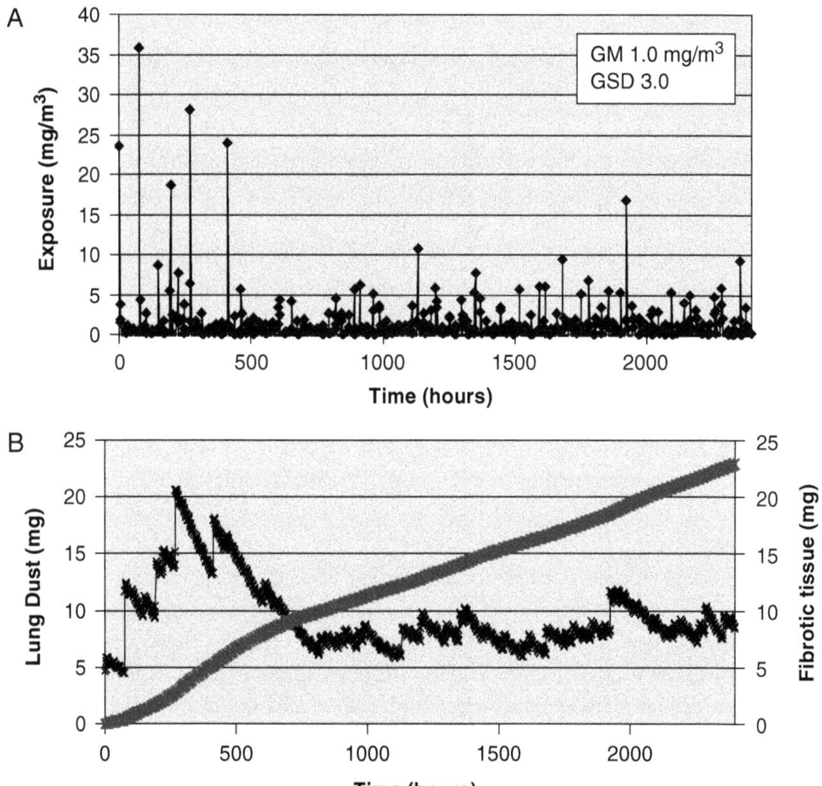

Figure 10.5 Calculations from the difference equation model of fibrogenesis from silica dust exposure for the model shown in Figure 10.4. Part A: Exposure distribution of dust exposures: 5 days/wk. Part B: Lung dust accumulation and resulting fibrotic tissue.

permanent central nervous system damage, which, in turn, causes lower IQs and adverse behavioral effects.

Figure 10.5 shows the typical temporal pattern of irreversible effects. Usually there is a delay in the appearance of detectable effects. The duration of the delay depends on both the nature of the effect and the method used to detect it. Once it is observed, the effect can progressively worsen, as long as exposure continues and there is sufficient time after exposure for the effects to develop. The rate of worsening of the effect is proportional to the intensity of exposure (tissue concentration), which can be seen for the two exposure periods in Figure 10.5B: the early period produces a higher slope than the later period. The effect may continue to progress when exposure stops, because the pathologic processes continue to respond to the residue of the agent or to earlier damage. When the effects are complete, the quantity of damage reaches a static plateau; this is not a steady-state condition because it is not a dynamic balance of damage and repair—it is a static level of damage. For example, some fraction of brain cells have been killed, or the liver contains a certain quantity of scar tissue. If there is additional exposure, then the pathologic process begins again from where it stopped with a further accumulation of damage. Additional or late pathologic processes may modify the effects or cause new ones. When high levels of damage occur, other secondary responses are often induced, such as compensatory processes or organ failure, with their own consequences. The irreversible proportional type of response is not a progressive response such as is seen with acute silicosis, in which pathogenic processes, once initiated, continue and may even increase after the introduction of the agent has stopped, because when that happens the proportionality with dose is lost. In Chapter 12 we explore in detail the irreversible proportional response.

11 Reversible Proportional Disease Processes: Effects of Ammonia and Ozone on Respiratory Symptoms

11.1. INTRODUCTION

In this chapter the modeling of reversible proportional disease processes is illustrated using two examples. We first continue to use the simple case of ammonia leading to eye irritation that was presented in Chapters 6 and 10. Then, a more complex case is explored: the respiratory effects of ozone. A historical perspective on ozone dose-response modeling is useful because the literature followed a path of increasingly complex and biologically relevant models as more and more data were obtained. As we lay out this history, we hope to illustrate the limitations of the empirical approach and the advantages of biologically based models.

As noted earlier, the key features of reversibility are defined by the damage coming to a steady-state level with a steady exposure and the damage diminishing or disappearing when the exposure diminishes or stops. In many cases the latter condition is self-evident. To be precise, the effect dissipates because of repair when the tissue concentration declines. However, there may be a lag between the end of exposure and the decline of the adverse effect as the body (dynamic biological system) adjusts to the absence of the agent. We must define other key features in order to develop a model as summarized in Table 11.1.

Table 11.1 Characteristics of the Reversible Proportional Disease Processes

Key Features	Characteristics	Data Sources
Target Tissue	Tissue(s) with adverse effects	Clinical reports; published findings; interviews with affected individuals
Evidence of reversibility	Time course shows a plateau in response with steady exposure and response declines when exposure stops	Clinical reports and interviews with affected individuals indicating recovery after exposure
Type of a proportional response	Magnitude of the response is linear or nonlinear with tissue dose	Scientific papers
Half-time of recovery	Time for half of the damage to be repaired	Clinical reports; interviews with affected individuals
ETI	Interval of time contributing to the observed effect; equals the time to complete recovery when exposure stops, ~5 half-times of repair, ETI $= N_{ETI}\Delta t$	Clinical reports and interviews with affected individuals indicating recovery after exposure
Disease process model	Effect is the sum of the new damage from exposure minus the damage repaired over ETI $$D = \sum_{i=1}^{N_{ETI}} K_{dam}\Delta t C[i]\left(1 - K_{rep}\Delta t\right)^{i-1}$$ Note: i counts backwards from the present into the past.	Begin with general model if descriptive model appears to fit; estimate K_{rep} from observations; K_{dam} is a scaling factor to adjust D to units of effects, i.e. the amount of damage per unit of exposure.
Dose Metric	$$DM = \sum_{i=1}^{N_{ETI}} C[i]\left(1 - K_{rep}\Delta t\right)^{i-1}$$	Use K_{rep} estimated from DP model. Or fit DM across subjects to get mean and SD.

11.2. MODELING THE IRRITANT EFFECTS OF AMMONIA

Our strategy in developing a model of effects is to decide on a physiologic model of the target tissue and its interaction with the environment—the exposure biology. A conceptual model of the eye irritation is critical to guide our exposure assessment and to link exposure to the target tissue dose, as was shown in Chapter 10. We can go through the development of the model presented earlier for the case of eye irritation from gaseous ammonia. First, the eye is in direct contact with atmosphere. Therefore, we do not need a pharmacokinetic model to describe the movement of the ammonia into the body and to its site of action.

Second, we know that ammonia is highly soluble in water and physiological fluids. There are tables of chemical data on materials in which we can look up their water solubility and the pH of solutions (Verschueren 2000; Lide 2007); some of them are also online. Third, we can read about the physiology of the eye and irritant responses in a basic physiology textbook (Ganong 2005) or by searching the scientific literature. This literature indicates that the cornea has a layer of watery fluid (tears) on its surface and that it also has irritant sensors on its surface. These have very rapid responses to protect the cornea from damage from airborne materials. Based on that information, and given that the cornea is the target tissue, we do not need to be concerned about other aspects of the eye's anatomy or physiology. Taken together, that information is sufficient to develop our descriptive model shown in Figure 11.1 (this is the same as Figure 10.2).

The next step is to formulate a model of the environmental interaction. We will hypothesize that it is the concentration of ammonia and pH of the eye fluid that stimulate the irritant sensors of the cornea. Therefore, we need to have a

Descriptive Diagram of Interaction

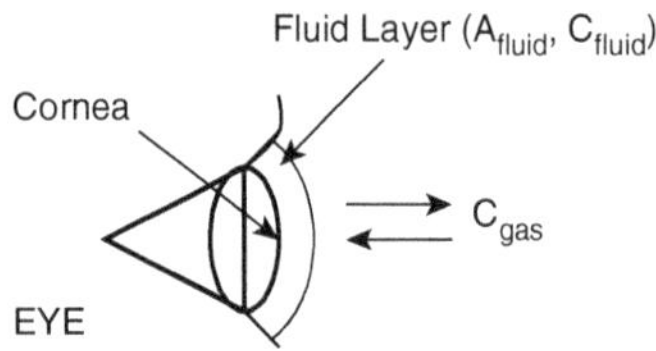

Cornea is the target tissue, with sensory nerves on the surface.

Fluid Layer Compartment

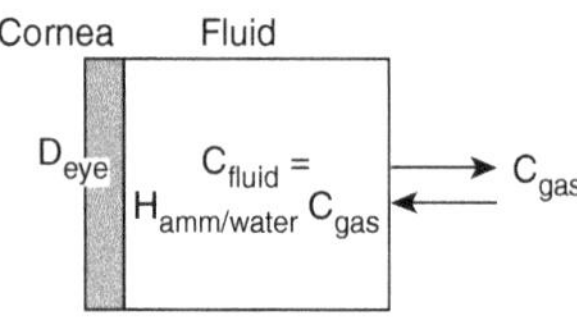

$$C_{fluid}[t+\Delta t] = C_{fluid}[t] + \frac{K_{dif\text{-}amm}\Delta t(H_{amm/water}C_{air}[\Delta t] - C_{fluid}[\Delta t]}{A_{fluid}}$$

Difference equation for estimating the fluid concentration

$$D_{eye}[t+\Delta t] = D_{eye}[t]\ (1-K_{rep}\Delta t) + K_{dam}\Delta t C_{fluid}[\Delta t]$$

Difference equation for estimating the damage level

Spreadsheet Model: In a box above the columns for the model calculations, place the values of the parameters: Δt, $K_{dif\text{-}amm}$, K_{rep}, $H_{amm/water}$, A_{fluid}, K_{dam} and K_{rep} (see text) plus the exposure distribution (GM, GSD) values. Four columns are defined as

A: Time ($i^*\Delta t$),
B: Exposure ($C_{air}[i]$) from random number generator using GM, GSD,
C: Fluid Concentration calculated from the difference equation above using $C_{air}[i]$ and parameters,
D: Damage Level calculated from the difference equation above using $C_{fluid}[i]$ and parameters.

Figure 11.1 Simple one-compartment PBTK model for tissues in contact with ammonia in the external environment. (See the text for the sources of the $K_{dif\text{-}amm}$, $H_{amm/water}$, and A_{fluid}, values.)

compartment that represents the eye fluid. We will initially hypothesize that the sensors are in direct contact with the fluid. Someone might reasonably ask "Why does the model represent the surface fluid and not the eye tissue, such as the cornea itself?" We want the model to represent the location of the response. If the irritant sensors were internal in the cornea, then there might be a lag between exposure and response to allow time for the ammonia to diffuse into the cornea. However, the sensors are on the surface to protect the cornea from environmental agents. Given that the observed response is very fast, if there is a lag it must be very small, so it can be ignored in our initial model. The result is a very simple model shown in Figure 11.1.

If the agent diffuses into the tissues and is directly toxic, we should model the tissues (an alternative hypothesis). For example, triethylamine (TEA) is highly water soluble and forms caustic solutions, too, but airborne exposures also produce visual effects (a blue haze and halos around lights). Those effects are more likely caused by TEA diffusing into the cornea and light-sensing tissues (Reilly et al. 1995; National Institute for Occupational Safety [NIOSH] and Health and World Health Organization [WHO] 1998).

As noted in Chapter 4, some target tissues are in direct contact with materials from the environment: the moist surfaces of the eyes, nose, and throat; the moist surfaces of the airways; and the skin. In those cases, a one-compartment model is often useful for describing the kinetics of contact. If, on the other hand, internal tissues remote from the site of initial contact are affected, then a more complex model may be needed. The reversible neurological effects of solvents are an example.

Getting Data on Physical and Chemical Properties

If an exposure assessor wished to estimate the temporal variation of the actual concentration of ammonia in the eye fluid, she or he would need values for the three parameters in the model (Figure 11.1): ammonia's diffusion rate, $K_{dif\text{-}amm}$, Henry's Law coefficient for partitioning from air to fluid, $H_{amm/water}$, and the surface area of the exposed cornea, A_{fluid}. However, for epidemiologic purposes, we may not need to model the very rapid changes in concentration of ammonia on the surface of the eye because the eyes are directly exposed to the environment and they rapidly respond to changes in the air concentration. One could make the simplifying assumption that the eye fluid instantaneously comes to steady-state concentration with the ammonia in the air. Then all that would be needed to use this model is the $H_{amm/water}$. For a given concentration in the air, Henry's Law defines the steady-state concentration in a liquid, such as water (Sander 1999). Thus we can use the Henry's Law constant, $H_{amm/water} = 5.8 \times 10^{-5}$ mol/(L*ppm), times the air concentration expressed as ppm, to define the fluid concentration for that air concentration (Manahan 2004). For example, 100 ppm ammonia will produce a steady-state eye concentration of 5.8×10^{-3} M and a pH of about 10 in the eye, which is caustic. This ignores buffering capacity of the tears.

The eye concentration will rapidly reach steady state with the ambient concentration because there is only a small volume of fluid. If the air concentration later falls, then the ammonia concentration in the fluid will also fall by off-gassing or back diffusion of ammonia out of the eye fluid. The diffusion rate defines how fast that will happen. We can check that this process is fast by calculating the ratio of the diffusion coefficient for ammonia in air, $K_{dif\text{-}amm}$, = 0.198 cm²/sec (Chen and Othmer 1962), and the exposed surface area of the eye, A_{fluid}. = ~2 cm² (roughly an ellipse 1 cm by 2.5 cm). The half-time is $\ln[2]*A_{fluid}/K_{dif\text{-}amm}$ = ~7 sec. This small half-time value clearly shows that the thin layer of fluid on the eye will very rapidly reach high values and equally rapidly be cleared of ammonia postexposure. Because the neurological response rate is also very fast, we will experience nearly instantaneous irritation from high environmental concentrations, and when the external concentration falls, the ammonia will quickly diffuse back out of the eye fluid. So it is a good assumption that the eye concentration will closely follow the changes in air concentrations.

Getting Data on the Damage and Repair Processes

Poison control data reported on the web in eMedicine (http://www.emedicine.com) and in journals on clinical toxicology are a useful source of temporal information on the development and dissipation of adverse effects, although they obviously focus on the more severe effects, and, aside from the very common hazardous materials, they rarely have data on exposures associated with the effects. Current textbooks on internal medicine, such as *Harrison's Internal Medicine* (Braunwald et al. 2002) can also have useful information. Textbooks on toxicology have some information, but they are usually not as detailed as the clinical works. Published case reports also will sometimes have sufficiently detailed data to be useful in developing models.

Personal symptom reports by the workers at industrial operations can also be helpful. For example, workers at operations using large amounts of ammonia, such as refrigerated warehouses or skating rinks, indicated that they became acclimatized to ammonia exposures that newcomers found very irritating (Hunter 1969). Such places also often have local legends about the "curative powers" of regular exposures to high concentrations of ammonia (perhaps in the range of 50–100 ppm). These exposures have been reported to reduce the frequency of colds and sinus conditions and even to quickly cure hangovers (Ferguson et al. 1977). Although these "war stories" are entertaining, they are probably not representative of the likely experience of the general population, because working groups are usually highly selected (the most resistant will remain in irritating or hazardous conditions). At most these stories may be useful for defining the upper limit of normal responses (and credibility!).

To make our reversible proportional DP model in Figure 11.1 useful, we need an estimate of how much damage is formed per unit of exposure. Later we can fit the data to estimate this, but to start we need a rough estimate. The eMedicine article on ammonia toxicity by Isley and Lang (2007) identified symptoms that were associated with selected ammonia concentrations. We can use those data to define a preliminary value for the damage rate constant, K_{dam}. For example, if we assign a score of 1,000 (arbitrary units) to mild and 10,000 to severe irritation, then our damage constant is 1,000/1,000 μM, or 1.0 per μM on the surface of the eye. Based on Henry's Law, 17.2 ppm should produce 1,000 μM on the cornea at steady state at 25°C.

Next we need estimates of the time scales for recovery or repair for target tissue, that is, the repair half-time. For the sensation of irritation, we know that within a few minutes at the most the irritation is gone, once the ammonia diffuses away, probably aided by tears flushing the high pH fluid from the eyes. Our present model does not include tears flushing the eyes. We may wish to add that to the model if we find the model's recovery time scale is too slow for high exposures. So the half-time for recovery from eye irritation is about 30 sec. A check of the clinical literature on caustic damage to the eyes and upper respiratory tract from intense prolonged exposures indicates that when a person develops more severe symptoms of significant eye inflammation or corneal damage without tissue destruction, it will take 24 to 48 hours for the symptoms to resolve with treatment (Isley and Lang 2007). We will assume that the half-time of repair for those inflammatory responses is approximately 8 hours beginning after the exposure stops. These values can be used to complete our model of damage and repair.

Behavior of the Ammonia Model

We are interested in the temporal variation in the response given an ammonia exposure. Ammonia air concentrations can be simulated by assuming a lognormal exposure distribution, such as GM = 25 ppm and GSD = 3.0, and drawing a series of random values to simulate the variation over time shown in Figure 11.2. This time profile of exposure can then be used with the difference equation in Figure 11.1 and the parameters developed earlier ($K_{dif\text{-}amm}$ = 0.198 cm^2/sec; A_{fluid}. = ~2 cm^2; $H_{amm/water}$ = 5.8×10^{-5} M/ppm; and K_{dam}. = 1 unit/μM); to estimate the stepwise eye concentrations and damage-repair responses at each time point (shown in Figure 11.2).

The time scale of Figure 11.2 is set to half of the shortest half-time for the effects, 15 seconds, and shows an hour of exposure with rapid variation in ammonia concentration, with several brief periods of high concentrations, >100 ppm. The time line for tissue concentration follows the air concentration with no lag,

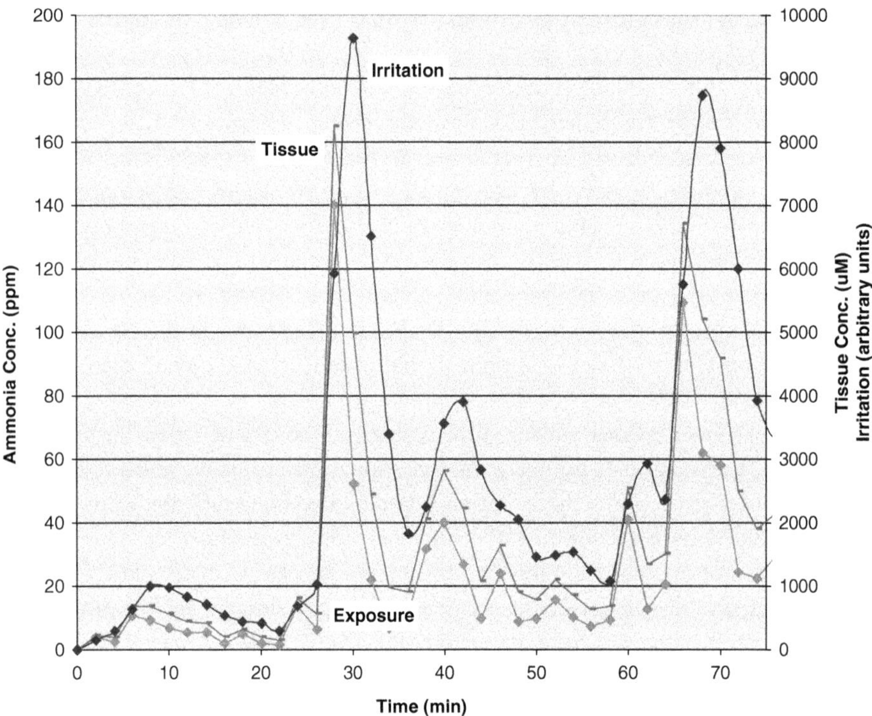

Figure 11.2 Temporal behavior of the ammonia concentration and sensory irritation in the eye from an airborne exposure using the difference equation model (Figure 11.1).

as expected from the rapid equilibration between the air and eyes. The eye concentration goes up and down rapidly, as does the irritation responses. The time line for irritation lags behind the exposure by 3–4 seconds and has a much smoother response, but it rapidly follows the short peaks of exposure, producing irritant responses in the 8,000 to nearly 10,000 range, which would be strong irritation.

Designing a Field Study of Eye Effects of Ammonia Exposure

Given an understanding of human responses to ammonia from the literature and a reasonable model of the damage-repair process, we could now use this information in an epidemiologic study. Let's assume that we wish to investigate complaints of eye irritation by residents of homes near a refrigeration plant that uses large quantities of ammonia in its operations. To design the exposure assessment, the investigator needs to determine the temporal pattern of the exposure and factors affecting exposure intensity. For example, presumably the ammonia emissions are being blown or drifting from the plant into the neighborhood where residents are reporting irritation. There are several things we would need to know: Are some emissions more concentrated or in a larger volume than others? Is there a continuous output, or is it intermittent? If the emissions are intermittent, can the emission times be identified? What are the prevailing wind patterns, and can real-time wind data be obtained from weather stations or nearby airports? Answers to some or all of those questions will guide the design of our study. Data from the subjects can also answer some of those questions. For example, a pilot survey of the neighborhood might ask such questions as who has experienced symptoms, what time of day, whether the symptoms vary in intensity, and whether wind direction affects the frequency or intensity of symptoms. If people are exposed repeatedly, do some notice that they become acclimatized and stop responding, that is, stop reporting irritation? Do they report eye inflammation, or have they had other, more serious responses, such as symptoms of upper respiratory tract irritation? In a real sense, we are using the reports of irritation and inflammation as biomarkers of exposure, in addition to broadening our understanding of the nature of the responses. Note that if we wish to model the inflammation as the response, we will have to change the parameters of the model, because the half-time of repair is much longer than that for irritation.

The exposure assessment must also be done on a time scale that is consistent with the time scale of the responses of interest. The etiologic time interval (ETI) for reversible proportional sensory responses is usually very short (see Chapter 6). For ammonia eye irritation, the nearly instantaneous diffusion of the compound into the target tissue means that the ETI is probably less than 1 minute. The rapid response suggests that ideally we would use continuous monitors positioned to measure instantaneous air concentrations near the subjects' eyes, so that we could calculate a 15-second average for association with reports or observations of eye irritation.

Next we need to decide how to measure the responses. First, there is no direct measurement for sensory irritation in humans. However, we can have them indicate whether they are experiencing irritation by having them keep a diary or press a button connected to a data logger that records a different voltage when the button is pressed. We could use a personal organizer or cell phone to remind the subjects when they need to note whether they have eye irritation (red eyes).

One concern is that if the exposure is sufficiently intense and sustained, it may produce a more prolonged response, such as inflammation, which could confuse the relationship with sensory irritation because the ETI for inflammation also produces irritation. We might have the subjects check for eye inflammation by giving them mirrors and asking them to observe whether their eyes are red at periodic intervals, especially after exposure stops. The apparent irritation associated with an inflammatory response will have a much longer recovery half-time than sensory irritation alone. Clearly, we must be precise in our choice of response and its association with the exposure. Multiple responses produced by a single exposure must be resolvable and not interdependent for our model of the damage-repair process to be appropriate. A poor fit with the data can indicate that our model is not appropriate. Remember: all models are provisional, and some will not be useful because they do not fit the data.

11.3. MODELING THE RESPIRATORY EFFECTS OF OZONE

Our next example of a reversible proportional response is the effect of ozone on pulmonary function seen in lab and epidemiologic studies. Ozone was discovered by Schonbein in 1851. He produced it with electrical discharges and also provided the first detailed description of the adverse effects of inhaling it. Ozone has been formally studied in both human and animal laboratory studies since the early 1900s. A 1939 literature review (Witheridge and Yaglou 1939) reported on data collected from human volunteers who were exposed to low levels of ozone to define the odor threshold (~0.01 ppm) and to define exposure levels "objectionable" to everyone exposed (~0.1 ppm). The latter level was also chosen as the American Conference of Governmental Industrial Hygienists (ACGIH) Threshold Limit Value in 1961 because it appeared to be tolerable for an entire day. Interestingly, these early human tests clearly showed that airborne ozone did not remove odors but deadened the sense of smell through a neurotoxic effect. Unfortunately, ozone generators are still being sold as odor removal devices.

Ozone Experimental and Epidemiologic Studies: A Short History

The history of ozone research is unusual because laboratory studies of humans preceded the epidemiologic studies, which did not begin until the 1950s (Environmental Protection Agency [EPA] 1996). Consistent epidemiologic evidence of acute eye and respiratory symptoms associated with ozone and oxidant pollutants in photochemical smog exposures was found beginning in the 1950s. Numerous studies of human responses in lab studies were conducted through the 1990s. Some investigators also found associations with pulmonary function decrements, although these effects were inconsistent across studies. All of the pulmonary function effects from exposures of a few hours were reversible within 24 hours. Epidemiologic studies of chronic effects through the 1970s showed little clear evidence of long-term effects, partly because of methodologic limitations and the relatively limited long-term duration of intense community-wide smog with high ozone levels.

One of the few laboratory studies to expose humans for more than a few days was conducted by Bennet and coworkers in 1962 (Bennet 1962). These human tests were performed by Air Force researchers because measurements in high altitude aircraft had found relatively high ozone levels, which raised concern about wartime exposures, when airmen might be flying large numbers of high-altitude missions. The goal of Bennet's study was to determine whether long-term exposure to ozone would harm the lungs of the airmen. Panels of 6 healthy males were exposed to 0.2 or 0.5 ppm ozone in the lab 3 hours per day, 6 days per week for 12 weeks. No effects on FEV_1 or upper respiratory infections were seen at 0.2 ppm exposure. However, slowly developing decrements in FEV_1 were observed at 0.5 ppm exposure. At this level, the investigators observed a significant decline in FEV_1 to 80% of its baseline level after 10 weeks of exposure. When the exposures stopped, the effects slowly dissipated, and by 6 weeks postexposure they were undetectable.

Some important complexities have been observed in the biological effects of ozone. First, effects on FEV_1 have been produced by short-term exposures (a few hours per day). However, after a few days of repeated short-term exposures, the magnitude of the response declines to near normal, a result that was taken as evidence of adaptation or acclimatization. Second, Bennet (1962) found that longer term repeated exposures (3 hours per day for 6 days/week, for 12 weeks) cause steady declines in FEV_1 with no evidence of adaptation, followed by a slow recovery postexposure function. The latter time course of effects is consistent with epidemiologic studies that have found reduced average FEV_1 in residents of Los Angeles at the end of a summer of high ozone, followed by slow recovery to normal airflow by the spring of the next year (Linn, 1988). Thus there appear to be effects occurring on two different time scales: brief exposures of 1–2 hours per day produce apparent adaptation, and longer exposures > 2 hours per day do not. This represents a significant disease process (DP) modeling challenge.

Modeling Ozone's Responses

Empirical Models

The earliest lab studies of ozone looked for empirical mathematical dose functions based on Haber's Law, which states that various combinations of intensity multiplied by duration that produce the same value ($C \times T = k$) will produce the same level of effects from a gas inhalation. Therefore, $C \times T$ can be used as the dose metric, such as ppm × min for ozone. The rationale was that the same amount of agent would be inhaled if everyone inhaled at about the same rate. However, for most of the short-term exposure experiments, the subjects were exercising at various levels, and exercise increases the amount of ozone entering the respiratory tract per unit of time by increasing pulmonary ventilation (Q). Empirical models developed by the investigators added ventilation to the dose metric, $C \times T \times Q$, also called the "effective dose," which improved the fit. Note that effective dose has some implicit assumptions: the same fraction of inhaled ozone is absorbed at the target locations in the lungs, and those are unaffected by C, T, or Q. Further investigations showed that nonlinear models actually fit better, such as $C^n \times T^m \times Q$ (Gerrity and McDonnell, 1986).

In the 1990s, investigators at the U.S. EPA conducted a set of lab experiments exposing subjects to a series of fixed concentrations applied for varying lengths of time (EPA 1996). Studies with exposures less than 3 hours found response curves that were concave upward, which was interpreted as an increase in the relative strength of the response at higher exposures, but studies lasting longer than 3 hours observed instead that there seemed to be a leveling off of the response for each concentration, an apparent reduction in the response per unit dose ($C \times T \times Q$).

Analyzing the temporal data more carefully, McDonnell and Smith (1994) found that a sigmoid curve fit best for a range of time points across several sets of exposure experiments. They observed that the response at different time points after the initiation of exposure grew slowly at first, then steeply, and finally tended to level off as time progressed. Although the investigators obtained excellent empirical fits with a sigmoid curve and defined the curve as a dose metric, in fact the curve is tracking the development of the effect over time for a given exposure level. As with all empirical models, this dose metric will produce good predictions under the conditions for which it was developed: starting with no exposure and no prior effect, using a steady exposure level, and measuring the effect at the end of the exposure period. If any of these conditions are not met, then the model will not give good predictions. The sigmoid response curve shows that the effects do not develop in a simple linear growth curve, nor do they start rapidly and then slow down and plateau, as is expected for a one-compartment model. The development is more complex. The initial cell damage does not immediately translate into stimulation of irritant receptors.

Time as a Dose Variable

The dose analysis using $C \times T \times Q$ as the dose metric contains an implicit assumption that T is important only for the magnitude of the administered dose. However, T is also important for the development of the effect. When T is a significant fraction (> 40%) of the ETI for a reversible effect, then the response will plateau as T increases because the effect is reaching steady state. This is clearly shown in Figure 2A of the 1994 McDonnell and Smith paper, which maps the change in response over time as a function of different exposure intensities (McDonnell and Smith 1994). Different concentrations gave different plateau heights but similar time courses—just what we would expect for a reversible damage-repair process that has run long enough to reach steady state.

Similar problems have been seen in a variety of other studies of toxic effects, in which attempts were made to rationalize and reformulate Haber's Law

($C \times T$ = constant) to form a universal dose metric (Miller, Schlosser, and Janszen 2000; Shusterman, Matovinovic, and Salmon 2006). A problem arises because duration of exposure (time) is treated as an independent experimental variable contributing to the total applied dose. However, time is also the framework within which causal processes occur. An agent at some concentration enters a biological system and, while we watch, causes what we observe as an outcome or response as the result of damage and repair processes. Time alone does not cause anything; it is our observational framework. Haber's Law can apply only when the delivery time is small relative to the kinetics of the agent and its effects (Rozman 2000). In environmental studies, the simple dose-response concept in which the dose is just given at a point in time is rarely appropriate. Usually the agent is being delivered concurrently with the response. Thus a process model approach is more biologically appropriate than an empirical statistical analysis.

Dose Model to Estimate Tissue Ozone

The target tissue for ozone is the small airways of the lungs, which have the largest surface area and slowest air flow rate of all airways (Kriebel and Smith 1990). Animal studies also have observed that ozone's effects are predominantly seen in the small airways. The amount of ozone deposited in the small airways is proportional to the air concentration. Ozone is highly reactive and does not require metabolic activation, but it is partially metabolically deactivated by superoxide dismutase (SOD), glutathione peroxidase (GPx), and catalase (CAT) (EPA 1996). This simple view suggests that there are four physiologic factors that affect the tissue dose: the pulmonary ventilation rate, the air concentration, the area of the small airways in which the ozone is absorbed, and metabolic deactivation. The area of the small airways is roughly proportional to body size, as is baseline ventilation. The ventilation rate is affected by physical activity, so we must verify that all subjects have about the same level of activity, or we must estimate their activity levels. A simple one-compartment model of the small airways should be adequate for the dosimetry, as shown:

$$C_{area}[t+\Delta t] = C_{area}[t](1 - K_{remv}\Delta t) + \frac{Q_{vent} f_{rem} \Delta t C_{air}[\Delta t]}{A_{resp}}$$

Equation 11.1 Concentration of ozone in the terminal airways tissue

Where C_{area} is the ozone concentration in the lung, Q_{vent} is the minute ventilation, f_{rem} is the fraction absorbed from the air, and A_{resp} is the area of the target tissue. Two of the parameters, Q_{vent} and A_{resp}, may need to be adjusted depending on the individual. K_{remv} represents the removal of ozone from the tissues by all mechanisms, including reactions with membrane lipids and glutathione that may become depleted and catalytic removal by SOD, GPx, and CAT that may vary with genetic factors. If we can identify that there is primarily one enzyme, for example, SOD, associated with ozone removal and there is information on the differences in functional abilities of genetically different forms, single nucleotide polymorphisms (SNPs) for SOD, then we can modify the K_{remv}. Q_{vent} will vary with physical activity by the subject. By itself, the tissue concentration model does not help us estimate dose because we do not know what the etiologic time should be.

Alternative Tissue Response DP Models

Kriebel and Smith (1990) developed a model of the initial responses to ozone based on the findings of the human exposure chamber studies and those of animal studies. Box 11.1 reviews the structure of this model. This model was developed from the literature on lab studies of ozone's effects on the respiratory epithelium of test animals. The amount of ozone in the tissue is a simple one-compartment model, shown earlier in Equation 11.1. The key descriptive steps from the literature were: (1) ozone damage to the airway epithelium is rapid (ozone forms

Box 11.1 Kriebel and Smith (1990) Damage-repair Model of Ozone Effects

This model was based on the earlier work of McDonnell, who studied effects of short ozone exposures on pulmonary function indicated by declines in FEV_1 during exposure. The descriptive model is shown in the diagram.

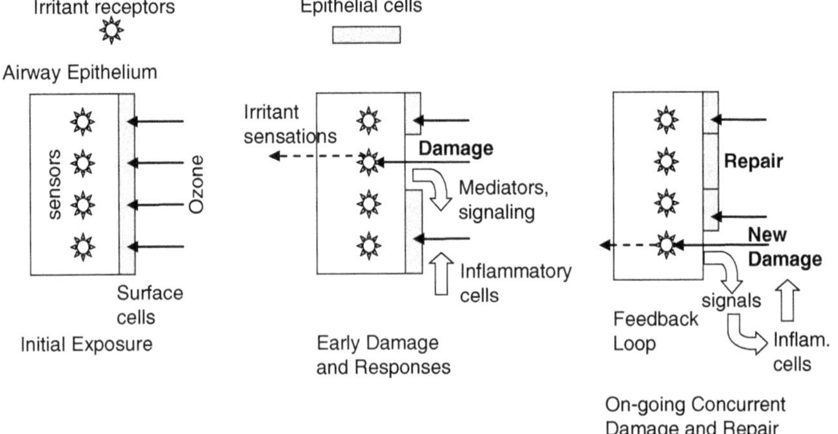

Different areas are independently undergoing damage and repair at the same time. There are two processes: the initial cell damage and the inflammation produced by recruitment of inflammatory cells. These two processes act as a positive feedback loop, stimulating further damage and responses. When exposure stops, then new cell damage stops, signaling goes down, and repair removes the damage. Increased exposure will increase the prevalence of damage.

membrane-bound unsaturated fatty acid peroxides) and allows ozone access to subsurface receptors; (2) the damaged epithelium releases inflammatory mediators and signaling compounds (relatively slower); (3) ozone reaching the receptors produces responses (fast) that cause bronchospasm, which reduced FEV_1 in sensitive individuals; (4) although there is damage, mediators and cell signaling produce recruitment of inflammatory cells and increased inflammatory responses (relatively slow). If exposure is high enough, severe damage can result from a feedback or circular process: damage → signals → inflammation → damage, and so forth—that amplifies the response. The model has two components, a fast receptor response and a relatively slow inflammatory response, similar to what we saw with the ammonia example.

The model worked fairly well for predicting the airway effects of short-term acute exposures administered in human exposure experiments. Figure 11.3, reprinted from the original Kriebel and Smith paper, shows that the predicted values of a standard measure of airway resistance, sR_{aw}, agreed well ($r = 0.75$) with the mean observed values from 36 human exposure experiments in seven different published studies. However, this model did not capture the attenuation or acclimatization to ozone exposures over several days seen with repeated daily

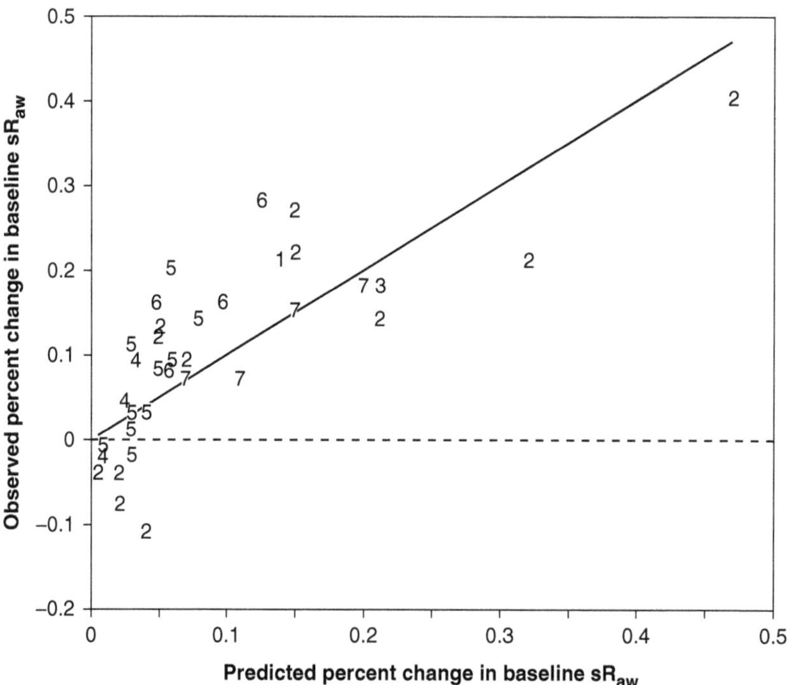

Figure 11.3 Results of Kriebel and Smith model for ozone effects compared with observed airway resistance in 36 human exposures in 7 studies.
(Reprinted with permission from Kriebel and Smith, 1990.)

exposures described previously. Figure 11.4A shows the observed responses (stars) from a protocol of daily 2-hour exposures to 3.8 ppm repeated on 5 days in a row (Devlin et al. 1996).

Attenuation or acclimatization implies that there is a secondary, slowly developing process that blocks or damps out the initial acute sensor responses affecting decline in FEV_1 (ΔFEV_1). This blocking process could be related to any of several factors: a reduction in sensor sensitivity; edema from the inflammatory effects; or

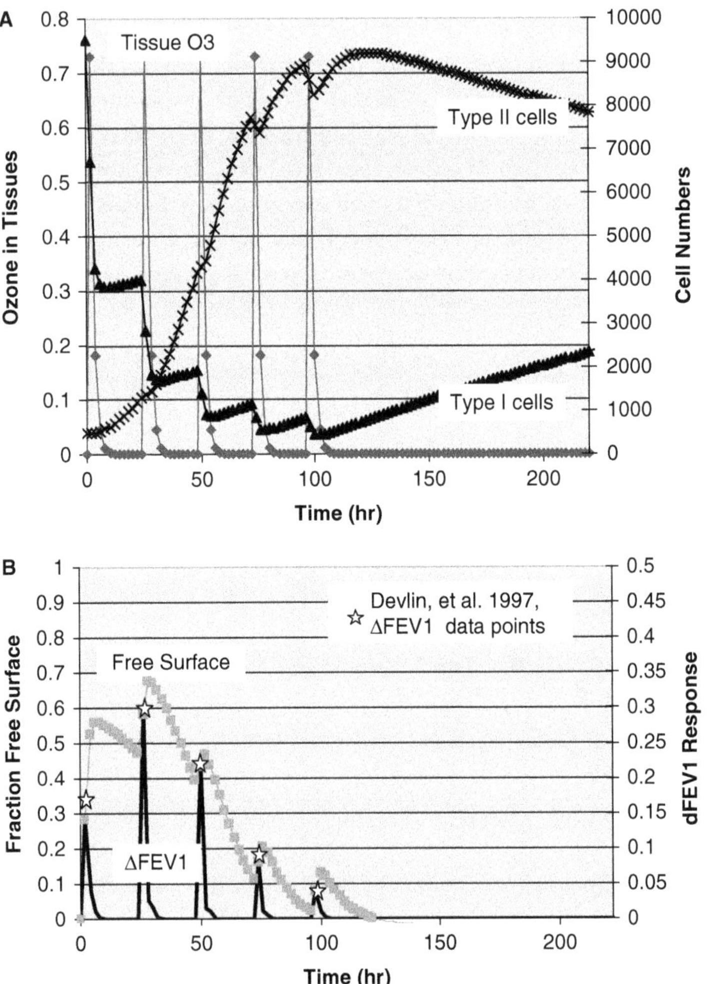

Figure 11.4 Box 11.3 simulation of a laboratory study of repeated 2-hr exposures to 3.8 ppm of ozone, and the Devlin et al. (1993) data points. Part A: Model estimated numbers of Type I and II cells in response to the exposures. Part B: Modeled FEV_1 responses with observed ΔFEV_1 points and modified Freijer model estimated free lung surface.

recruitment of new cells into the area. Given data on the time course, such as in Figure 11.4A, we can define the temporal characteristics of the blocking process, even without knowing what the process is.

Freijer and coworkers developed an new model using animal data to identify a cellular response involving changes in the types of epithelial cells (Freijer, van Eijkeren, and van Bree 2002). Based on laboratory data, they hypothesized that there were two different cell populations, one sensitive and one resistant, and that the relative surface coverage between them could explain the longer term changes in responsiveness to ozone that have been observed in repeated exposure studies. Box 11.2 diagrams their model.

The Freijer and coworkers (2002) model used a similar mathematical approach to that of Kriebel and Smith (1990), but also incorporated the laboratory finding that sensitive Type I airway cells were killed by ozone and replaced by resistant Type II cells. Type II cells also replicate faster than Type I cells. This shift in cell types means that soon after the onset of exposure, ozone starts killing the Type I cells, which exposes the irritant receptors to ozone, so there is a strong early response. Then the Type II cells start growing to cover and block access to the irritant receptors, reducing the response (as diagrammed in Box 11.2). Lab studies showed that there was a lag of approximately 24 hours before the killed cells are

Box 11.2 Damage-repair Model of Freijer and Coworkers

This model is similar to the Kriebel-Smith model, but it has two types of cells with different sensitivity to ozone, and the FEV_1 response is proportional to the amount of open surface area that exposes underlying irritant sensors. The ozone quickly kills sensitive surface cells. More resistant cells grow in to replace them, which changes the responsiveness of the system. After exposure stops, the sensitive cells regrow and replace the resistant ones. The FEV_1 response was assumed mediated by the exposed receptors, which are quickly covered when the exposure stops.

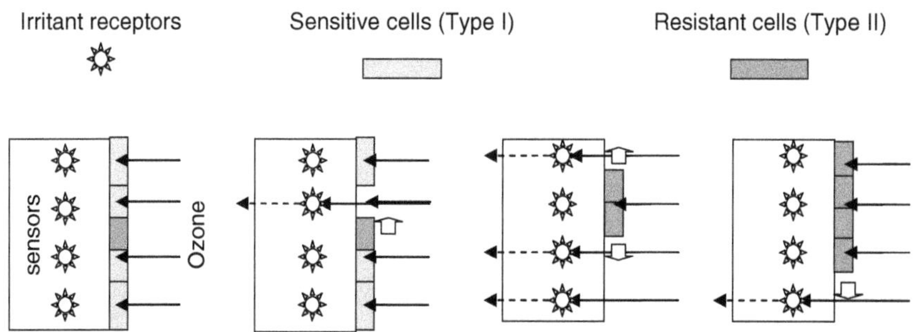

Diagram of the surface airway tissues in the terminal bronchioles and the effects of ozone on the cells. Starting on the left, initial exposure to ozone kills cells to expose receptors, then resistant cells cover the receptors, reducing the response. After exposure stops, Type I cells regrow, displacing Type II cells.

replaced by resistant cells (Type II), which Freijer modeled as a 24-hour offset for the Type II response. When they applied their model to the available data from a broad range of lab experiments with acute exposures, they found a very high degree of fit ($R^2 = 0.95$).

One check on this type of mathematical model is to look at the steady state that it settles into when exposure conditions are fixed. The original Freijer model did not include a baseline value for Type II cells, the source of replacement cells for ozone sensitivity in Type I cells, which is needed for biological plausibility. They also wished to avoid either cell type overgrowing the other when exposure stops, which was not inherent in their model but was achieved by requiring that the sum of the two types of cells do not exceed N_0 a reasonable mathematical solution, but not biologically meaningful. When exposure stops, the longer lived Type I cells gradually replace the Type II cells because the latter leave open surface during normal turnover that can be occupied by Type I cells. This type of biological competition among cell types is common when there is damage and cell injury.

In comparing the Freijer model to the Kriebel and Smith model, we see that they represent different cellular processes and time scales. The Kriebel and Smith model was developed to represent short, single exposures on the scale of hours. The Freijer model was developed to represent multiple exposures over several days. Within their designated time scales, they both work well. The Freijer model focuses on a different dimension of the biological response, the cell kinetics, and drops the short time scale inflammatory response. This results in a more useful model for longer time scales because it covers both single and multiple exposures. This is more useful because it correctly predicts the response to ozone over a wider range of lab exposure patterns and durations. However, neither model is "true," because neither covers the full complexity of the biological response. Recall that our goal is not to represent the full complexity; it is to develop a useful model for the time scale of the particular response we are studying. We want a model that has complexity that is "just right" to fit the time course of the specific response under investigation the Goldilocks Principle. And if we shift to a different response or effect, we may need a different model. There is no model that fits every response.

Fitting Models from Chamber Studies to Epidemiologic Data

All of the human lab studies conducted by the EPA and other investigators noted earlier hoped to predict human responses to air pollution using the inhaled concentration × duration × ventilation ($C \times T \times Q$) as the dose metric. However, the epidemiologic findings in air pollution studies are not consistent with the simple dose-response relationships seen in the laboratory. The human chamber studies did not see FEV_1 effects below 0.08 ppm. More important, with repeated exposures they found that the FEV_1 effects of ozone appeared to dissipate after about a week of repeated environmental exposures. In air pollution exposures, they do not. One possible reason is that people with air pollution exposures do not have short periods of high exposure separated by 18- to 22-hour periods without exposure, as is the case in the chamber studies. Instead, the exposures are continuous but variable over 24 hours each day in the summertime. In that context, what is the right $C \times T \times Q$ dose metric? Should it be based on 24-hour exposure or some other time period? There is no obvious answer to this question coming from the empirical lab studies.

How does the Freijer model behave when it is applied to ambient patterns of continuous ozone exposure? Ozone exposures in Los Angeles show a diurnal cycle, with the highest one 1-hour values in the summer: 0.04 to 0.09 ppm. This range is high for environmental exposures in the United States but low for the chamber study exposures, which were 0.08–0.4 ppm for up to 6 hours in the McDonnell studies and as high as 4 ppm for some of the other human studies (EPA 1996). Los Angeles ozone concentrations typically increase from the morning to afternoon on sunny days and decline during the evening and nighttime hours. With such a pattern of continuous ozone exposure, the original Freijer model predicts that even at 0.04 ppm, only 1% of the Type I cells will remain, and the Type II cells will expand to cover nearly all of the epithelial surface, blocking all responses. This is implicit in the structure of the original Freijer model.

This is a good example of an implicit limitation in a model developed for a specific set of conditions. This type of limitation may not become apparent until the model is applied in other circumstances. Freijer and colleagues' goal was to fit the available data, and they were very successful. However, it still may not be useful if extrapolated outside the data used to develop and test it. Without considering the nature of the exposures from the epidemiologic setting, it is difficult to determine whether the model will be relevant. But as we have said before, the model now projects how the biological system will behave in epidemiologic settings, and therefore it is testable. The model is a hypothesis for the new conditions—it may or may not be useful.

Modified Freijer Model

We can modify the Freijer model in several ways to make it more compatible with the biology and longer term exposures. First, a baseline level of Type II cells needs to be set: 5% worked well. If we do this, we do not need to set the 24-hour lag, because there is a natural lag while the small number of Type II cells starts growing (see Figure 11.4A). Second, the growth behavior of Type I and II cells is fundamentally different: Type I cells are quickly killed by ozone; $N_1[i]$ drops down from its baseline N_{01} and then grows back to N_{01} when exposure stops. The model of

Type I growth makes the rate proportional to the distance from the baseline, $K_{n1}(N_{01} - N_1[i])$. Whereas the Type II cells need to respond when there is open surface and quickly grow the small number of cells until they cover the open surface, these cells need a different type of growth control. This can be achieved by hypothesizing that the rate of Type II cell growth is a response to the amount of damage as indicated by the fraction of open surface (f_{os}); then the rate is $f_{os}K_{n2}N_2[i]$, where $f_{os} = (N_0 - N_1[i] - N_2[i])/N_0$. This also will limit the amount that the cell types can overgrow each other; a small amount of overgrowth is biologically reasonable. Third, Type II cells can be given some susceptibility to ozone—for example, K_{dam2} can be set to 0.06 per hr per ppm × hr ozone—then, while there is exposure, some surface will always remain open, as shown in Figure 11.4B. The modified model will use the same simple response-recovery model for the ΔFEV_1 response that was used by both Kriebel and Smith (1990) and Freijer et al. (2002).

Box 11.3 shows the resulting modified Freijer model. The proof of the model's utility is its fit with the data. A reasonable fit can be achieved using parameter values within the literature identified range collected by Freijer for the Type I cells, and choosing a higher growth rate and lower sensitivity to ozone for Type II cells. As shown in Figure 11.4, the modified model's findings are a reasonable match for Devlin and coworkers' (1996) data. Freijer's relative change in FEV_1 response for ozone, K_{resp} = 0.8 per hr per ppm × hr, was set based on fitting the laboratory data. We would expect that the modified model would fit the lab data just as well as the original Freijer model did, but what does it do with a long-term lower environmental exposure?

The long-term epidemiologic data for Los Angeles show a seasonal pattern (EPA 1996). High ozone exposures in the summer lead to lower population levels of pulmonary function (FEV_1). During the low exposures in the winter, there is a slow recovery from effects caused by the summer exposures. This is analogous to the findings of Bennet's lab study of long-term effects. We can simulate a 20-day period of 24-hour cyclic exposure and see how the modified Freijer model responds, which is shown in Figure 11.5. First, there is an interesting pattern of cell type changes, suggesting a period of heightened sensitivity when the ozone exposure starts again in the early summer, shown in Figure 11.5B. This sensitivity passes when the fraction of sensitive Type I cells is substantially reduced to approximately 30%, and the amount of open surface is reduced by the growth of Type II cells (Figure 11.5A). Second and more important, the model-predicted FEV_1 response to ambient ozone levels is very small; they will not produce detectable FEV_1 effects (Figure 11.5B). The predicted FEV_1 reductions are less than 0.1%, which is undetectable. Because K_{resp} in the model (Box 11.3) is a proportionality factor, we could adjust it upward to make the model match the observed response level for a given level of actual ozone exposure. However, this change shows that the continuous exposures produce a heightened responsiveness compared with the

Box 11.3 Model of Cell Kinetics and FEV_1 Response: Modified Freijer Model

Model of Cell Kinetics: Two difference equations, based on the Freijer model, representing the two cell types are shown in the following diagrams. Simple first-order kinetics are assumed, such as we have used for the ammonia model. The formation of new cells of both types depends on the amount of open surface, N_0 - (N_1+N_2), where N_0 represents the number of cells needed for full coverage.

Type I Cells: K_{n1} is the new cell rate per minute; K_{dam1} is the cells killed per ppm ozone per hour.

$$N_1[i+\Delta t] = N_1[i](1-K_{dam1}\Delta t C_{oz}[i]) + K_{n1}\Delta t(N_{01} - N_1[i])$$

Type II cells, ~500, which grow into open surface. K_{n2} is the new cell rate per hr with a multiplier sensitive to free surface; K_{dam2} is the cells killed per ppm ozone per hr.

$$N_2[i+\Delta t] = N_2[i](1-K_{dam2}\Delta t\, C_{oz}[i]) + K_{n2}\left(\tfrac{(N_0-N_1[i]-N_2[i])}{N_0}\right)\Delta t\, N_2[i]$$

Parameter values: A variety of values were manually explored to fit the lab and epidemiologic findings simultaneously, without formal fitting. Total cells N_0 = 10,000.

		Type I	Type II
Number of starting cells	N_{0i}	9,500	500
Cell growth rate, per hr	K_{ni}	0.002	0.1*
Cell death rate, per ppm per hr	K_{dami}	0.4	0.06

* The K_{n2} growth rate is a response to cell injury or death; this is not the normal turnover of uninjured cells, which is expected to be slower.

Model of FEV_1 Response: The receptor response is assumed to be a simple response-recovery type of system, which is analogous to the damage-repair model. The response is proportional to the fraction of uncovered surface, N_0-N_1-N_2, multiplied by the ozone concentration and the recovery has a half-time of ~2 hr.

FEV_1 Deficit: K_{resp}, 0.8, is the response rate per ppm ozone per hour; K_{rec} 0.2 is the recovery rate per hour; and $C_{oz}[i]$ is the tissue ozone concentration.

$$D_{FEV1}[i+1] = D_{FEV1}[i](1-K_{rec}\Delta t) + K_{resp}\Delta t C_{oz}[i]\underbrace{(N_0 - N_1[i] - N_2[i])}_{\text{Open surface}}$$

Spreadsheet Model: A block of cells define the model parameters as noted above, plus the exposure distribution (GM, GSD) and Δt values. Six columns are defined:

- **A** - Time ($i*\Delta t$);
- **B** - Exposure ($C_{exp}[i]$) fixed values or from a random number generator using GM, GSD. A diurnal variation in exposure can be added using a sine function as a multiplier with a 24-hr cycle and a baseline offset;
- **C** - Tissue ozone concentration ($C_{oz}[i]$) from the one-compartment model;
- **D** - Type I cells calculated from the difference equation above using $C_{oz}[i]$ and parameters;
- **E** - Type II cells difference equation above using $C_{oz}[i]$ and parameters;
- **F** - FEV_1 deficit ($D_{FEV1}[i]$) calculated from the difference equation above and $C_{oz}[i]$ and K_{rec} and K_{resp} parameters.

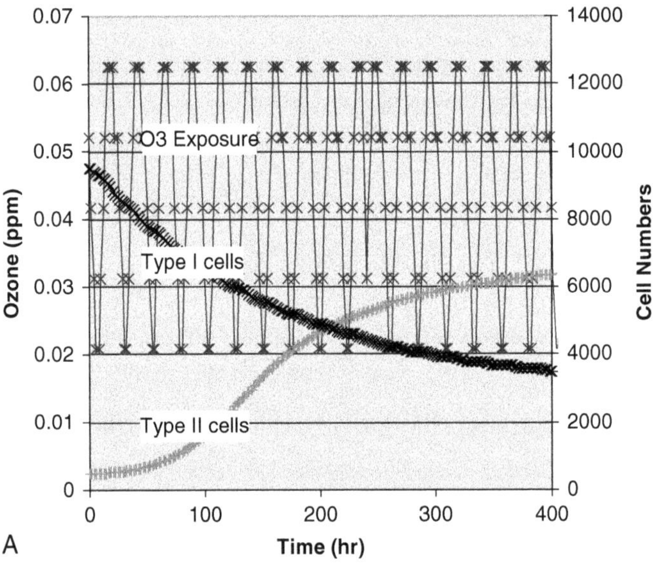

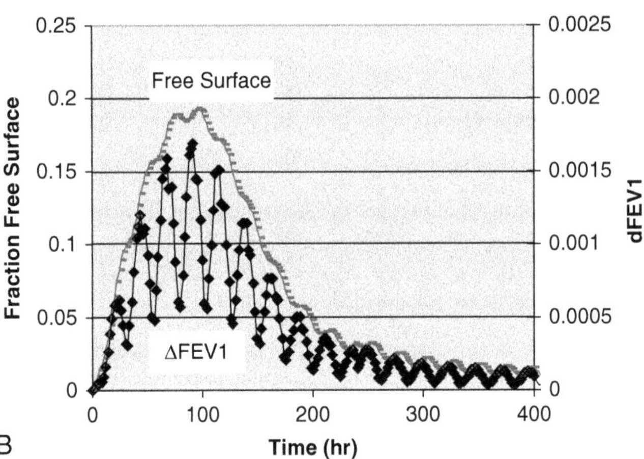

Figure 11.5 Simulated environmental ozone exposure application of the modified Freijer model (see text) to the exposure time profile using the same parameters as in Figure 11.4. Part A: Model estimated numbers of Type I and II cells in response to the exposures. Part B: FEV_1 responses and modified Freijer model estimated free lung surface.

lab studies, which also suggests that there may be additional pathological processes operating on a longer time scale that are not seen in the lab studies.

Etiologic Time Interval

The ETI for this system has two dimensions—a long time dimension for the cell kinetics and a short one for the irritant response—and both must be considered. The epithelial cell kinetics define the amount of open surface that will allow ozone to reach the irritant sensors. The acute decline in FEV_1 responds to the current tissue concentration of ozone, which varies on an hourly time scale. Therefore the long-term mean ozone defines the amount of open surface and the hourly ozone defines the immediate reduction in FEV_1. We can see both of these effects in Figure 11.5. In Figure 11.5A we see the slow changing of numbers of different cell types, and in Figure 11.5B we see how the amount of open surface and FEV_1 response have begun to stabilize after about 20 days. The 10-day period immediately after exposure began had the largest amount of open surface because many of the Type I cells were killed off and the Type II cells had not yet grown up to cover most of the open area. With continuing exposure and no long periods without exposure, after about five half-lives of Type I growth (70 days), the levels of cells will stabilize, as does the amount of open surface; higher ozone levels leave more surface open. Therefore the ETI_{os} for open surface is ~70 days. The ETI_{FEV1} for FEV1 response is five times the half-time for recovery (7 hr) for the FEV_1 response, which is ~35 hr. Because the open surface area changes relatively slowly compared with the FEV_1 response, we only need to estimate f_{os} at a time point to calculate the response.

Dose Metric for Ozone's Pulmonary Effects

A dose metric (DM) for ozone effects on FEV_1 can be based on the modified Freijer model. The DM_{FEV1} uses the assumption that ozone tissue concentration in the small airways is directly proportional to the inhaled concentration as long as protective enzymes are not saturating and antioxidants are not being depleted. The dose model has two steps: calculating the open surface area and calculating the weighted response to hourly concentration of ozone. First, we need the ozone time profile for the previous 70 days weekly averages are suitable to calculate the fraction of open surface. If, during that time, there have been some periods of very low or no exposure lasting at least two half-times of Type I cell growth, 4 weeks, so that the Type I cells could expand and increase the epithelial sensitivity to ozone, then the spreadsheet model in Box 11.3 should be used to calculate the open surface.

The DM_{FEV1} is defined in the following equations, and the parameters are taken from the model in Box 11.3.

$$DM_{FEV1} = \sum_{i}^{N_{ETI-FEV1}} C_{oz}[i] f_{os}[i] (1 - K_{rec}\Delta t)^{i-1}$$
$$\textit{where } f_{os} = 1 - f_1[i] - f_2[i] \quad \textit{and} \quad N_{ETI-FEV1} = ETI_{FEV1} / \Delta t$$

Equation 11.2 Dose metric for ΔFEV_1 responses to ozone

If the ozone has been relatively stable over the previous 70 days, then amount of open surface can be calculated with the overall average ozone using the model parameters and the following steady-state equation:

$$f_{1,ss} = \frac{K_{n1}}{K_{n1} + K_{dam1}\bar{C}_{oz}} \quad \textit{and} \quad f_{2,ss} = 1 - f_{1,ss} - \frac{K_{dam2}\bar{C}_{oz}}{K_{n2}}$$

Equation 11.3 Estimators for steady-state fractions of Types I and II cells

In an epidemiologic application, it is rare to have the personal ozone time profiles for each of the subjects. Alternatively, we have residential groups in relatively small areas in which it is reasonable to assume that the subjects have all experienced the same ozone time profiles. This greatly simplifies the DM_{FEV1} calculations and matches them to the time when the subjects' ΔFEV_1s were measured.

You could justifiably ask the question: Do we really need all this complexity to represent the FEV_1 effects of ozone? The answer is yes, with the simplification we observed when the ozone levels are relatively stable. However, we would not be aware of the potential simplification if we had not developed the modified model. The fundamental problem is that the biology is complex, and that has implications for how we need to formulate the dose metric.

11.4. DISCUSSION OF REVERSIBLE PROPORTIONAL RESPONSES

Reversible proportional responses are the result of tissue damage processes that cause injury and repair processes that fix the damage. Unless the damage is severe, we assume that damage and repair operate concurrently and independently. These processes operating together produce two observable effects: (1) the level of damage will plateau with a fixed level of exposure, and (2) when the exposure stops, the damage will decline as it is repaired. The level of the plateau depends on the intensity of exposure, but once achieved it becomes independent of duration. Sustained damage can lead to irreversible effects by other mechanisms and is common when damage is significant. Therefore, we have to be alert to the possibility that reversible damage may also be the first step in an irreversible process when the damage is intense.

Reversible proportional responses apply to tissue damage-repair situations and to sensory response-recovery processes. These are very different mechanisms, but they have the same type of temporal behavior. Sensory responses are much faster than cellular damage, as we saw in the ammonia example. The same disease process model and dose metric will apply to both. However, the more complex cellular response caused by ozone does not fit the simple model because two types of cells are affected. Further examination shows that on a long time scale the overall behavior affecting FEV_1 fits a simple damage-repair model.

When a disease process model is fit to data from a group of individuals, the structure of the model is assumed to be the same for each subject, but the parameters will vary across subjects. In the case of the reversible proportional model, the amount of damage at the steady-state plateau and the time required for damage to decline by half after exposure stops (half-time of repair) will not be the same for each individual. Considering the ammonia example presented earlier, we would not expect each subject to stabilize at the same level of irritation for a given exposure because of interindividual differences in the intensity of irritation produced per unit of exposure, although we would expect a plateau to exist for all subjects.

What are the implications of this variation for an epidemiologic study of irritation by ammonia in an exposed population? Consider some examples. Suppose that we divide a large population into subgroups using a crude exposure measure, such as potential contact with ammonia (yes or no), and we examine the intensity of symptoms reported by people in each of the two subgroups. If the exposed group has high potential for contact with intense ammonia concentrations and the group with no ammonia exposure has no contact with other airborne irritants, we probably will see a large difference in the intensity of irritant symptoms between the groups. However, if we find that the mean exposure differs only between none and a "low" level, then the concentration may not exceed the threshold for many of the subjects, and we might observe only a small difference in symptom intensity between the two groups, or possibly only a small average difference that we could not confidently say was not due to chance. This is a common scenario in many epidemiologic studies. At relatively low exposures, the fraction of the population that reports low-intensity symptoms is determined by the intersection or product of the right tail of the exposure distribution and the upper tail of the population sensitivity distribution. That is, those who are responding are those who are both more sensitive *and* more heavily exposed. The responders are typically not just the most sensitive or the most highly exposed but both.

This model clearly shows that we must consider the biology of the effect when we design the measurement of exposures. There are three points to note. First, air concentration variations shorter than the half-time of response, a fraction of a second, will not affect the eye concentration unless they are very large. Second, if we measure the air concentration with an averaging time much longer than the response time, we may miss peaks that will cause a response. Third, the ozone example of cell kinetic effects shows the importance of exposure history for the individuals' responsiveness, especially periods of no exposure, which may increase populations of sensitive cells. Exposure monitoring must be matched to the temporal behavior of the effect.

This book's basic premise is that informative models of biological processes can be built around the temporal behavior of the limiting processes. If we had more data, for example on the temporal variations in signaling proteins or genetic controls, we might further increase the complexity of the model and be able to test additional aspects of the processes. But there is no real gain from adding complexity that cannot be tested. Remember, too, that biological processes operating on time scales shorter than those on which the epidemiologic outcomes are measured are not likely to be detectable. During the process of developing a model, we will often rule out biologically plausible processes because they cannot fit the observed time course. For example, the Kriebel and Smith (1990) model for ozone was satisfactory for a single exposure but did not capture the dynamics of responses to multiple exposures over a period of days. The power of the DP model is that it can provide an incremental way to move our understanding ahead without having to have the full mechanism before we can make progress. We can make reasonable inferences that are testable and that accommodate the growth in our understanding of mechanistic processes.

Skeptics about this modeling approach will often complain that a complex model has so many parameters that it can fit anything. This criticism is often made by statisticians, who are thinking of an unconstrained empirical model. Consider the example of the modified Freijer model for ozone, Box 11.3, in which there are four parameters for the two cell types and two for the response and a relatively complex model. This criticism ignores the model's many constraints. First, the model's structure is not arbitrary; it reflects a biologically plausible pathophysiologic process based on the available data. Second, the kinetics of cell turnover are known from in vitro studies of epithelial cell kinetics, so they cannot be set to any values. Third, the human acclimatization behavior has been characterized in laboratory studies, and only a limited combination of parameters will correctly capture this complexity of the response. Fourth, the cell-killing effects of ozone have been characterized in vitro, and these data should be used to define the range of plausible damage rates. These constraints exert powerful limiting effects on the structure of the model and its parameters; they cannot take any arbitrary values. Finally, a disease process model is not acceptable if it is not consistent with *all* of the preexisting data on the processes.

We can learn about the nature of the effect by noticing when there are systematic changes in the fit of dose metric (DM) with changes in exposure features. If the fit changes with different exposure profiles, then the dose metric provides a means to examine how the biological processes are changing. For example, the response ratio per unit of exposure may not be constant across the range of exposure intensities. Epidemiologists have long recognized that short, intense exposures may produce more adverse effects than long-duration low exposures. This effect can be explored by observing that low-level DMs produce little or no response and that intense exposures produce large response ratios. Intense exposures may affect the repair process, causing the recovery half-time to increase because of slowed recovery, so the estimated K_{rep} values decrease (less repair per unit of time). We can learn a lot about the nature of the response by looking for heterogeneity of responses relative to the exposure intensities. This is implicit in the biologically based dose metric, even for the simple one-step model for linked damage-repair processes.

In some cases, the actual mechanism causing the response is obscure and poorly understood or the result of a complex and shifting mixture of different agents, as in the case of respiratory symptoms during cigarette smoking. When the effects show a typical reversible proportional time course, we can use a reversible proportional dose metric to adequately describe the exposure-response relationship, even without a clear understanding of the specific causal agent. The severity of respiratory symptoms, cough and phlegm, in long-term cigarette smokers is proportional to the packs per day they currently smoke, largely independent of how long they have smoked, as long as they have been smoking for a year or two (Department of Health and Human Services [DHHS] 1990). This suggests that the processes causing the symptoms are being balanced by repair processes. When smokers stop smoking, their symptoms will decline with an apparent half-time of 1 or 2 months. This does not negate the likely occurrence of *other* irreversible effects that have difference DPs, such as formation of fibrosis around some of the damage.

The quantitative dose-response relationship for a reversible proportional effect depends strongly on the time between the receipt of the dose and the time the effect is observed. The investigator must consider the speed of the development and repair of the damage. However, these processes in groups of individuals do not necessarily move at the same rates, just as there are interindividual variations in pharmacokinetic processes. Consequently, responses to bolus exposures in a population will not peak at the same time. If the exposure is relatively steady, then the opposing damage and repair processes may come to a steady state, and the prevailing level of damage will largely depend on the intensity of exposure. Thus reversible processes are best studied in situations in which there has been enough exposure time for the effects to reach steady state. If there has not been much time since the onset of exposure, then the investigator should be alert for nonlinear behavior and the possibility of a poor fit for simple dose metrics.

12 Irreversible Proportional Diseases Processes: Neurobehavioral Effects of Mercury, Popcorn Workers' Lung

12.1. INTRODUCTION

In this chapter we present two examples of irreversible proportional disease processes. The first example, the neurobehavioral effects of mercury, is relatively well studied, with a long and tragic history. The second example is much more recent, and there remain many uncertainties about the causal mechanisms by which a recently identified lung toxin—the artificial butter flavoring in microwave popcorn—leads to severe lung disease. We hope to show how, in both a familiar and an emerging hazard, there are benefits for researchers in applying a disease process model.

In Chapters 6 and 10, we presented the basic principles of irreversible proportional disease processes, summarized in Table 12.1. Here we apply this model and show how it can be used in exposure assessment and epidemiology. We have noted before that there is often no clear line between reversible and irreversible effects; some agents will cause largely reversible effects unless the exposures are very high, whereas agents that are known to cause irreversible damage may appear to have reversible effects at low concentrations or when exposures are brief. For example, exposure to organic solvents when stripping paint from old furniture during a home renovation project can make you dizzy and cause a headache. However, you will usually recover fully after a period of no exposure. On the other hand, working 8 hours per day stripping furniture in a used-furniture shop for many years, which involves prolonged exposure to high levels of solvents, can kill neurons and lead to irreversible dementia.

Table 12.1 Characteristics of the Irreversible Proportional Disease Processes

Key Features	Characteristics	Data Sources
Target tissue	Tissue(s) with irreversible adverse effects	Clinical reports; published findings; interviews with affected individuals
Evidence of irreversibility	Time course shows a plateau in response after a lag, when exposure stops; no recovery from effects	Clinical reports and interviews with affected individuals indicating no recovery after exposure
Type of a proportional response	After a lag, magnitude of the response is cumulative linear or nonlinear with tissue dose	Scientific papers
ETI	Interval of time contributing to the observed effect; equals the time of exposure minus the lag. $\text{ETI} = T_{expo} - L$ $L = n_{lag}\Delta t$	Clinical reports and interviews with affected individuals indicating a lag before detectable effects and progressive worsening during exposure. Lag may be difficult to quantify.
Disease process model	Effect is the sum of the new damage from exposure over the ETI period, excluding the lag period. $D\left[N_{ETI} - n_{lag}\right] = \sum_{i=1}^{N_{ETI}-n_{lag}} K_{dam}\Delta t C_{tis}[i]$ (Note: i counts backward from the time of observation into the past.)	Model covers the full period of exposure; K_{dam} is a scaling factor to adjust D to units of effects, i.e., the amount of damage per unit of exposure; high exposures may produce a nonlinear process altering the value of K_{dam}.
Dose Metric	DM is analogous to cumulative exposure except for the lag. $DM = \sum_{i=1}^{N_{ETI}-n_{lag}} C[i]\Delta t = \overline{C}\left(N_{ETI} - n_{lag}\right)\Delta t$	Use K_{rep} estimated from DP model.Or fit DM across subjects to get mean and SD

12.2. MODELING THE NEUROBEHAVIORAL EFFECTS OF MERCURY

The Minimata Bay Tragedy

The town of Kumamoto in Japan was a small fishing village in the early 1900s. It is located about 550 miles southwest of Tokyo on the edge of Minamata Bay, which is part of the Shiranui Sea.[1] At the request of the town fathers, in 1908 the Nippon Nitrate Fertilizer Company, later renamed Chisso Corporation, constructed a factory for fertilizer and carbide production. The company was commercially successful, and in 1932 it expanded into petroleum and plastics products. Reopening after World War II, Chisso's work in plastics and related chemicals left it well positioned to expand with Japan's rapid economic development during the years 1952 through 1960.

One of the plant's major products was acetaldehyde, which was first produced in 1932. The production process catalytically produced acetaldehyde from acetylene, using mercury sulfate as the catalyst. Methyl mercury was a by-product waste from the process. The factory's wastewater was initially dumped directly into Minamata Bay without treatment. In 1945, 1958, and 1959 ineffective control technologies were applied to the wastewater. Finally, in 1960 some effective treatment and partial recycling were installed. Total recycling was started in 1966, which stopped the releases of mercury. During the period from 1954 to 1960, the production of acetaldehyde was rapidly ramped up to meet the growing demand. It is estimated that more than 5,000 Kg of mercury wastes (salts and methylated) per year were released into the Bay, quickly producing widespread massive contamination of water and sea life.

The fishermen began complaining of reduced catches and noticing dead fish in the waters near the plant outfall in the 1920s. At that time it was the company's policy to compensate the fishermen because it was cheaper than cleaning up the wastewater. This policy was not unique to Japan in this period; many American companies behaved similarly. In the American west, around the large copper and lead smelters were "smoke farmers," who grew crops and sold them to the company each year when the sulfur dioxide or heavy metals in the smelter smoke damaged and destroyed them. Economic damage to crops and fish came first; the more subtle health effects came later.

The released elemental mercury was converted to methyl mercury (MeHg) by microoganisms. The MeHg then entered the food chain, where it was taken up by marine organisms and transported into fat; MeHg is highly fat soluble. Through a process of bioaccumulation (discussed in Chapter 2) the MeHg became more and more concentrated as it moved up the food chain into the fish. Japanese fishermen and their families living on the Bay caught and ate the contaminated fish.

Starting in the mid-1950s, local residents began to be afflicted with degeneration of the nervous system, including numbness in the limbs, slurred speech, and loss of normal vision. Severe brain damage occurred in some. The cause was not immediately identified, and dumping continued. Severe fetal malformations were also found to be occurring at an alarmingly high rate in this population. Clinical and epidemiologic studies followed (Harada 1995). The studies found high rates of certain birth defects in the fishing villages compared with national rates and with rates in nearby nonfishing villages. At first, the conclusion was that something about the fishing lifestyle in the Kumamoto area was associated with a large increase of normally rare neurological fetal malformations. By 1959, organic mercury was strongly linked to what had by then become known as Minamata Disease (Harada, 1995). Unfortunately, the mining company and Japanese government were very slow to correct the problem and compensate the victims. Political and legal fights over apportioning the blame and the provision of compensation lasted into the 1990s (International Center for Environmental Technology Transfer 1998) (ICETT 1998), more than 40 years after the damage was first identified.

Epidemiologic Studies

Minamata Disease was a tragedy of terrible proportions because of the intensely high exposures of the fishing populations and because of the slow response of scientists and the government to the first sentinel cases. Although epidemics as severe as Minamata do not happen often, there are undoubtedly many instances of toxic damage from mercury and other neurotoxins that were never detected because the exposures are much lower. At low levels, the effects of neurotoxins are typically manifested as subtle modifications of normal development and behavior that may not appear until many years after the causal exposure.

Recently, large population studies have been conducted in the Faroe and Seychelles Islands, where there are unusually high human exposures to MeHg because of diets rich in seafood. The Faroe Islanders obtain most of their MeHg intake from eating pilot whale meat and blubber, but they also consume contaminated cod fish (Weihe, Grandjean et al. 1996). Whale fishing is seasonal, but the meat is available year round because it is well preserved by freezing. Based on 1987 questionnaire data collected at 990 births, pregnant women consumed a daily average of 72 g of fish, 12 g of whale meat, and 7 g of blubber (Weihe, Grandjean et al. 1996). MeHg averages ~0.07 μg/g in the fish and ~1.7 μg/g (total Hg was ~3.3 μg/g) in the whale meat. However, there was wide variation in the number of meals of whale meat they ate per month, ranging from 0 (21% of the subjects, who had a median 11 μg/g Hg in maternal hair) up to > 3 per month (27% of the population, who had a median 35 μg/g Hg in maternal hair). Infant cord blood levels were also closely related to maternal meals of whale meat. The investigators also examined other potentially toxic components of whale meat. The meat contains selenium, which was increased in cord blood in proportion to meals of whale meat. However, the selenium levels were only slightly elevated relative to normal Danish levels. Pilot whale blubber contains high levels of PCBs, which can pass the placental blood barrier and are present in mothers' milk. Faroe Islander intakes of PCBs with blubber have been estimated to be ~200 μg/d with wide variation; this is much higher than the estimated Scandinavian intake of 15–20 μg/d. Because PCBs are reportedly neurotoxic in children and because the intake is correlated with meals of whale meat, there is potential for confounding from this exposure.

The Seychelles Islanders have a diet with high intake of fish, but no marine mammals (Myers, Davidson et al. 1995; Shamlaye, Marsh et al. 1995). A cohort study was initiated in 1989–1990 to follow the development of infants born to mothers with "high" intake of MeHg through their diets. They reportedly had, on average, 12 meals containing fish each week. There was a range of MeHg in the species of fish consumed, but they averaged 0.3 μg/g, considerably less than in the whale meat eaten by the Faroe Islanders. The mothers in the Seychelles had hair levels averaging 6.9 μg/g, which was less than the mothers with the lowest exposures that consumed no whale meat in the Faroe Islands. The Seychelles mothers had much higher exposure than the U.S. general public or pregnant women who eat similar kinds of fish, but based on hair levels, the Faroe Islanders have much higher MeHg and inorganic mercury intake because of their ingestion of pilot whale meat and blubber.

The subtle effects of prenatal low-level exposures via maternal diet have been a topic of considerable interest. Repeated prospective evaluations of developmental effects have been conducted. There were marked differences in neurological effects between the children from the Faroe and Seychelles Islands that were consistent with the differences in the maternal exposures.

The studies of children born in the Seychelles Islands found no adverse effects associated with maternal hair mercury (National Academy of Science (NAS) 2000). Cord blood tests were not conducted. The number of abnormal and questionable findings was very low for the cohort, which precluded much data analysis.

The studies in the Faroe Islands looked at children at birth, at 1 year, and at 7 years of age. The early tests showed that infants tested at 2 weeks after birth showed a significant decline in their neurological optimality score (NOS), which looks at "an infant's functional abilities, reflexes, responsiveness, and stability of state" associated with increasing cord blood mercury level (NAS 2000). At ages 1 and 7, additional neurological examinations were administered to determine motor coordination and perceptual-motor performance (Grandjean, Budtz-Jorgensen et al. 1999). These showed clear reductions in recall, language, and attention that were best associated with cord blood mercury. A recent follow up of 14-year-olds using the same tests found that cord blood MeHg was most strongly associated with reduced performance in finger-tapping speed, reaction time on a continued performance task, and cued naming (Debes, Budtz-Jorgensen et al. 2006). Later postnatal measurements of MeHg showed no relationship with effects. The predictive power and magnitude of the mercury exposure contribution at age 14 was similar to that at age 7. The investigators performed an analysis of the differences in results at 7 and 14 years and concluded that the effects had not changed in the 7 years between the two examinations. Thus the MeHg effects were proportional to late fetal exposure and irreversible.

Mechanism of Effects

Methyl mercury is a potent nerve toxin and teratogen (National Research Council (NRC) 2000). Several decades of research have begun to unravel the tangled causal mechanism of these effects. The severity of the effects in adults and children depends on the dose received, but the fetus is much more sensitive. The key effects are that methyl mercury kills neurons and disrupts their normal growth in the fetus. Normal fetal development requires cell migration, cell-to-cell signaling, differentiation, organogenesis, and apoptosis in selected parts of tissues. MeHg can bind to sulfhydryl groups in key proteins, damaging neurons and modifying their behavior during the development of the neural system, which can severely affect normal growth and linkage within the nervous system. This damage, when severe, causes mental retardation and neuromuscular problems that are irreversible. The timing of various developmental activities in the fetal central neurological system (CNS) also varies. There is evidence that third-trimester exposures to MeHg can produce more severe damage (Grandjean and White 1999). Consequently, there can be both accumulative toxicity from general CNS effects and specific developmental toxic effects that are not cumulative beyond the gestational period. These will depend on the intensity and timing of the exposure. An early high exposure can damage or kill neurons that are scheduled to develop into key parts of the CNS. Additionally, MeHg is slowly converted to inorganic Hg in the brain, which may be more toxic than the methylated form, and this may account for the progression of toxic effects after exposure has ceased.

A physiologically based toxicokinetic (PBTK) model for MeHg ingestion in adults and children and transport into the fetus has been developed that permits estimation of the tissue concentrations based on dietary exposure data (Clewell, Gearhart et al. 1999). MeHg is readily absorbed (90%) into the blood from the gastrointestinal (GI) system, where it strongly but reversibly binds with hemoglobin in red blood cells. It also binds reversibly with sulfhydryl groups on glutathione and with proteins, such as albumin in blood and keratin in hair. The red blood cells (RBC) and kidneys have the highest binding coefficients, and they serve as medium-term repositories for MeHg. The RBC/plasma partition coefficient for MeHg is a critical determinant of both the fetal exposure and hair absorption because it determines the amount of free MeHg in the plasma. Because of the large RBC binding coefficient, less than 10% of the total MeHg in the blood exists in unbound form in the plasma, and this is the only form that can penetrate into the tissues. Consequently, MeHg is only slowly transferred from the RBC into the other tissues. Free MeHg easily passes through the blood/brain and fetal/maternal placental barriers. Therefore, even though peak exposures (e.g., from ingestion of large amounts of MeHg in highly contaminated food) could produce a significant step increase in the maternal blood level, the slowness of the transfer process smooths out these short-term exposure peaks.

Removal of mercury and MeHg from the maternal body and fetus is very slow. The primary route of exit from the body is binding in the liver and excretion with the bile into the feces. In adults, there is an approximately 50-day half-life for removal of MeHg from the blood; it is much longer in the fetus because of the necessary back-diffusion into maternal blood before clearance. Daily or weekly intake of MeHg from contaminated fish will produce a gradual increase in maternal blood levels that will stabilize after about six half-lives of clearance, ~300 days of exposure. If the mother makes a large change in her intake of fish—for example if she stops eating fish—her blood level will only decline by 50% after 2 months. Therefore, modest day-to-day variation in input will have little effect on her blood level. The fetus absorbs free MeHg from maternal plasma. An amount is absorbed each day proportional to the concentration in circulating plasma pool.

During the 9-month gestation period there is no clear susceptibility window for fetal CNS development. Development of the cerebellum during the third trimester is a potential target for motor neuron problems later in life. Although there are likely periods of increased sensitivity, the integrating effect of the long half-time of MeHg and Hg in the brain make it unlikely that periods of increased sensitivity can be associated with exposures from dietary intake unless there are large differences in intake. As the fetus grows, its placental blood supply also grows, and the ready exchange of MeHg through the placenta is likely to keep fetal blood in steady-state equilibrium with the mother's plasma level. MeHg entering the fetus is rapidly taken up by fetal hemoglobin, which keeps the free MeHg low. Free MeHg is transferred from the fetal plasma into the brain. MeHg moves along the concentration gradients between mother and fetus plasma and between fetal plasma and fetal brain. Because most of the MeHg is bound in the blood, the transfer rates into the tissues are relatively slow.

There are probably at least two mechanisms of MeHg neurotoxicity: first, MeHg binds with critical proteins, blocking or misdirecting normal development of neuronal migration and formation of CNS connections; second, at higher levels MeHg directly kills neurons, including some that are critical for nervous system development (Murata, Grandjean et al. 2007). Even at low MeHg levels, there may be subtle effects on neurological development and subsequently on an individual's psychosocial functioning. On the positive side, the adult human brain is capable of impressive feats of compensation when affected by neurotoxins. However, if the fetal brain does not properly develop in the first place, it may not be able to compensate and achieve "normal" behaviors because critical connections are not made or neurons are killed that would develop into a necessary brain structure.

Assuming that each unit of MeHg arriving in brain tissue produces an equal amount of damage per unit time by binding to critical proteins, then the total amount of damage will be proportional to the total dose, which would be represented by the area under the curve (AUC) for the brain concentration. Note that this is different from the total mass of MeHg in the fetal brain, which will constantly increase during gestation as the brain grows. This proportionality may hold overall, whereas the precise adverse effects in an individual may vary depending on which brain structures and connections are developing and susceptible at the moment that the MeHg arrives. But usually these detailed individual effects are too subtle to be precisely identified, and so it is reasonable to assume that effects are simply proportional to the AUC within broad time windows, and especially the third trimester of pregnancy.

Etiologic Time Interval

For pregnant women and their developing fetuses, the most sensitive target tissue would appear to be the developing fetal brain. One would expect there to be a time window of increased vulnerability, but there is only limited evidence to suggest when this might be. Because motor neuron problems are associated with the cerebellum and because they are common MeHg toxic effects in children, the last trimester is possibly a sensitive period, as the development in the cerebellum peaks just after birth. During the third trimester and infancy, the brain grows extensively and undergoes extensive remodeling of neuronal connections that could be affected by MeHg or Hg. Therefore, even though the effects—neuronal killing and misdirecting neuronal connections—are irreversible and cumulative, the precise nature of these effects is not clear. It is known that loss of growing neurons can lead to undersized brains, and we can infer that an accumulation of faulty neuron connections will progressively affect brain function, but the degree of proportionality of these effects with increasing dose is not known. A reasonable first choice for the etiologic time interval (ETI), then, would be the whole gestation period, 9 months, or perhaps the last trimester plus a few months after birth, so as to include motor neuron effects in the cerebellum observed during childhood.

Continuing MeHg exposure during infancy while nursing may amplify the damage received as a fetus. Even though a mother's milk may have relatively high MeHg, proportional to her plasma level, the relative level and quantity of intake are small. Infant blood samples collected during the first few years of life, while children are nursing, show that their blood levels decline by more than 50% (Bjornberg, Vahter et al. 2005). Exposure during nursing does not appear to be important, as shown by this study of Swedish women, in which the infant blood MeHg was only modestly associated with breast milk total mercury (Spearman R = 0.55). The infants' blood MeHg declined by more than 50% by 13 weeks post-birth, whereas the mothers' blood levels remained elevated. It was not clear why, but infant clearance processes are not known. Therefore, for effects observed in infants, tissue levels as a fetus may be most relevant. However, during childhood the nervous system continues to develop. If a weaning child is fed fish, continuing exposure may add to adverse effects. In that case, for childhood studies the etiologic period should include both fetal and childhood exposures.

Hair Mercury as a Biomarker

The mother's hair will absorb MeHg from the free MeHg in her plasma. Hair absorption is a diffusion exchange between the hair's sulfhydryl binding sites on keratin and the plasma until the hair shaft grows away from its blood supply, at which time the mercury content becomes fixed. Thus the mercury content of hair is defined by the plasma level at the time the hair is formed, which will follow the plasma level. Large changes in maternal dietary Hg intake affect hair levels significantly during pregnancy (Bjornberg, Vahter et al. 2003). Analysis of total mercury in maternal hair samples has been widely used as a biomarker of exposure, but uncertainties about maternal hair growth rate (~1 cm/mo) make it difficult to localize exposures in time (NRC 2000). Nevertheless, a number of studies of mercury's effects on children have used mercury in mother's hair collected at the child's birth as a biomarker to quantify fetal MeHg exposures. If the hair collected is at least 9 cm long, then the hair concentration should represent the time-weighted average of the maternal blood level. However, if the hair is much shorter than 5 cm, it will represent only the most recent exposures, and these may not be representative of the full ETI, especially if the mother changed her diet during pregnancy to reduce Hg intake. Investigators using hair concentration in a cross-sectional study must assume that all women have about the same rate of hair growth, that their hair has the same MeHg binding constant, and that their dietary intake of MeHg is stable over time. Random variation in these values will add noise to the dose metric. It is not known how rapidly fetal levels will decline in response to a decline in maternal concentrations; however, when infant intake is relatively low from breast milk, infant blood levels decline by more than 50% in about 80 days.

Dose Metrics

As noted in Chapter 6, cumulative exposure is often an adequate dose metric for irreversible proportional disease processes, and this guidance should apply to studies of neurologic effects of mercury in adults or children. If diet is the main exposure, as in the Faroe Islands and Seychelles studies, then food intake and food Hg concentration data could be combined to estimate the microgram-days of cumulative exposure over the ETI for each study subject. A more complex metric, such as that proposed in Equation 10.9, could be used if detailed time-varying exposure data (quantitative weekly dietary intake, for example) were available.

12.3. EXAMPLE OF POPCORN WORKERS' LUNG

A second example of a irreversible proportional disease process is the outbreak of bronchiolitis obliterans (BO) in workers packaging microwave popcorn (CDC 2002) (Kreiss, Gomaa et al. 2002). This example was briefly presented in Chapter 6 to illustrate the utility of disease process models for new and emerging diseases. There have now been several clusters of cases of BO among workers in microwave popcorn factories in the midwestern United States and in other parts of the country, as well as abroad. BO involves acute destruction of the air exchange region of the lungs and irreversible loss of lung function. Fortunately, the disease is rare, and few environmental causes are known. Starting in 2002, small clusters of workers in microwave popcorn packaging plants have been diagnosed with BO. Investigations of these outbreaks were summarized in a report from the National Institute for Occupational Safety and Health (NIOSH 2002).

History of Popcorn Workers' Lung

As early as 1986, a NIOSH Health Hazard Evaluation reported that substantial lung damage might be associated with flavorings used in the baking industry. Flavorings are concentrated organic chemicals added in small amounts to food to impart a desired taste. The chemicals are all food grade, which means they are designated by the U.S. Food and Drug Administration (FDA) as "generally recognized as safe" (GRAS) for ingestion in small amounts. Although GRAS chemicals are safe for ingestion, they are not necessarily safe for inhalation; chili pepper is a good example. Many people find this anti-intuitive: If it's safe to eat, shouldn't it be safe to breathe? However, the sensitivities of your gastrointestinal tract are very different from those of your respiratory system. For example, phosphoric acid is added to many well-known soft drinks to give them an acid bite, but it can cause severe lung damage if inhaled at the same high concentrations.

Microwave popcorn is a popular inexpensive snack food; large amounts are produced. The popcorn is packaged in special containers that will expand as the corn pops in the microwave oven, and flavor ingredients will melt with the heat of popping and cover the kernels. Because buttered-flavored popcorn is highly desirable, the manufacturers have developed proprietary flavorings with the proper taste and smell that are also stable during storage (real butter is not).

In the factory at which the principal investigation was performed, the disease was concentrated in the area of the plant in which artificial butter flavoring was mixed (Kreiss, Gomaa et al. 2002). Onset of the disease sometimes occurred within only a few months after first employment in the factory. Subsequently, the disease has been diagnosed among workers in several other factories processing the same flavorings.

One of the challenges for this type of investigation is determining what is in the proprietary flavor mixture; this information is covered by trade-secret laws. There are many potentially toxic flavoring ingredients, including alpha- and beta-unsaturated aldehydes and ketones, aliphatic aldehydes, aliphatic carboxylic acids, aliphatic amines, and aliphatic aromatic thiols and sulfides. One or more of these could be agents of respiratory effects. NIOSH researchers performed experiments in which rats were exposed to vapors from the flavoring mixture and found multifocal, necrotizing bronchitis at concentrations in the range found in the mixing areas of popcorn factories (Hubbs, Battelli et al. 2002). The pathologic findings in rats were sufficiently similar to BO that the findings were considered relevant to the human disease outbreak. Chemical analysis of the vapors revealed a complex mixture of diacetyl (2,3-butanedione), acetic acid, acetoin (3-hydroxy-2-butanone), butyric acid, acetoin dimers, 2-nonanone, and δ-alkyl lactones. Several lines of evidence suggest that diacetyl is the most important toxic compound in the mixture (Kullman, Boylstein et al. 2005).

The earliest case cluster was identified by a NIOSH Health Hazard Evaluation reported in 1986 that examined workers at a manufacturing plant making butter-flavored additives for the bakery industry (NIOSH 2002). The composition of the flavorings was not reported. Four workers had developed mild to severe symptoms of airway obstruction after working in the mixing room. After only 2 months on the job, the two most severely affected young, nonsmoking workers had severe symptoms of irreversible airway obstruction: shortness of breath with exercise and persistent cough. Pulmonary function tests by spirometry showed severe fixed airway obstruction. Chest radiographs and diffusing capacity for carbon monoxide (lung-diffusing capacity DL_{CO} reflects thickening of alveolar walls) were both within the normal range. Treatment with bronchodilator or corticosteroid drugs did not produce any clinical improvement for either subject. Even after months away from the workplace, neither individual had any improvement; they both had severe symptoms of shortness of breath with exertion.

Three case series have been investigated in microwave popcorn production plants (Akpinar-Elci, Travis et al. 2004). In each of these, the most severely affected workers were in the mixing room where large batches of the flavorings were mixed with heated oil in tanks. A small amount of this mixture is added to each package. At the time of the evaluations, typically the mixing room was unventilated and the concentrated flavorings were moved manually in open buckets and poured into open tanks. Others less severely affected worked on the production line close to the mixing room, or in the quality control lab where large numbers of test bags were popped. Few of the individuals with symptoms were smokers. For most, the onset of the symptoms occurred within months to a few years after starting work in the plants. None of the workers reported that their symptoms were associated with acute overexposure events. In general, symptoms began with shortness of breath on exertion, cough, and wheezing. Later, when they were tested by spirometry, the findings included severe airway obstruction and normal chest X-rays; most had normal DL_{CO}, and symptoms were unrelieved by bronchodilators or steroids. Four individuals had very severe symptoms and were placed on lung transplant lists.

The clinical picture is consistent with BO caused by exposure to an airborne chemical irritant that destroys tissues in the airways without affecting the alveoli, similar to the effects of high-intensity exposure to nitrogen dioxide, chlorine, or phosgene (King 1998). There is no evidence of recovery from the damage that causes shortness of breath. Bronchodilators and corticosteroids would be expected to modify symptoms associated with asthma and some types of immunologic conditions. The severity of the effects appears related to the degree of enclosure of the mixing area and was reduced by ventilation of the mixer or use of high-efficiency respiratory protection by the workers.

Application of a Disease Process Model

Assuming that the case information was the only data available, how could we proceed to design an epidemiologic study to follow up on this problem? Using the process model approach we need to identify a tentative process model, identify one or more markers of the key effect, choose an ETI, select the hypothetical agent(s) and a measurable marker for exposure, and finally define a summary dose metric that represents our mechanism to be tested in the epidemiologic study.

1. Based on the preceding clinical summary, shortness of breath (SOB) and reduced pulmonary function that do not dissipate long after exposure stops are distinctive features of bronchiolitis obliterans. We assume that the severity of SOB and reduced pulmonary function are proportional to the total damage and can be our markers of the effect. Thus we use the model for an irreversible proportional effect.
2. Given an irreversible effect, the ETI should be the whole period of exposure plus a lag time for the development of the effect(s).
3. We assume that the agent is diacetyl, although there may be other hazardous components of the flavoring mixture, as well.
4. Consistent with the irreversible proportional process model, our dose metric will be cumulative exposure for our measure of exposure over the whole period of exposure for each subject. Based on the clinical observation that adverse effects require a minimum of only a few months following first exposure to appear, we might choose a very short lag of 1 or 2 months, or perhaps no lag period at all.

This analysis provides the rationale and basic parameters for conducting a cross-sectional study of pulmonary symptoms and pulmonary function, with a retrospective determination of exposure for each subject based on his or her job history.

Epidemiologic Study of Popcorn Workers

We can check how our proposed study compares with a study of one of the popcorn plants. A cross-sectional study of workers at a microwave popcorn plant was conducted in 2000 to assess the association between exposures to the flavorings and losses of pulmonary function (Kreiss and Gomas, et al. 2002). A total of 117 out of 135 individuals in the workforce were recruited to participate in the study. A standardized questionnaire was used to obtain data on SOB, wheeze, cough, and chest tightness and their severity and duration. Additional questions obtained data on smoking history, work history, exposures outside work, and demographics. Spirometry and lung-diffusing capacity (DL_{CO}) measurements were made, chest X-rays were taken, and all were evaluated with standard methods and compared with normal values. In November 2000, an exposure survey was conducted, with measurements throughout the plant that identified more than 100 chemicals associated with the butter-flavoring additive. The most common was diacetyl (2,3-butanedione), which was chosen as a marker for the intensity of flavoring exposure. Samples were collected to characterize diacetyl exposures by jobs and work areas. A diacetyl exposure metric was obtained from each subject's job history by calculating cumulative exposure using the currently measured air level for each job times the time worked in that job, summed over all jobs reportedly held by the subject. Exposures to several other substances possibly of interest were also measured, including nitrogen dioxide, respirable particulate, endotoxin, bacteria and fungi, acetoin, nonanone, methyl ethyl ketone, acetaldehyde, and acetic acid. All of the vapors were at lower levels than diacetyl and were correlated with diacetyl. None of the compounds in the flavoring had been previously identified as causes of bronchiolitis obliterans.

External Comparison

Overall, 117 out of 135 subjects were evaluated. Most chest X-rays and DL_{CO} values were in the normal range. This population of workers showed a prevalence ratio 2.6 times higher for SOB and 3.3 times higher for airway obstruction than expected from national rates (Third National Health and Nutrition Evaluation Survey [NHANES III]) after adjusting for age and smoking. Young people who never smoked (< 40 years of age) had the highest prevalence ratio for SOB, 4.8, and older smokers had the lowest ratio, 1.8. Excess prevalences were also seen for wheezing, chronic cough, and chronic bronchitis compared with the general population.

Internal Comparison

Comparing the unexposed with the exposed workers within the plant, the exposed had significantly more "exertional shortness of breath" ($p < 0.01$), "regular trouble breathing" ($p < 0.004$), and "unusual fatigue" ($p < 0.003$). Smaller nonsignificant differences were seen for chronic cough, wheeze, and chest tightness.

A categorical analysis was conducted for pulmonary function reductions associated with quartiles of diacetyl cumulative exposure (ppm-years). The cumulative exposure distribution was highly skewed, as expected from the large differences in the diacetyl exposure in the work areas. Even with the small number of subjects, there were strong trends for FEV_1, airway obstruction, and abnormal spirometry in the expected directions across the quartiles of cumulative exposure (Table 12.2). Thus subjects with longer work in the mixing room or quality control testing area, and so with the highest cumulative exposures, were the most likely to have clinically

Table 12.2 Results of a cross-sectional study in a microwave popcorn factory: pulmonary function among workers categorized into quartiles of cumulative exposure to diacetyl

Diacetyl cumulative exposure Quartiles (ppm-years)	0–0.65	0.65–4.5	4.5–11	>11
Average FEV_1 as percent of predicted	97%	92.5%	88.1%	84.5%
Airway obstruction[a]	10.3%	10.3%	24.1%	27.6%
Abnormal spirometry[b]	13.8%	24.1%	31.0%	37.9%

a. Defined by as "a low ratio of FEV_1 to FVC in the presence of a low FEV_1 value," where a low value is below the 95% confidence interval from the NHANES III data by age and gender; test for trend $p < 0.04$.

b. Defined as "airway obstruction or low FVC value"; test for trend $p < 0.02$.

Note: Table taken from Kreiss and Gomas (2002).

reduced spirometric function. These findings are consistent with our proposed disease process hypothesis.

Comments on the Popcorn Worker Example

The epidemiologic investigation of Kreiss and colleagues added a critical dimension to the investigation: it is clear that BO, a terrible and disabling lung disease, is not the only consequence of breathing diacetyl or other flavoring chemicals. There appears to be a continuity of effects, with mild, asymptomatic lung disease detectable in workers with relatively short-term and low-level exposures. This is what is meant by a proportional disease process. At the highest exposures, a shockingly high percentage of individuals were irreversibly affected—five out of six workers in the quality control room. Lung biopsy data in the index case "revealed scattered, non-necrotizing granulomas; focal bronchiolar fibrosis; fibroblast proliferation compressing one bronchiolar lumen; and no interstitial pneumonia," (Kreiss and Gomas, et al. 2002). This type of diffuse fibrotic response is consistent with long-term permanent loss of conducting airways, that is, airway obstruction. Even with this small study, important data on the dose-response relationship can be obtained when an appropriate dose metric is available.

If diacetyl itself is not the agent (or not the only agent), these findings provide several clues about the characteristics of the actual causal agent(s). They are likely to produce significant amounts of vapor with heating to 54°C, based on descriptions of the production process. When concentrated vapors contact cool air, they will condense into very small droplets, because these materials are liquids or solids at room temperature. The agents have low to moderate water solubility, and they are moderate sensory irritants, consistent with the modest upper airway irritation and cough; but they can produce severe deep lung damage. A good example of another such agent is nitrogen dioxide. Respirable particles (PM_4) are unlikely to have contributed much because the levels were low and had little variation across the lowest and highest exposure areas. Also particle deposition would not be concentrated in the small airways, as it can be for some gases and vapors. Some of the concentrated vapor agents may have condensed into droplets when the vapors were dispersed into the cooler air above the tanks, but there is little evidence in the concentration gradient between the mixing room and other areas to support this.

We have classified this as an irreversible proportional disease process; consequently, we would assume—at least as a first approximation—that there is no threshold for a safe exposure that will protect the entire population. What are the effects of exposure to microwave popcorn vapors for consumers? There are no studies to provide data on this question, but we would hypothesize that there would be effects on the airways and that they would likely be very mild—roughly proportional to the exposure intensity. Thus we would not expect to observe BO or serious lung disease in consumers, but sensitive tests of frequent users might detect small decrements in lung function.

Prevention strategies are quite feasible. First and foremost, the toxic agents can be eliminated. Because no essential human need is being provided for by artificial butter flavor, a prudent approach would be to eliminate the chemicals involved, at least until safe alternatives can be developed. Failing this direct approach, manufacturing workers can be protected through a fully enclosed manufacturing system, such as a hood for small amounts, or piping with pumps should be used to handle concentrated flavoring added to batch tanks; manual handling of open containers should be prohibited (NIOSH, 2003). When maintenance is done on these systems, inside of exposure controls, workers must wear suitable respiratory protection. Obviously, these controls will only protect the workers; it is not clear how large the risk might be to consumers, but we would recommend that they minimize and avoid exposure however they can.

12.4. DISCUSSION OF IRREVERSIBLE, PROPORTIONAL DISEASES

This disease process model is one of the simplest that we explore. However, as you have seen from the two examples, methyl mercury and popcorn flavorings, the biology of an irreversible disease can be very complex and involve widely varying mechanisms. When we want to apply a disease model, we need a clear justification that the case under study fits the model's three principal assumptions. First, each unit of the agent causes damage at a rate independent of other units arriving in the tissue before or after. Second, damage is (mostly) irreversible. Third, there is typically a lag between the tissue dose and the irreversible response. The two examples illustrate different aspects of irreversible processes.

The effects of irreversible damage in a fetus are very different from those in a child or adult. Damage in a fetus can affect a development process and the resulting organs and life functions. In the case of the nervous system, developmental effects may be difficult to detect if there is a pattern of random interference with the programming of the neural network rather than in a target region that is linked to specific neuronal functions. Exposures can have specific effects only when they coincide with development of specific structures or pathways. In addition, the definition of the etiologic period for the fetus is limited at most to gestation. The agent must be able to pass the maternal-fetal and blood-brain barriers to have a neurological effect on the fetus, setting significant limits on the types of candidate agents.

For irreversible cumulative effects, the temporal sequence of exposure intensities may be important. This dependence on temporal sequence will occur if vulnerability of biological systems varies through time for individual cells, tissues, or whole organisms or if there is nonlinearity with increasing tissue concentration. When this happens, the most common dose index for irreversible effects, cumulative exposure, $C_{tis} \times T$, will not be correct because it implies that short periods of high exposure followed by prolonged low carry the same risk as long, low exposures followed by a short, intense exposure. To account for this, the more general dose metric in Chapter 10 (Equation 10.9) can be used for this type of effect, as shown here. It has a weight attached to each interval that can vary over time or depending on the intensity of the exposure, which depends on the mechanism.

$$DM[ETI-L] = \sum_{i}^{N_{ETI}-n_{lag}} w[i] C_{exp}[i]$$

Equation 12.1 General dose metric for irreversible proportional effects

The time intervals in this expression are intergerized, ETI = $N_{ETI}\Delta t$ and $L = n_{lag}\Delta t$, and it is assumed that the subjects are currently exposed.

NOTE

1 Material on this case was taken from "Approaches to Water Pollution Control (Case Study-2), Minamata City, Kumatmoto Prefecture," prepared by the International Center for Environmental Technology Transfer, Yokkaichi City, Japan, 1998 (http://www.icett.or.jp). ICETT (1998). Approaches to Water Pollution Control (Case Study-2). Minimata City, Kumamoto Prefecture Yokkaichi, International Center for Environmental Technology Transfer.

13 Modeling Discrete Disease Processes

"God does not play at dice."

Albert Einstein

"Oh, but she does Albert. See who lives and who dies during the pandemic," chuckled the Joker.

TJ Smith

13.1. INTRODUCTION

The remaining two disease process models differ from the first two in that for these we typically observe an "all or nothing" response to the exposure. We call these *discrete* responses rather than *proportional* ones, and in general the degree of response is independent of the magnitude of the stimulus. As exposure increases, the *probability* of response increases, rather than the severity of response, or amount of damage or loss of function. Perhaps the simplest example is death (a very discrete outcome), in which the degree or severity of the outcome has no meaning. Less dramatic examples include asthma attacks, allergic responses, autoimmune diseases, and the onset of cancer. In these cases, it is useful to view the risk or probability of the discrete outcome as some function of the exposure, whereas the *severity* of the outcome will either be similar with each response and among subjects or may have other determinants than the immediately preceding exposure.

The stochastic or probabilistic nature of discrete effects such as the onset of asthma or the clinical presentation of a tumor suggests that they are not simply the accumulation of effects or damage building up, but rather that something has triggered an altered state. Such events include mutations, a heightened immune response, or an altered biochemical signal during fetal development. Another way a trigger occurs is when a cellular metabolic system fails, through loss or repair of a normal defense or possibly by direct cell killing.

Whether a response is proportional or discrete may depend to some extent on the time scale of observation. In many of the examples we cite, a very close view of the evolving disease process on a very short time scale (minutes or even seconds) might reveal a "gradual" development of a response. But when viewed from a time scale of clinical relevance—hours or days for an allergy or years for a cancer—the response will appear "instantaneous": one moment the subject was healthy, and the next she was sick. Investigations of delayed conception and fecundity (the probability of conception) provide a good illustration of this point (Davies, deLacey, and Norman 2005). It would seem reasonable to think of pregnancy as a discrete outcome, which is put nicely in the old adage, "you can't be a little bit pregnant." But it turns out that if one looks very closely, this is not completely true. Prospective studies of sexually active women who provide daily urine samples can detect very early conceptions by monitoring the concentration of chorionic gonadotropin, an early marker of implantation (Tingen, Stanford, and Dunson 2004). But this test requires an arbitrary cutoff indicating the level of chorionic gonadotropin that will be considered a positive indication of implantation. Many early implantations are lost and do not lead to pregnancies. How much chorionic gonadotropin means a woman is "truly pregnant"? It is an arbitrary decision. Within a fairly short time—weeks—this condition will become clearer, as a fetus will either develop or not, but in the hours immediately following intercourse, the picture is fuzzier. Thus one could say that a woman with a somewhat elevated level of urinary chorionic gonadotropin really *is* "a little bit pregnant."

Disease processes with important immunologic components are frequently discrete. Because the immune system functions in prevention of infections, it is capable of rapid response and can also amplify small signals into strong responses, presumably for "getting ahead of" a developing infection. Many disease processes result from what appears to be an overreaction to a foreign protein; allergies are an example. Thus many types of "hyperresponsiveness" of an immune function can lead to rapid and seemingly discrete ("all or nothing", "on/off") responses to a toxic exposure. The key characteristics of immune responses are briefly summarized in an accompanying box.

13.2. REVERSIBLE DISCRETE DISEASE PROCESSES

Discrete responses can be either reversible or irreversible, and this distinction closely parallels the preceding explanations for the reversible and irreversible proportional disease processes. We start with the reversible discrete process, of which the onset and resolution of an asthma attack is an excellent example.

Box 13.1 Immunological System and Its Responses

The immune system is a complex interactive system of different types of cells and proteins that is critical for defense against microbiological invaders, cancers, and toxic chemicals and particles. There are two main system components: innate immunity and acquired immunity. Both of these are the result of actions by cellular components, such as lymphocytes and macrophages. *Innate immunity* is a nonspecific defensive system that attacks microorganisms entering the body through inhalation, ingestion, or skin penetration. "Natural killer" (NK) cells, polymorphonuclear (PMN, leukocytes) cells, and macrophages are all part of this system. NK cells can recognize cells that are infected with viruses or are cancerous by changes in their surface markers. They attach to altered cells and release cytolytic granules onto the surface of the target cell, which cause membrane disintegration and apoptosis (programmed cell death). PMNs and macrophages are phagocytic and will engulf and kill invading microorganisms. In this process, they also release mediators that attract other PMNs and macrophages and cause inflammation. In the tissues, macrophages will ingest free particles and cellular debris, in addition to attacking invading microorganisms. Beyond the killer cells, the innate immunity system also includes soluble materials, proteins and complement. When there is an infection, macrophages release signaling proteins called cytokines that produce systemic responses such as fever and a global acute-phase response. The complement system is also triggered by microbial surface proteins, which results in a cascade of protein production that can disrupt membranes and bind to microorganisms, marking them for destruction by PMNs and macrophages. *Acquired immunity* is a signaling process whereby specialized cells, T-lymphocytes, recognize specific pathogens and their products, leading B-lymphocytes to produce signaling proteins, antibodies, that will attach to specific receptors, proteins, carbohydrates (often bacterial), nucleic acids, lipids, and other foreign substances. There are four types of hypersensitivity responses. Type I immediate response hypersensitivity is produced by IgE antibodies that attach to mast cells in circulation or tissues. When an antigen binds with its antibodies, mast cells degranulate, releasing mediators that cause vasodilation, constriction of bronchial airways, and inflammation. This response is rapid, occurring within minutes of exposure. Type II cellular hypersensitivity results from IgG antibodies binding to cell surface antigens forming complexes that attract killer cells, or cell destruction by complement. Type III circulating antigen hypersensitivity involves IgM (sometimes IgG) antibodies binding to circulating antigens, which are then bound to tissues where platelets aggregate, complement is activated, and inflammatory cells are attracted. Type IV delayed response hypersensitivity is caused by memory T-lymphocytes that respond to antigens by producing a delayed cellular cytotoxic or inflammatory response, usually in the skin by direct antigen contact; but it also may occur in the lungs by contact with antigenic dusts. An individual may show more than one type of response to a given exposure, such as both immediate (Type I) and delayed (Type IV) responses. (For a more detailed discussion see Klaassen, *Casarett and Doull's Toxicology*, 6th edition, 2001.)

Example: Allergens and Asthma Attacks

Asthma is a disease characterized by discrete episodes or attacks of shortness of breath, cough, and wheezing (Braunwald E, Fauci AS et al. 2002). The shortness of breath is caused by acute airway constriction, bronchospasm (spasms of muscles in the walls of the large airways, the bronchi), mucus production, and inflammation. Asthmatics usually have developed specific sensitivity to one or more agents—usually allergens—and in addition often have "twitchy airways." This means that their airways may constrict in response to a range of inhaled irritants or even cold air that nonasthmatics will generally find innocuous or only mildly irritating. We discuss the risk of *becoming* asthmatic as a result of environmental exposures in the next section on irreversible discrete responses. Here, we discuss the triggering of an asthma attack among people (asthmatics) who have already acquired the tendency to react strongly to one or more airborne exposures.

Typical Time Course

Exposure of sensitized individuals, either to the agent that sensitized them or to nonspecific irritants, can precipitate a rapidly developing asthmatic response. The initiating stimulus only needs to last a short time—minutes, perhaps. For this reason, exposure assessments often focus on short-term "peak" exposures during which the airborne concentration is much higher than the normal or background level. Removal from exposure may only gradually lead to a reduction in the intensity of the response. Shortness of breath, cough, wheezing, and other responses can continue for many hours, even days for a highly sensitive individual. With medication, the attack may be prevented from developing or may be limited in its duration. After a period of recovery, an individual becomes susceptible to having another attack. When attacks occur on consecutive days, it can be hard to say whether each day represents a new response or is a carryover from the previous day.

Mechanisms

Asthma is a complex immunologic disease with two parts, sensitization and acute attacks (Klaassen 2001). Intense exposures to irritants and sensitizers may produce no response in unsensitized individuals, but once sensitized, even a tiny exposure may produce a large response. An asthmatic's airways are more reactive than normal and prone to spasms of the bronchial smooth muscles, which produce the characteristic shortness of breath and difficulty breathing. Unfortunately, once an asthmatic's airways are hyperreactive, he or she will often react to a variety of nonspecific irritants, in addition to the initial sensitizing agent. Some individuals develop cross-reactivity to other sensitizers.

Once an attack is triggered, the magnitude and duration of the response are usually unrelated to the concentration of the agent, as long as it exceeds the individual's threshold of personal sensitivity. The threshold may be higher for nonspecific triggers. Clinical challenge testing can define an individual's threshold for response to an agent, but it must be used with caution because the challenge testing can itself cause or enhance sensitization.

Etiologic Time Interval (ETI)

How long must an exposure last to provoke an attack? For some receptor responses, once the tissue concentration threshold is reached, there may be a response. On the other hand, if tissue inflammation or other tissue responses are needed, then an exposure of minutes to hours may be needed. Fundamentally that depends on the mechanism of the response. However, we can use clinical reports or data on exposed workers to identify approximate time scales of tissue responses. For asthma, the response can be very rapid—tens of seconds or minutes. For example, asthmatics frequently report that the onset of symptoms occurs within a few minutes of entering an exposure setting, which implies a time scale of minutes. For dermatitis, the time scale is usually hours.

Process Model for Discrete Reversible Responses

A discrete response can be thought of as behaving like a switch: if you push on it hard enough it switches to "on," and once it is on, the amount of input (exposure) is not predictive of how strong the response will be or how long it will last. For the example of asthma, when the agent's tissue concentration is high enough in an asthmatic's airways, it triggers a rapid cascade of cellular responses that result in an attack. For the proportional disease process models, the amount of response could be calculated in each time interval as a fairly simple function of the amount of exposure or dose in that interval. For discrete responses, we need to modify this modeling approach. Now the presence/absence of a "switch response" in each time interval corresponds to a branching relationship, where the branch taken depends on whether the tissue concentration C_{tis} exceeds some critical value C_{Th} (Figure 13.1). However, a more realistic model would recognize that a given individual will not always respond at the same threshold and that thresholds vary across individuals. Thus a more realistic model is one in which the exposure-response probability function is an *S*-curve (Figure 13.1). You can think of this as flipping a coin within each time interval to determine whether there is a response, and the probability of the coin coming up heads (an attack) is not fixed but instead is a function of the exposure or tissue concentration. Therefore, to define an exposure-response relationship, we measure the exposure in each exposure interval and observe when responses occurred. For better precision, we could put the subject in an exposure chamber and observe how many times he or she responds at each level of exposure. Although this can and has been done, it is cumbersome and must be done very carefully, because the exposure may trigger an overwhelming

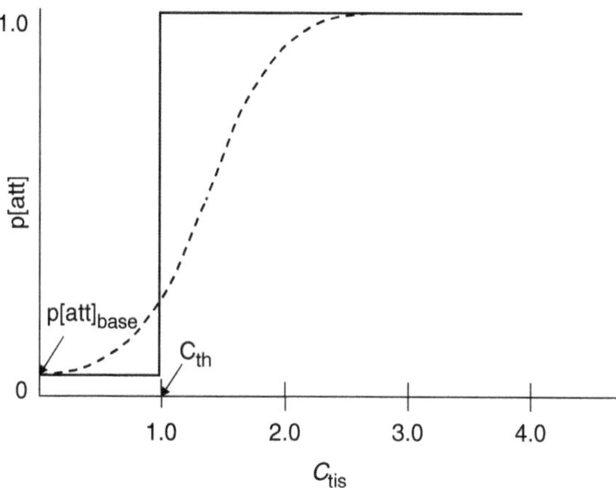

Figure 13.1 A hypothetical function relating the tissue concentration of a toxin (C_{tis}) to the probability of a response in a reversible discrete disease process model. The x-axis scale is in arbitrary units. The threshold, C_{th}, is set at an arbitrary level when a "response is likely."

response in some individuals. Additionally, a large number of measurement points would be needed to define the shape of the curve precisely for a single individual.

For these reasons, it is often not realistic to use this approach to study populations. We would need to measure all environmental exposures during a sequence of Δt intervals for each subject and match them to the intervals when each subject responded in an epidemiologic study. To make this problem more tractable, we reframe the question: How is the overall probability of attacks in an exposed group related to the probability distribution of exposures during some time period? First, we choose groups of subjects for whom it is reasonable to assume that everyone's exposures are random during Δt intervals but drawn from the same distribution and independent of time (some temporal dependence may be present, such as a diurnal cycle of sensitivity, and this could be addressed either through design—studying everyone at the same time of day—or in the analysis). If we are comfortable assuming that exposures vary randomly in intensity during Δt intervals, then we do not have to measure every time interval during a period of exposure, nor every subject. We can measure a set of intervals for a random subgroup of the population to determine the frequency distribution of exposure concentrations. We can also observe the number of attacks in the group during the measured period of exposure, which gives us the overall probability of an attack associated with the distribution of exposures. Those data reflect the underlying group exposure-response curve.

The group response is the superposition of all of the personal exposure-response curves. It strongly reflects the distribution of sensitivities of the members. At low levels of exposure, only the most sensitive will respond, and at high levels nearly all who can respond will. The group curve will be broader than the individual curves and probably will be more stable than the individual curves. When a group is exposed over a long period of time, self-selection will tend to remove those with high sensitivity, reducing the overall responsiveness of the remaining population and shifting the response curve.

Exposure Response Curve

Binary responses have been extensively studied in toxicology and pharmacology, and the *S*-shaped curve described by the probit or logit functions is commonly used. In epidemiologic studies, the logit or logistic function is often used.

An epidemiologic study of discrete responses may use a logistic regression analysis to determine the exposure-response relationships for the proportion of a group responding. This regression has the form shown in Equation 13.1.

$$\ln\left(\frac{p[att]}{1-p[att]}\right) = \text{logit}(p[att]) = a + \beta\,(C_{\text{expo}}\Delta t_{ETI}) + \sum \gamma_j X_j$$

Equation 13.1 Logistic model of discrete risk from exposure dose

Slightly different models will be used depending on the measure of disease frequency being estimated, but the differences are not important for this illustration. The parameters are α the background risk of a response (p[att] = $1/(1+e^{-\beta})$), β the sensitivity of the response to the dose, and a series of covariates, X_j, for age, gender, and so forth. Examples of the curves for the simplest relationship with only α and β are shown in Figure 13.2. More parameters can be added to adjust the shape of the curve, such as bounding the total response by making the numerator γ instead of 1.0 to represent the fraction of responders in the population. For small values of C_{expo} the p(att) increases approximately exponentially until it gets to about 10% of α/β. When C_{expo} equals α/β, then the p(att) equals 0.5, and when C_{expo} is larger than α/β the curve becomes concave downward and approximately logarithmic. Because $1/(1+e^{-\alpha})$ is the background prevalence of the response without exposure, we can use the local prevalence of the response, such as asthma attacks, to set α. There is a problem with the logistic function when the baseline rate is zero because α becomes infinite, but most immunologic responses have some prevalence in the population without exposure. The relative simplicity of this function allows us to make reasonable assumptions about the underlying form of the exposure-response curve when we develop dose metrics. For our purposes the most important feature of this relationship is the effect of different values of α on the shape of the curve at low dose levels. Examining Figure 13.2 shows that when α is less than about -4 (background risk of 1.8%), the lower end of the curve is very concave, and when it is greater than about -2 (background risk of 12%),

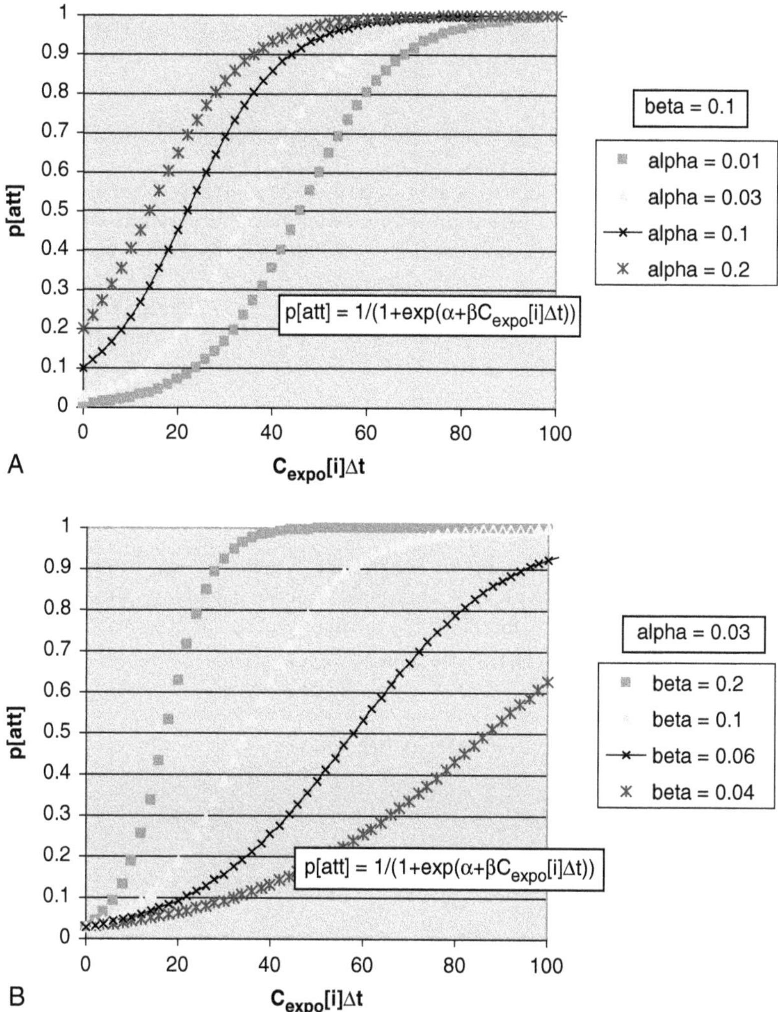

Figure 13.2 Simulated probability of response curves using the logistic relationships, $p[\text{att}] = 1/(1+\exp(-\alpha-\beta C_{\text{expo}}[i]\Delta t))$. Example 13.2A shows how changing the baseline rate, α, affects the shape of the bottom end of the curve. Example 13.2B shows how changing the slope, β, affects the steepness of the curve.

then the curve is nearly linear at low doses. This has important implications for the choice of dose metric that will have the best predictive value.

An example of the underlying relationship between the exposure distribution and the probability of an attack is shown in Box 13.2, the reversible discrete model. The exposure distribution tells us how likely specific exposures are when measuring fixed time intervals, Δt, and the population logistic exposure dose-response curve tells how likely a response is for a given exposure. (Note: the Δt

Box 13.2 Reversible Discrete Effect Model: Calculation of the Probability of an Asthma Attack for an Individual during Δt

- The exposures during Δt intervals have a lognormal distribution, which is defined in this example by the GM = 0.1 and GSD = 2.0.
- The exposure distribution is a set of probabilities, $p[C_{expo}[i]]$, for all possible exposures in the time period (in the following table, 49 intervals were observed, and in four instances the exposure was within 0.02–0.04; therefore $p[C_{expo}[i]] = 4/49 = 0.08$).
- A logistic model defines the exposure-response curve indicating the probability of an attack given an exposure intensity, $p[\text{att} \mid C_{expo}[i]]$. In Table 13.1, for 0.02–0.04 the probability of an attack is 0.008, assuming a logistic curve:

$$p\left[\text{att} \mid C_{expo}\right]_{\Delta t} = 1/(1+\exp(-\alpha - \beta * C_{expo})) \text{ where } \alpha = -5 \text{ and } \beta = 5.$$

The α defines the background risk, and β defines the risk per unit exposure (see text).

- The joint probability is the probability that a given exposure will occur and an attack will occur with that exposure: joint probability is $p[C_{expo}[i]] \times p[\text{att} \mid C_{expo}[i]]$. In the table, for 0.02–0.04 the joint probability of an attack is 0.0006.
- The overall probability of an attack in a given Δt is the sum of the joint probabilities,

$$p[\text{att}]_{\Delta t} = \sum p[C_{expo}[i]] \times p[\text{att} \mid C_{expo}[i]],$$

which in this case is $p[\text{att}]_{\Delta t} = 0.024$ for an exposed individual.

Table 13.1 Calculation of the overall probability of a response during a random Δt Period

Exposure Distribution				Exposure-Response Curve	Joint Distribution
Exposure Ranges	Median Exposures	Number of Observations	$p[C_{expo}[i]]$	$p[\text{att} \mid C_{expo}[i]]$[a]	Joint Probability
0 – 0.02	0.01	1	0.020	0.007	0.0001
0.02 – 0.04	0.03	4	0.082	0.008	0.0006
0.04 – 0.08	0.06	19	0.388	0.009	0.0035
0.08 – 0.16	0.12	15	0.306	0.012	0.0037
0.16 – 0.32	0.24	6	0.122	0.022	0.0027
0.32 – 0.64	0.48	3	0.061	0.069	0.0042
0.64 – 1.28	0.96	1	0.020	0.450	0.0092
		49	Sum = 1		Sum = 0.024

[a] The exposure-response curve is given above in the third bullet point.

Note: all exposure measurements are made during Δt, so $p[C_{expo}[i]] = p[C_{expo}[i]\Delta t]$.

has been chosen to match the ETI needed to trigger a response.) This relationship is asymmetric across exposure because the exposure distribution is usually lognormal and highly skewed to large values, as shown in the box. When the background rate of the response is low ($p[\text{att}]_{\Delta t} < 0.05$), the relative contribution of the upper tail of the exposure distribution is disproportionately higher than the lower tail because the shape of the response curve at low response probabilities is concave upward. For example, in Table 13.1 the joint probabilities of the two highest exposure ranges, 0.32–0.64 and 0.64–1.28, contain 8.1% of the exposures, but they account for 56% of the joint probability of an effect. This is a common combination of conditions when the background risk of a response is low, less than ~0.01. This also may be the cause of our impression that discrete responses are associated with peak exposures. It also has important implications for sampling strategies, which we discuss in Chapter 14.

Assuming an underlying logistic relationship, the overall attack probability can be calculated by the equation

$$p[att]_{\Delta t} = \sum_{i=1}^{m} p\left(C_{expo}[i]\right) \times 1 / \left(1 + \exp\left(-\alpha - \beta * C_{expo}[i]\right)\right)$$

Equation 13.2 Logistic relationship for the probability of an attack during Δt

There are m intervals covering the range of exposures, and α and β are risk parameters. Equation 13.1 represents the mean probability of a response given the exposure distribution. The rate variable has a bionomial distribution with a mean value of $p[\text{att}]_{\Delta t}$ and a standard deviation of $\text{sqrt}(p[\text{att}]_{\Delta t} \times (1\text{-}p[\text{att}]_{\Delta t}))$. When the mean $p[\text{att}]_{\Delta t}$ is small, it has a highly skewed distribution. So we must have a long period of observation to obtain stable estimates. The mean number of attacks during a period of observation, $T_{obs} = N_{obs}\Delta t$, is given by

$$N_{att} = N_{obs} p(\text{att})_{\Delta t}$$

Equation 13.3 The expected number of attacks during a period of observation

The standard deviation of N_{att} is $N_{obs}\text{sqrt}(p(\text{att})_{\Delta t} \times (1\text{-}p(\text{att})_{\Delta t}))$. Thus the longer our observation time, the greater the chance of observing a reasonable number of attacks.

Another useful parameter to estimate is the probability of at least one attack during an observation period. During a period of N_{obs} one or more attacks can occur in a wide variety of ways, so it is easier to estimate the probability that there will be no attacks and then subtract that from 1 to estimate the possibility of one or more, as is shown in Box 13.3. The expression for at least one attack is

$$p(\geq 1\ \text{att}) N_{obs} = 1 - \left(1 - p(\text{att})_{\Delta t}\right)^{N_{obs}}$$

Equation 13.4 The probability of at least one attack during a period of observation

Box 13.3 Reversible Discrete Effects Model: Calculating the Probability of One or More Attacks

During each Δt interval in the Box 13.2 example, we observe that the overall probability of an attack is 0.024 for the joint distribution of exposures and exposure-response curve for an individual. Therefore, the probability of no attack is:

$$p[\text{no att}]_{\Delta t} = 1 - p[\text{att}]_{\Delta t} = 0.976.$$

- The probability of no attacks during N_{obs} observation period is the product of the probability of no attack during each interval multiplied together, or

$$p[\text{no att}]_{Nobs} = p[\text{no att}]_{\Delta t}^{Nobs}.$$

- Therefore, the probability of at least one attack during the time period of $N\Delta t$ is

$$p[\geq 1 \text{ att}]_{Nobs} = 1 - p[\text{no att}]_{\Delta t}^{Nobs},$$

which is not linear with large numbers of intervals, as shown.

Table 13.2 Probability of at least one attack with increasing duration of observation

Duration in $\Delta ts(N_{obs})$	Probability of no attacks in $Np[\text{no att}]_{\Delta t}^{Nobs}$	$p[\geq 1 \text{ att}]Nobs$
1	0.976	0.024
5	0.89	0.11
10	0.78	0.22
50	0.30	0.70

- The expected number of individuals in a group of N_{subj}, with ≥1 attack:

$$\text{Expected individuals} = N_{subj}\, p[\geq 1 \text{ att}]_{Nobs}$$

- We can develop a dose metric from this:

$$DM_j = N_{subj}\, p[\geq 1 \text{ att}]_{Nobs}$$
$$DM_j = N_{subj}(1 - (1 - \sum p[C_{expo}[ij]] \times p[\text{att} \mid C_{expo}[ij]])^{Nobs})$$

This metric will be proportional to the number of individuals having at least one attack in the *j*th exposure group, given the hypothesized dose-response relationship.

N_{obs} is the number of Δt intervals observed, and the total period of observation is $T_{obs} = N_{obs}\,\Delta t$. As expected, the probability of observing at least one attack increases strongly with the number of intervals observed. Notice that the expected or mean number of attacks with 100 intervals is 2.4, but there is still a 9% chance of no attacks. This also has important implications for sampling strategies, which we discuss in Chapter 14.

Key Points

1. To be able to quantify the risk relationship, the frequency distribution for the exposures must be the same for each random time interval, that is, the exposure distribution must be "stationary," having a fixed mean and standard deviation through time. This is critical because we cannot measure a given time interval more than once. Note that this does not mean that the exposure concentrations are constant. This assumption can be supported by observations that show that exposure determinants for the setting are stable for the time period under study, as noted in Chapter 2.
2. When exposures are lognormal, which is common, the upper tail of the skewness of the distribution increases the frequency of high-intensity exposures relative to the median, for example, two GSD values above the GM commonly may be four- to tenfold larger than the GM.
3. At low response frequencies (p[att] < 0.5) and low background response rates (p[att] < 0.01) the logistic response curves tend to be concave upward, which gives disproportionately higher risk to increasing exposures. When this is combined with a lognormal exposure distribution, which also tends to have disproportionately frequent high exposures, those high exposures can account for a major part of the overall risk, such as seen in Box 13.2.
4. For simplicity, we often assume that the target tissue concentration is directly proportional to the exposure intensity, such as the concentration on the surface of the airways. However, that assumption is not required. Pharmacokinetic modeling (Chapter 4) can be used to estimate the tissue concentration when simple linearity is inappropriate.
5. As a result of points 2 and 3, the exposure characterization should emphasize a good assessment of the upper tail of the distribution. Our goal is not just to obtain a good estimate of the geometric mean (GM) or mean exposure. A large enough sample size must be collected to obtain a stable estimate of the geometric standard deviation (GSD), such as $N > 30$, so the frequency of high-intensity exposures can also be estimated with reasonable precision.

Although this probabilistic model is relatively simple, it can capture some of the complexity of the biologic responses. Individuals will not respond every time a particular "threshold" (the theoretical minimum concentration at which that individual will respond) is exceeded, but the probability increases with increasing exposure, and sometimes they respond at exposures below the threshold. However, this model does not reflect all of the important temporal dynamics of the response. An individual cannot respond with a new attack in the period of time immediately following an asthma attack, for example? If the subject is removed from exposure when an attack begins, then a period of symptoms is likely to persist for some time without exposure, but then gradually subside. If exposure continues, however, the severity of the attack may increase or the duration may be prolonged. After an attack finishes, there may also be a refractory period, possibly amplified by medication, during which a new attack cannot be easily initiated. If individuals vary substantially in their exposure-response curves, it will substantially affect how the composite curve looks. These phenomena are not accounted for in the preceding simple model. To capture these dynamics, we could add another component to our model that defines what happens after a reversible discrete response occurs.

Dose Metrics for Discrete Reversible Responses

Simple Dose Metrics

In some cases it is not possible to estimate the approximate exposure-response curve. However, given the relationship in Equation 13.1, several simple metrics may be chosen. First, duration of exposure has often been used in the past. This is still a reasonable metric because the risk of an attack is a strong function of duration of exposure: the longer the exposure, the more chance for a response. Second, risk is also a strong function of the exposure intensity distribution, especially the upper tail. Another simple metric is to simply count the number of "peak" exposures during a period of exposure, such as the number of values larger than the 90th or 95th percentiles. However, the definition of a causal peak also depends on the response curve. Large values of exposure relative to the mean or median may not be large enough to produce a detectable increase in risk. Third, if there is anecdotal or clinical information that some types of exposures are associated with attacks, we could estimate the exposure intensities and use them as our cutoff for peak values. For example, asthmatics have "twitchy" airways that can respond to irritants. If we know the rough threshold for irritation for a chemical, we can begin by setting the minimum value for a peak at that level. Fourth, if there is no exposure data, we could count the number of tasks, or exposure situations, in which peaks are likely during a time period. This metric is likely to have some misclassification because all of the tasks may not have a peak sufficient to cause attacks. There is also the problem of variations in the sensitivity of the subjects, which will also introduce misclassification for the sensitive individuals.

Another problem with the simple dose metrics is the availability of relevant exposure data. What if all we have are data on monthly or annual average exposure intensities, but the ETI is 10 minutes? Should we abandon the study? As it happens, when we compare agents and settings with similar temporal variability, that is, similar distributions over time, then the frequency of peaks will affect the mean; more frequent and higher peaks cause larger means. Consequently, comparing the risk of response across settings with different means can reflect the underlying peaks, too, but only if they all have similar dispersion (spread; e.g., the GSD). In that case, we may detect differences in risk, but we may also be misled about what constitutes a peak exposure. A more rational alternative dose metric is one based on the probabilistic model we developed earlier.

Probabilistic Dose Metric

With proportional effects we used the sum of exposure intensities multiplied by weighting factors to obtain a dose metric, such as cumulative exposure, that is proportional to the effect. However, probabilistic effects are different in two important ways: their response intensity is not proportional to the intensity of exposure, and each subject can only respond or not. Therefore we need to adopt another strategy to develop a dose metric that is proportional to the risk of the effect. In this case, we look at groups of subjects with the same exposure distribution and observe the overall proportion of subjects who respond. The proportion responding is proportional to the group exposures, which is defined by the mean exposure and the dispersion of the distribution, the peak exposures.

The overall risk of responses is the joint distribution of exposures and the dose-response curve given exposures. The exposure distribution specifies how likely a given exposure is, and the response curve shows how likely a response is when the group receives a given intensity of exposure. When multiplied together they specify how likely a response is from that exposure. Finally, if we sum across all possible exposures in the distribution, we obtain the overall likelihood of responses from the exposed group.

An example of this calculation is shown in Box 13.2. How can we use this to calculate a dose metric for the discrete response? We have defined a dose metric as a metric calculated from the measured exposures that is hypothesized to be proportional to the effect. In this case, the "effect" is a group risk of an adverse response, such as asthma. If we could estimate a dose-response relationship that was proportional to the true relationship, then, given the measured exposure distribution, we could calculate a joint probability estimate for $p[\text{att}]_{\Delta t}$, and this would be proportional to the true $p[\text{att}]_{\Delta t}$. If we multiply the joint probability by the number of subjects in the group, we would have a metric proportional to the expected number of people responding to that exposure distribution. Note that this metric is not intended to be an estimate of the number of subjects responding, but an indicator calculated from the exposure that increases or decreases in proportion to the group risk.

To construct this dose metric, we could assume that a logistic relationship defined by summary data on the exposed population (baseline rate of attacks and overall average attack rate of exposed) was approximately the correct form. Therefore, the baseline rate of attacks and overall attack rate could be used to estimate the logistic parameters, α and β. The observed baseline rate of the response, $p[\text{att}]_{\text{base}}$ with no exposure can be used with Equation 13.1 to estimate $\tilde{\alpha} = \text{logit}(p[\text{att}]_{\text{base}})$. Then given an estimate of the mean exposure $\overline{C}_{\text{expo}}$ and an overall attack rate $p[\text{att}]$ associated with that set of exposures, we can estimate the slope from Equation 13.1, $\tilde{\beta} = \dfrac{(\text{logit}(p[\text{att}])}{C_{\text{expo}}}$. Given those parameter estimates, we can use this logistic relationship with the group (indicated by *j*) exposure distributions (elements indicated by $C_{\text{expo}}[ij]$) to calculate the following dose metric (DM) for each exposure group, as was done in Box 13.2. The DM_j is assigned to each member of the group.

$$DM_j = N_j N_{\text{obs}} \left(\sum_{i=1}^{m} \frac{p[C_{\text{expo}}[ij]]}{1 + \exp(-\tilde{\alpha} - \tilde{\beta}[C_{\text{expo}}[ij]]\Delta t_{ETI})} \right)$$

Equation 13.5 Dose metric for discrete reversible processes

This dose metric is intended to be proportional to the expected number of attacks during the period of observation, $N_{\text{obs}}\Delta t_{\text{ETI}}$, for the subjects N_j in the group. The DM_j will give a metric proportional to the expected number of attacks for the *j*th exposed group. Because the overall probability is sensitive to the joint distribution of the exposure distribution and the exposure-response curve, small changes in either may have a large effect on DM_j. The utility of this DM requires that enough exposure samples be collected to make a good estimate of the exposure distribution, including the upper tail. In essence this metric is a weighted sum across the exposure distribution for the group.

Our dose metric has several useful intuitive characteristics. First, duration of observation ($N_{\text{obs}}\Delta t$) during exposure is an important part of the metric; longer observation is associated with increased frequency of responses. Second, although it is very difficult to measure the concentration in the exposure interval before an attack, we can measure the exposure *distribution* ($p[C_{\text{expo}}[ij]]$) with a desired precision for an exposure period in a given setting, assuming that the distribution is stationary and not changing over time, which will have the same distribution for each subject's periods before the attacks. Even if it is not possible to observe average concentrations on very short time intervals, as noted earlier, longer Δt averages may still be useful because averages are sensitive to peak values. Third, although each person may have a somewhat different exposure-response curve ($p[\text{att} \mid C_{\text{expo}}]$), the population curve is an "average response," which we can estimate based on clinical data from the population, as was suggested earlier.

One of the common observations about asthma attacks and other immune responses is that they seem to be associated with short-term intense exposures, so-called "peak" exposures. Examination of the data in Figure 13.2 shows that there can be an important interaction between the shape of the exposure distribution and the exposure-response curve. In general, most exposures are low enough that probabilities of an attack are well below 0.5 on the exposure-response curve. If the baseline response rate is low, $p[\text{att}]_{\text{base}} < 0.05$, then the probability values increase nonlinearly with increasing exposure. Consequently, the larger exposures in the upper tail can contribute disproportionately to the total probability of a response, as is shown in Box 13.2. When the exposure distribution has a larger GSD, its upper tail will extend further under the exposure-response curve, which increases the likelihood of a response for high exposures. A distribution with a large GSD, greater than 2.5 or so, has more "peak" exposures, that is, exposures much larger than the median (> tenfold). In that case, most of the observed risk comes from the highest exposures, which is implicit in the overlapping concave upwards shapes of the exposure distribution and the response curve. However, the response is not inherently associated with peaks, because lower values of exposure also cause responses, too; but because they have a low probability, they are rarely observed.

13.3. DISCUSSION OF REVERSIBLE DISCRETE RESPONSES

Some immunologic processes lead to reversible responses, such as dermatitis and immune suppression. Most hypersensitivity reactions to chemicals, such as isocyanates, nickel, and acid anhydrides, are reversible after exposure stops. But a severe response of this type can be life threatening and may require a long time to resolve. All four types of immune reactions can be produced by environmental or occupational exposure: Type I (mast-cell-mediated rapid responses—asthma, rhinitis, skin reactions), Type II (circulating antibody cytotoxicity—pulmonary disease, anemia), Type III (immune complex effects for specific tissues—hypersensitivity pneumonitis), and Type IV (delayed hypersensitivity—contact dermatitis). All agents do not produce all types of responses, and some may have more than a single type of response. Subjects who have prolonged responses may not have fully reversible effects. For example, severe asthma may develop into continuous bronchospasm, inflammation, and overproduction of mucus that may become unresponsive to treatment, with fatal consequences.

An important part of the process models is the time course of the recovery process. You cannot have an identifiable new asthma attack or dermatitis if you have not recovered from the previous one. Thus determining the incidence of attacks caused by environmental conditions requires knowledge of the recovery process. In the absence of data, subject reports and medical specialists' observations can be very useful to define the time course. Individuals do not necessarily have similar recovery dynamics, a fact that must also be considered because it has implications for the occurrence of false positives. Recovery can be handled in an analogous fashion to the recovery from reversible continuous responses. In essence, the physiologic system needs to reset. We explore this in more detail in Chapter 14.

Our dose metric for reversible discrete responses is a combination of the duration of exposure and the joint probability of a given exposure and the likelihood of a response. This metric is a population parameter, not an individual measure. As a result, it avoids the problem of needing to make very challenging short time interval measurements of individual exposures and responses.

An important aspect of the DM is its disproportionate emphasis on increasing exposures. The joint probability is asymmetric because the dose-response curve is approximately exponential going from low to high exposures, and as a result, the high exposures—the apparent peaks—may contribute disproportionately to the overall risk. The population distribution for sensitivity to the agent is also important. As the GM (approximate median) and GSD (indicator of skewness) of exposures increase, a greater fraction of the sensitive individuals will respond. Importantly, a population with a history of exposures may have a much lower frequency of highly sensitive individuals because of self selection out of the population.

13.4. IRREVERSIBLE DISCRETE DISEASE PROCESSES

The fourth disease process model covers discrete outcomes of mechanisms that involve several ordered steps where the final step is not to be repaired or not reversible. There are two important groups of diseases of this type: acquired sensitivities, such as asthma, and carcinogenesis. This model applies when there are at least two steps to a process that must occur in sequence and the last step is irreversible; these conditions impose some important characteristics on the temporal patterns that we observe. For example, it can appear that time is an independent predictor of risk or that the length of time for which someone is exposed to a toxin is more important than the intensity of the exposure. Although these simplifications may not be accurate representation of the details of molecular mechanism, there is often a kernel of truth in them.

Example: Air Pollution and Development of Asthma

Respiratory sensitization to air pollutants is a good example of an irreversible change in a person's health. McConnell and coworkers have shown that air pollution exposures can cause an increase in the prevalence of asthmatics in groups of children who participate in vigorous exercise in areas with high concentrations of smog (McConnell, Berhane et al. 2002). They followed groups of children living in 12 communities with different levels of air pollution (particulate and gaseous) in the Los Angeles area during a 4-year period. A group of 3,535 initially nonasthmatic children was enrolled. Each year a questionnaire was administered, and the number of children who had been told by a doctor that they had asthma was recorded. Monitoring stations measured the air pollution levels (ozone, nitrogen dioxide, and PM_{10} [particulate matter less than 10 μm dia.]) at a central site in each community. Overall, children living in high ozone areas, 73.5 ppb average daily 1-hour maximum, showed a small increase in new cases of asthma compared with low-ozone communities that had an average 44.6 ppb hourly maximum. However, children who engaged regularly in three or more active sports had an annual incidence rate of 0.05 new cases per year in the high ozone areas compared with 0.02 cases per year in the low-ozone or nonexercising groups. No association was found between asthma incidence and community levels of nitrogen dioxide or PM_{10}. Because of its prospective design, this study was able to make a clear distinction between the incidence of new cases and the occurrence of asthma attacks.

Typical Time Course

The ETI for developing asthma or other sensitization is difficult to define because the sequence of etiologic events is not known. In an exposed population, only a fraction of the total is susceptible and able to develop asthma. A person susceptible to developing asthma has airways that can become more reactive than normal and that are prone to bronchospasm when provoked by a variety of stimuli. Animal studies of sensitization suggest that it typically occurs after two episodes of a sufficiently intense exposure to the sensitizing agent within a relative short period of time, approximately 20 days (Gregus and Klaassen 2001). The duration required can vary considerably and depends on the agent. After that, the individual is sensitized and will respond accordingly. Sensitization typically appears to be a permanent change in immunological status, but some individuals apparently do slowly recover and lose their sensitization if they are not exposed for a long period of time.

Mechanisms

Given the right combination of high inhalation or skin exposure to a sensitizing agent and an individual with the right disposition, that person will become sensitized. Sensitization results when the immune system can recognize an antigen and responds by producing either specific protein antibodies, IgE, IgM, or IgG, or memory T cells. An example time course for animal sensitization for a humoral immune response is based on the finding that production of antigen-specific IgM requires 3–5 days following the first sensitizing exposure (Gregus and

Klaassen 2001). After a second sensitizing event, B cells switch to producing IgG antibodies, which have a higher affinity and a higher serum level (titer). The second response takes about 5–10 days. The antibodies may be bound to tissues or circulating, and this distinction defines the type and locations of responses. Sensitized individuals can develop cross-reactivity to other agents and may respond to nonspecific irritants. In some cases, the responses become progressively stronger with each exposure because of increased production of antibodies and other cellular components. There also are genetic factors that increase the risk of sensitization. For example, a family history of allergies may increase the likelihood of becoming sensitized.

The likelihood of sensitization also depends on the intensity of exposure. Short-duration, high-intensity exposures appear to be more likely to sensitize than long-term low levels. Exposures less than needed for sensitization can produce measurable levels of antibodies in unsensitized individuals. Some investigators have used antibodies as a biomarker of exposure. Exposures via inhalation, ingestion, or skin contact can lead to sensitization, both local and systemic, depending on the type of immunologic response. An individual's susceptibility to sensitization can vary with age, becoming either more or less susceptible. The dose-response for sensitization cannot be studied in an individual, because an individual is usually sensitized only once. Progressive responses can also occur, in which an exposure triggers a response that continues to develop after the initiating event has ended and no further exposure occurs, as in acute silicosis and beryllium disease.

Disease Process Model for Irreversible Discrete Responses

Irreversible discrete responses result from changes in an internal state that, once made, are irreversible or nearly so. As noted earlier, two important classes of diseases involving such mechanisms are acquired sensitivities, such as asthma, and cancer. We briefly discuss the development of models for these processes.

Sensitization Process

A simplified model that is based on animal experiments is a useful point of departure in developing a rational disease process model. An initial intense exposure to an agent forms a protein antigen, whose concentration must exceed some threshold to produce an immune system priming response. After a period sufficient to allow the development of the immune response, a second intense exposure produces a systemic response that results in an irreversible sensitization. The probability of these events most likely depends on the antigen concentration exceeding some minimum value. We assume that the animal model is relevant to the human response. However, it is rare that we can identify two well-defined exposure events that clearly led to sensitization. More commonly, individuals work or live in settings in which there may be regular exposures of varying intensities for periods of weeks, months, or even years before they become sensitized. Sometimes they report events with high exposure that preceded becoming sensitized, such as a chemical spill, but often they do not.

For many human chemical sensitizers, there is a reactive agent (hapten), such as an isocyanate, which covalently binds with a protein in the blood or skin to form the antigen. The antigen is cleared by normal protein recycling processes, so an intense exposure may be needed to produce sufficient antigen to provoke the response. It is common that exposed individuals may have antibodies in their blood without being sensitized. How high the antigen concentration must be, and how long it must be maintained, are not known, but with appropriate temporal data they may be estimated. This may be modeled as shown in Figure 13.4 with a branch switch that is keyed to the subject's sensitivity, or an assumption of general

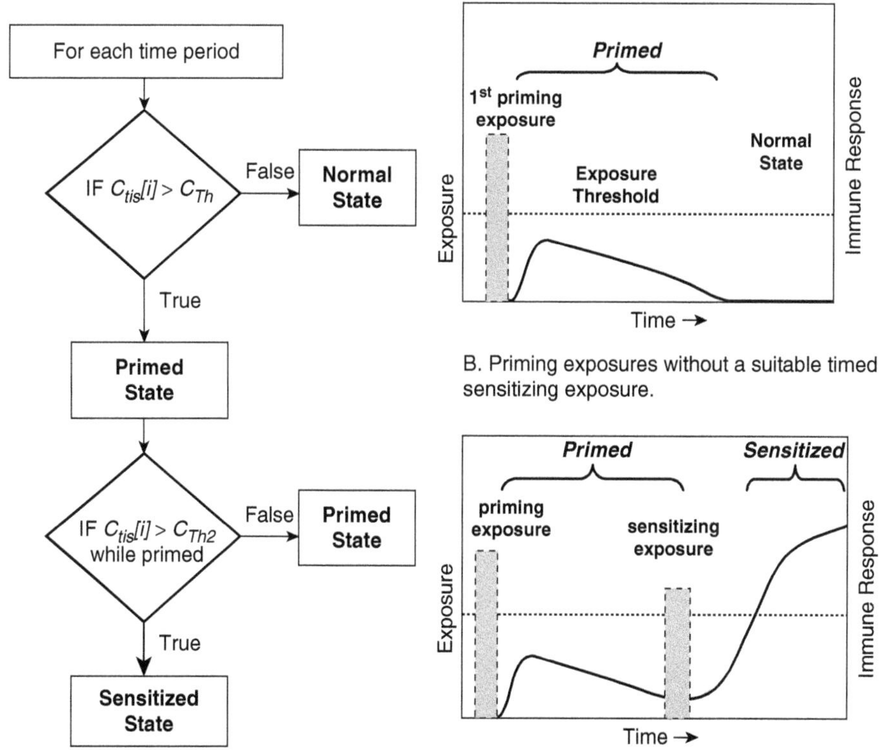

A. Diagram of discrete events associated with the development of sensitization.

B. Priming exposures without a suitable timed sensitizing exposure.

C. Priming exposures without a sensitizing exposure.

Figure 13.3 Processes associated with sensitization and their timing.

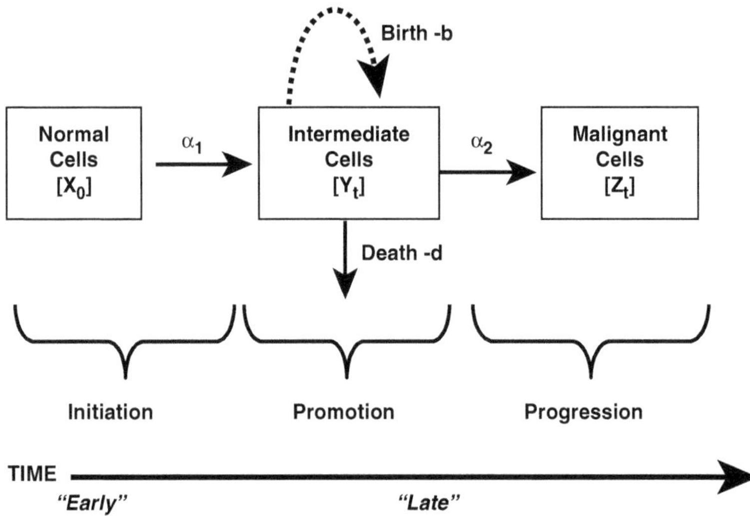

Figure 13.4 Moolgavkar-Knudson model of cell kinetics for cancer. (The model is described in more detail in Box 13.6.)

sensitivity. When a person has had a priming sensitivity response, there is a required minimum lag during which the immune response must develop before a secondary response can develop. Based on animal data, it is hypothesized that when a person has had a priming response, there is a lag during which the immune response develops before a sensitizing response can occur. If the priming response is not sustained, it may fade over a period of time (Lesley 2005). Once the minimum time has passed, a second exposure and immune response produce permanent sensitivity. We do not know whether the second antigen exposure must exceed some minimum intensity (C_{Th2} in the diagram), but this seems like a reasonable assumption, based on experimental and clinical evidence.

The sensitization step is assumed to have approximately the same temporal structure as the first step. However, clinical or other observational data may suggest otherwise. In that case, the parameters can be changed, as can the structure of the model. Because this model is based on the response to an antibody formed by a protein reacting with a reactive agent, we must consider the dynamics of that process, as it may affect the temporal characteristics of the sensitization outcome. Importantly, the sensitization step is dependent on the completion of the priming step. If the time is too short or too long relative to the first step, then sensitization will not occur.

Even without any knowledge of the immunologic processes, we can still model the overall process as a series of linked or conditional probabilities for each of the two steps. As we saw earlier for reversible discrete events, the probability of the priming step is a function of both the exposure distribution and the subject's exposure-response curve. Assuming that the animal model is correct, the occurrence of the sensitization step must follow the priming step within a period of time, $T_{prim} = N_{prim}\Delta t$, to have a sensitization. We can estimate the probability of sensitization by considering the joint probability that both will happen in sequence. If we were following the risk of sensitization in an individual, we could represent the risk for each time period as the chance of the priming step times the probability that the sensitization step would occur within the next N_{prim} time intervals. For the priming step, each time interval is a separate chance to start the process, and the total probability is the probability at each interval times the number of intervals observed. Therefore, during a period of observation, we can calculate the likelihood that at least one priming step will occur, which must be followed by at least one sensitization step within T_{sens} interval. This is presented in more detail in Box 13.4.

The quantitative probability relationships shown in Box 13.4 are more complex than the earlier relationships we have explored. There are two important features of these relationships. First, the sensitization exposure must follow a priming exposure within a fixed time period, which is defined by the biology of the response. As a result, if the high priming exposures are infrequent, there may be little risk of sensitivity for most people; only those with high sensitivity are at risk. Second, the investigator has control only of the duration of observation. Normally, increasing the duration of observation will increase the chance of observing a response; however, that will not change the probability of two high exposures occurring sufficiently close together to produce sensitization.

Clearly, if high exposures are required and if they must occur within defined time periods, then the chance of someone becoming sensitized in a setting with low exposures may be quite low. However, there also appears to be a wide range in personal exposure-response curves across individuals such that a subgroup of individuals may not require high exposures to become sensitized.

13.5. DOSE METRIC FOR IRREVERISIBLE DISCRETE PROCESSES

We take our dose metric from our expression for the probability of sensitization in Box 13.4. Assuming that the exposure-response curves for priming and sensitization are very small, we can simplify the relationship and develop a relatively simplified dose metric, shown in Box 13.5.

We can use the data from the McConnell et al. (2002) study of the incidence rate for new asthma cases in Los Angeles, mentioned earlier, to examine the possible values for $p[\text{prime}]_{\Delta t}$ and $p[\text{sens}]_{\Delta t}$. A simulation was performed using the expressions in Box 13.5 and the high and low ozone exposure distributions and asthma case incidence rates. Assuming that the minimum duration ozone exposure to cause a response is Δt = 60 min, and the duration within which a sensitization

Box 13.4 Probability Relationships for Irreversible Sensitization

- Two steps are required for sensitization: priming followed by sensitization within a defined period. We assume the probability of priming during Δt is analogous to a reversible immunologic response given the exposure distribution, discussed in Box 13.2:

$$p[\text{prime}]_{\Delta t_i} = \sum p[C_{\text{expo}}[i]\Delta t] \times p[\text{prime} \mid C_{\text{expo}}[i]\Delta t]$$

where $p[\text{prime} \mid C_{\text{expo}}[i]\Delta t]$ is a logistic relationship as it was in Box 13.2.

- The probability of the sensitization step during Δt has a similar probability relationship:

$$p\left[\text{sens} \mid \underset{i}{\text{prime}}\right]_{\Delta t} = \sum p[C_{\text{expo}}[i]\Delta t] \times p[\text{sens} \mid C_{\text{expo}}[i]\Delta t]$$

- However, to become sensitized, a sensitization event must follow priming within T_{sens} (where $T_{\text{sens}} = N_{\text{sens}}\Delta t$). For a given Δt, the probability of a priming event being followed by at least one sensitizing event during T_{sens} has the form:

$$p[\text{sens} \mid \text{prime}]_{\Delta t} = p[\text{prime}]_{\Delta t} \times \left(1 - p[\text{no sens} \mid \text{prime}]_{\Delta t}^{N\text{sens}}\right)$$

- During a period of observation, N_{obs}, each Δt is a separate opportunity for sensitization to follow priming. Therefore, the overall probability of at least one sensitization event following a priming event during a period of observation, N_{obs}, is

$$p[\geq 1 \text{ sens} \mid \text{prime}]_{N\text{obs}} = 1 - \left(1 - p[\text{sens} \mid \text{prime}]_{\Delta t}\right)^{N\text{obs}}$$

Equation 13.6

- This probability can be used to estimate the minimum expected number of new cases, $N_{\text{sens\_subj}}$, during observation by multiplying by the group size, N_{subj}:

$$\text{Expected } N_{\text{sens\_subj}} = N_{\text{subj}}\left(1 - \left(1 - p[\text{sens} \mid \text{prime}]_{\Delta t}\right)^{N\text{obs}}\right)$$

Equation 13.7

must occur after priming, $T_{\text{sens}} = 10$ days ($N_{\text{sens}} = 80$ midday hours), and an observation period of a year, $T_{\text{obs}} = 365$ days ($N_{\text{obs}} = 2{,}920$ midday hours), and assuming that the risk of priming and sensitization during a Δt are approximately the same, then if $p[\text{prime}]_{\Delta t}$ and $p[\text{sens}]_{\Delta t} = 0.0003$ the overall risk of sensitization is $p[\text{sens}]_{N\text{obs}} = 0.02$. The probabilities of priming and sensitization are so small

Box 13.5 Probabilistic Dose Metric for Irreversible Discrete Effects

Our goal is to develop a probabilistic dose metric that will be proportional to the risk of sensitization for a group of individuals with the same probability distribution of exposure intensities during Δt_{ETI} intervals. Simulations suggest that probabilities of both priming, p[prim] $_{j,\Delta t}$, and sensitization, p[sens | prime] $_{j,\Delta t}$, are very small. If so, then we can simplify the logistic weighting relationship to $e^{\alpha+\beta C_{expo}[ij]\Delta t}$ related to the logistic function (see text), where j is the exposed group, i is an interval of the exposure distribution, and α and β are the same parameters as in the logistic relationship. Both the priming and sensitization for the jth group will be weighted this way, but the α and β values will not be the same.

$$p[\text{prim}]_{j,\Delta t} = \sum_i p\{C_{expo}[ij]\Delta t\} \times \exp(\alpha_P + \beta_P C_{expo}[ij]\Delta t)$$

$$p[\text{no sens} \mid \text{prime}]_{j,\Delta t} = 1 - \sum_i p\{C_{expo}[ij]\Delta t\} \times \exp(\alpha_S + \beta_S C_{expo}[ij]\Delta t)$$

As expected, the overall probability will be strongly weighted by the upper tail of the exposure distribution. We will use equations from Box 13.4 to estimate the probability of sensitization within N_{obs} intervals of Δt. The dose metric for the jth exposure group is

$$\mathbf{DM}_j = N_{subj}\left(1 - \left\{1 - p[\text{prim}]_{j,\Delta t} \times \left(1 - p[\text{no sens} \mid \text{prime}]_{j,\Delta t}^{\,N sens}\right)\right\}^{N obs}\right)$$

where N_{sens} is the number of Δt intervals in T_{sens}, N_{obs} is the number of Δt intervals in T_{obs}, and N_{subj} is the number of subjects in the jth exposure group. This probabilistic dose metric will be proportional to the number of subjects who become sensitized in a period of observation, given the exposure distribution and a hypothesis about the exposure probability weighting function (in this case the logistic function).

because there are many opportunities to have a response. This information can be integrated with the exposure distributions to estimate the dose-response functions (see Chapter 14).

This metric has the same characteristics as the probability statement for the risk of sensitization. First, the metric increases with the duration of the observation period. However, the effect of increasing duration is limited by the number of time intervals associated with the period of heightened sensitivity after priming, N_{sens}. As a result, the risk does not increase directly with the period of observation, N_{obs}, but is more sensitive to the frequency of peak exposures that produce the priming step. Most important, the dose metric is a strong function of the overlap between the exposure distribution and the exposure-response curve, which together define the likelihood of a response during a given time interval. Second, because the probability of a sensitizing step is squared, this dose metric will be nonlinear and extremely sensitive to values in the upper tail of the exposure distribution, that is, the peak exposures. Finally, the step 1 and 2 exposure-response curves may not be the same, which will affect the likelihood of becoming sensitized.

Simple dose metrics for sensitization are related to those for allergic responses but must include some estimate of peak exposures to be meaningful. So the frequency of peaks greater than a biologically relevant value, such as the threshold for irritation, must be included. Also, the metric should consider the frequency of peaks greater than the threshold during some estimate of the critical biological time (T_{sens}) within which the second step should occur. For example, an exposure should have two peak exposures greater than the threshold within T_{sens} to produce sensitization.

13.6. CARCINOGENESIS

Cancers also occur through a multistep cellular process, and a number of researchers have argued that a two-step model based on two rate-limiting steps may capture the important dynamics, even if there are many more than two steps in actuality (Moolgavkar and Knudson 1981). According to the two-stage model, tumors develop after two sequential and irreversible steps, "initiation" and "transformation" (Box 13.6). In the first stage, this is usually envisioned as a mutation: the DNA in a normal stem cell is damaged and then fixed as a mutation by cell replication or misrepair or other processes, which produces an "intermediate" cell. This event is presumed to be rare, and we do not need to know what the event is to model it. The intermediate cell is affected by normal kinetics of replication and death, and most important, it may undergo a second mutation to form a malignant cancer cell. The second mutation to form a fully malignant cell is assumed to occur rarely and may or may not be affected by chemical promoters. This final step will be more likely to occur if there are many cells in the intermediate stage as a result of clonal expansion (promotion). The fully transformed malignant cell may be killed by immune surveillance, or it may begin a clonal expansion leading ultimately to a clinically diagnosed tumor.

We can model this two-step process with a simple stochastic model. The key processes are those affecting the initiated cells—their formation, replication, death, and transformation into malignant cells. The formation and transformation are probabilistic events, and the replication and death are cell kinetics. Turnover times vary widely for different tissues. You can see how there is an interaction between the level of DNA damage, the frequency of altered base pairs, and the likelihood that a cell replication will lead to a mutation. In fact, this is an oversimplification, but it may capture the most important events. More important, it is a testable hypothesis, which can be made more complex if the model does not fit the data.

Box 13.6 Disease Process Model for Cancer

The kinetics of initiated cells drive the cancer process (formation of a malignant cell). Once an initiated cell is formed, it is the relative rates of replication and death that will determine whether the initiated cells clonally expand or die off (Figure 13.6). If there is clonal expansion of the initiated cells, the probability of at least one cell being transformed increases with time. The probability of initiation and transformation are both very small, and one or both may be sensitive to the effects of toxins.

This process can be modeled by a single compartment model representing the initiated cells with four processes affecting their numbers, as shown:

Initiation: K_{ini} = probability of forming an initiated cell during Δt
Replication: K_{rep} = number of new cells formed per Δt
Death: K_{dth} = fraction cells dying per Δt
Transformation: K_{trn} = probability of a malignant cell from an initiated cell during Δt

A stochastic difference equation can calculate the number of initiated cells during each Δt interval:

$$N_{ini}[i+1] = N_{ini}[i] + \left\{ \underset{\text{New initiated cells}}{\text{Poisson}(K_{ini} C_{exp}[i]\Delta t)} + \underset{\text{Cell growth}}{\text{Poisson}(K_{rep})} - \underset{\text{Cell death}}{\text{Poisson}(K_{dth}\, N_{ini}[i])} - \underset{\text{Transformation}}{\text{Poisson}(K_{trn}\, N_{ini}[i])} \right\}$$

This equation is recalculated for each interval beginning from birth. By definition, when a fully transformed malignant cell occurs, the subject has cancer.

This is discussed further in Chapter 15, where a suitable dose metric will be developed.

13.7. DISCUSSION OF IRREVERSIBLE DISCRETE RESPONSES

Less is known about the precise exposure circumstances that lead to sensitization and cancer than about proportional responses such as lung fibrosis, because the discrete responses are difficult to study. Short-duration, high-intensity exposures are labor- and equipment-intensive to measure and are often not suited to personal monitoring. Low- probability biological events that are not externally evident are the most difficult to observe. As a result, we know little about the intermediate individual human responses except what we can glean from clinical and epidemiological studies, the former of which emphasize personal idiosyncratic responses to nearly unique situations and the latter aggregate responses across a population. Our understanding of these irreversible processes needs data from better tools that can follow personal exposures and intermediate responses—by using biomarkers of early effects, for example.

14 Reversible Discrete Disease Processes: Asthma and Indoor Air, Dermatitis and Metalworking Fluids

14.1. INTRODUCTION

A child from New York City goes to a special summer camp for asthmatic children in the Catskill Mountains in upstate New York to escape part of another hot summer in the city. During a warm afternoon playing softball the child starts to wheeze and cough, despite having taken a bronchodilator before the game. Researchers from New York University, who are studying the relationship between asthma attacks and air pollution, record the attack. During the 8 weeks the children are at the camp, the NYU investigators will record the number of children reporting attacks and other symptoms each day, what activities they were doing, and their medication use, and they will measure their pulmonary function (forced expiratory volume in 1 sec, FEV_1, and forced vital capacity, FVC) and the pollutant levels in the air: ozone, nitrogen oxides, particulate matter < 10 μm in diameter (PM_{10}), acid particles, and other pollutants. Later, reviewing their data, they observe that 6 of the 20 children playing softball had reported asthma attacks on that afternoon. They also note that in the morning, auto emissions from New York City had drifted northeast on the light prevailing southwest winds, and the ultraviolet (UV) rays from the sun produced ozone and other reactive components (photochemical smog) in the drifting air. The result was a high concentration of ozone when the air mass moved over the camp. A few days earlier, the same children had played softball in the afternoon with only one asthma attack, but the wind had been from the north and the ozone level was low.

This fictionalized anecdote was extrapolated from a study conducted by Thurston and coworkers in the 1990s (Thurston, Lippmann et al. 1997), which found a relationship between the frequency of asthma attacks and the concentrations of ozone in an upstate New York summer camp.

Reversible discrete (yes/no) responses, such as asthma and dermatitis, are consequences of an important group of immunologic processes that can be initiated by environmental challenges, which were discussed in Chapter 13. Their basic feature is a change in a person's responsive state, "sensitization," after which the person exhibits sensitivity to both specific and nonspecific agents. Atopic, or allergic, individuals are the most likely to develop immune reactions. Exposure (contact with the agent) produces a nonproportional response whose severity is often not determined by the magnitude of the exposure but by "host" factors. Constructing models to study the onset of sensitization and other irreversible stochastic changes is the subject of Chapter 15; here we focus on modeling acute reversible discrete events.

In this chapter we illustrate the application of the disease process model approach to discrete reversible disease processes. We apply the process model approach in five steps: (1) identifying and showing the relevance of a tentative process model; (2) identifying one or more markers of the key effect; (3) choosing an etiologic time interval; (4) selecting the hypothetical agent(s) and a measurable marker for exposure; and (5) defining a summary dose metric that represents our hypothesis about the agent and mechanism to be tested in an epidemiologic study.

14.2. AIR POLLUTION AND ASTHMA IN URBAN CHILDREN

Health surveys of children in large cities like New York, Baltimore, and Boston have found an epidemic of asthma among children, especially among African Americans and Latinos (Busse and Mitchell 2007). The frequency of emergency room visits for asthma is high in these communities, access to health care is limited, and the quality of housing is often inferior. Studies over the past 15 years have reported a variety of factors to be associated with increased risk of asthma. Environmental exposures include allergens in the home from molds, dust mites, cockroaches, mice, cats, and dogs; exposures to endotoxin; second hand cigarette smoke; air pollution from industries and traffic; stress from poverty and violence; and poor diet, limited physical exercise, and obesity (Gruchalla, Pongracic et al. 2005). The associations with these factors have not all been consistent, and important questions about etiology remain (Gold and Wright 2005).

Housing conditions in the poor neighborhoods of large U.S. cities are of notoriously low quality. Infestations by cockroaches, rats, and mice are common. Leaking roofs and windows have caused problems with mold growth. An evaluation of conditions in public housing in Boston found that many of the apartments had evidence of roach, mouse, and rat allergens in the kitchen, bedrooms, and other rooms (Peters, Levy et al. 2007). Indoor air concentrations of endotoxin from the cell walls of gram-negative bacteria have also been shown to be associated with increased risk of wheezing (Litonjua, Milton et al. 2002). Comparisons of allergen levels between inner-city and suburban homes typically show that levels are higher in inner-city homes (Simons, Curtin-Brosnan et al. 2007).

The clearest quantitative linkage between allergens and responses in children is found for cockroach allergens in bedding, bedroom floors, and kitchen floors, where children are most likely to be exposed (Busse and Mitchell 2007). Weaker relationships are found between other allergens and frequencies of asthmatic responses.

We use this example to show how we could apply the process model approach to answer some of the remaining questions about childhood asthma in the inner city. Our basic hypothesis is that some allergenic agent causes asthma attacks.

Tentative Disease Process Model: Responses of Sensitized Individuals

When cockroaches and rodents seek food in an apartment, they deposit allergenic proteins in urine and feces. Depending on the intensity and frequency of cleaning, the allergens are present on the floors, kitchen work surfaces, furniture, and bedding in the apartment to varying degrees all of the time. These can become aerosolized and inhaled by the children. Asthmatic children typically only have asthma attacks once per month or less. Some may be taking prophylactic drugs to damp their allergic responses, which decreases their sensitivity and raises the intensity of exposure needed to produce an asthma attack. Thus, although the allergens are present all the time, the intensity of the challenge is probably only strong enough to produce a response once in a while. Perhaps there are only a limited set of vigorous activities, such as household cleaning by dusting and sweeping or running and playing in an area with high levels of allergen, that can produce high enough levels of airborne allergens to provoke an attack.

We can use the discrete reversible disease process model to represent the asthma process, as we did earlier (Chapter 13). The initiating dose is some exposure intensity, C_{expo}, for a minimum duration, Δt_{ETI}. We can consider the process trigger as a logistic dose-response probability distribution, as shown in Equation 14.1, where α defines the baseline attack rate and β defines the response factor per unit dose:

$$\ln\left(\frac{p[att]}{1-p[att]}\right)=\alpha+\beta(C_{expo}\,\Delta t_{ETI})\text{ or }p[att]=\frac{1}{1+e^{-\alpha-\beta(C_{expo}\,\Delta t_{ETI})}}$$

Equation 14.1 Two forms of the logistic model of risk from exposure

The population will partly determine the baseline attack rate. Our goal is to define the parameter β. However, as noted earlier, we do not define individual dose-response distributions; we define the dose-response for our study population, which is a function of the individual baseline rates and their personal sensitivity factors. If everyone in the population has the same exposure distribution, then when the probabilistic dose-response is steep or our ability to precisely measure the exposure is poor, the process may appear indistinguishable from a threshold response. As a first approximation, the threshold assumption is a useful way to represent the process. Viewed this way, we can see that when an exposure exceeds the threshold exposure for a sufficiently long period of time, an attack is triggered.

Markers of Effects: Tracking Responses

Biomarkers such as allergen skin tests can determine whether the children have evidence of immunologic sensitivity to the cockroach or rodent allergens, which may be related to risk of asthma. Questionnaires can determine the frequency of asthma attacks, housecleaning, home conditions, and so forth. Daily diaries completed by the caregiver can track when the attacks occurred and gather data on the child's activities and conditions surrounding the attack.

Etiologic Time Interval: Duration of Exposure Causing Effects

Personal reports can define the likely range of the etiologic time interval (ETI). However, for some individuals it may not be clear when they experience a sufficient exposure to trigger a response. Alternatively, we may use clinical reports for similar responses to estimate the ETI. Typically, asthmatic patients report that when they have a noticeable exposure, the onset of the attack is rapid, within minutes at the most. However, it is possible that at low levels of exposure a longer time interval is required to allow sufficient allergen to accumulate on the respiratory surfaces. To begin the assessment process, we assume that the process is fast and choose a Δt_{ETI} of 10 minutes immediately before the attack.

Exposure Marker for Hypothetical Agent: Intensity of Exposure

For asthma in children in substandard housing, we hypothesize that the cockroach allergen is the primary agent because cockroach allergen has appeared to be a strong factor in several studies (Busse and Mitchell 2007). We can measure it in dust samples and use it in skin tests to detect sensitivity.

Given that our hypothesis is that a short-duration exposure to a sufficiently high allergen concentration is needed to trigger an attack, we need to make sequential brief measurements of the allergen in the child's breathing zone so that we can observe the concentration just before an attack. This would let us see the personal dose-response relationship. But those short-duration levels are rarely measurable with current technology. If we measure longer time intervals, we may miss the short exposure (peak) needed to provoke an attack. However, given an approximately lognormal distribution of short-term temporal exposures, the mean of longer time intervals also will be increased by the presence of peaks. A series of longer duration air measurements can be made at fixed locations in the living room, bedroom, and kitchen to define the temporal variation in those areas. They will be longer than the Δt_{ETI} for personal exposures, but we can assume that a relationship exists. An alternative strategy, noted in Chapter 13, is to use the 24-hour air level, although this would likely misclassify the relevant exposure to the ETI substantially.

Dose Metric: Weighted Probability of Exposure Intensity

In Chapter 13, we discussed a probabilistic dose metric based on the assumption that a logistic relationship defines the probability of an attack, and a rough estimate of that probability can be used to weight the probability of an exposure, as shown in Equation 14.2:

$$DM_j = N_{obs}\left(\sum_{i=1}^{m}\frac{p\{C_{expo}[ij]\Delta t_{ETI}\}}{1+\exp(-\tilde{\alpha}-\tilde{\beta}C_{expo}[ij]\Delta t_{ETI})}\right)$$

Equation 14.2 Probabilistic dose metric based on weighted exposure probabilities

The weighting function (the denominator) is the least complex form of the logistic relationship. The baseline asthma risk occurs when C_{expo} is 0. For example, if we know that the frequency of population risk of asthma is 8%, then the right-hand expression is $\tilde{\alpha} = \ln(0.08/1\text{–}0.08)$, or $\tilde{\alpha} = -2.44$ when there is no exposure. We can use the observed overall increase in the asthma attack rate in the exposed population to calculate our initial estimate of β. Given the measured or estimated exposures for subgroups of the study population who have estimated attack rates, we can calculate the metric, which can then be used in the full epidemiologic analysis.

Sampling Strategies

The accuracy of the probabilistic dose metric depends on having a good estimate of the exposure distribution, particularly the upper tail that contains the peak exposures. At least 30 exposure samples are needed to make good estimates of the geometric mean (GM) and geometric standard deviation (GSD), when the GSD is not extreme (> 3). When specific activities are known to cause high exposures, they should be measured specifically and time-activity records used to determine their relative frequency in the overall pattern of exposures. Real-time monitoring with direct reading instruments may be very useful when the Δt_{ETI} is relatively short, as in this case.

Application to Asthma in Children Exposed to Cockroach Allergen

A large study of childhood asthma has been conducted in the Boston area, the Epidemiology of Home Allergens and Asthma Study (Litonjua, Milton et al. 2002). As a part of this study, a group of 226 children less than 5 years of age was followed for 4 years. Their respiratory problems were tracked through a baseline home visit and an annual telephone follow-up with a questionnaire that covered respiratory symptoms and identified individuals with one, two, or more episodes of wheezing each year during the 4-year period. Each child's home was visited and vacuum cleaner samples collected with a standardized protocol to measure house dust in the baby's bedroom and bed, parent's bed, living room, and kitchen floor. For the older children in the study, the data for the living room and kitchen were used to indicate exposures. The samples were analyzed for allergens from cockroaches (Bla g 1 or g 2), dust mites (Der f 1 or p 1), and cats (Fel d 1), and when there was sufficient sample material, it was analyzed for endotoxin (total amount and concentration). These data represent the status of a relatively stable source of exposure. This is a necessary but not sufficient index of exposure. High contamination may not produce high exposure if it is not made airborne by home activities. If home activities are not too variable across the population, then home contamination may be a good surrogate for exposure. This variation was not evaluated in the study, however.

The fraction of children with wheeze during each of the 4 years was contrasted with those who did not wheeze. There is very likely to be a range in sensitivity across the population. As a result, nonresponders may either be unsensitized or have a high response threshold (low sensitivity). Overall, 47 (20.8%) children had two or more episodes of wheezing on different days each year during the follow-up period. A 2.2-fold difference in relative risk of wheeze was observed for subjects from homes with concentrations of cockroach antigen above the median (30% vs. 15% prevalence of wheeze, comparing those above and below 0.05 U/g). A decline in annual wheezing episodes was associated with a decline in median house dust content of endotoxin.

Using Available Data to Calculate the Dose Metric

The original data analysis compared wheezing between groups living in homes with above versus below the median vacuumed dust cockroach allergen concentrations during each of the 4 years. The responses were limited to the number of daily occurrences of wheezing (a person can probably have only one asthma attack or episode of wheezing in a day). This creates a challenge for our modeling approach because our exposure measure is an annual dust sample and the outcome measure is the annual frequency of wheezing attacks. Clearly, the temporal link is very weak. We have proposed that the true Δt_{ETI} is 10 minutes, but we cannot localize either the exposure or the response within 10-minute intervals. To make this tractable, we must assume that the daily exposure is proportional to the home contamination and that during each year there are 365 exposure periods during which an attack could occur. We have little choice, then, but to assume that the Δt_{ETI} is 24 hours, despite our a priori assumption that it is very likely to be much shorter.

The observations of wheeze risk with and without home contamination can be applied to estimate the dose metric curve, as shown in Figure 14.1. The dose metric is a quantity calculated from the contamination data that represents the potential for response. This is expected to be proportional to the response if our hypothesis about the response process is true. Calculation of the dose metric is described Box 14.1, "Estimation of Reversible Discrete Dose Metric Parameters."

Box 14.1 Estimation of Reversible Discrete Dose Metric Parameters

The logistic probability curve can be used to describe the probability of an attack for a given exposure intensity. The relationship from Equation 14.1 is shown:

$$\ln\left(\frac{p[att]}{1-p[att]}\right)=\alpha+\beta\left(\bar{C}_{\mathrm{expo}}\Delta t_{ETI}\right) \text{ and } p[att]=\frac{1}{1+\exp\left(-\alpha-\beta\left(\bar{C}_{\mathrm{expo}}\Delta t_{ETI}\right)\right)}$$

Equation 14.1 Logistic relationship for probability of an attack given an exposure

The Litonjua et al. (2002) hypothesis was that detectable cockroach contamination was associated with increased risk of asthma attacks. The biggest problem with application of our approach for the dose metric is the uncertainty in the levels of exposure associated with the observed 15% baseline and 30% response rates. The baseline response is assumed to be the population background rate without exposure. Many of the cockroach allergen measurements were below the limit of detection (LOD). The investigators stated that they used the LOD as the cutoff for the yes/no decision about a subject's exposure in the response analysis. Consequently, the actual value of the exposure might have been higher or lower.

(*Continued*)

Box 14.1 Estimation of Reversible Discrete Dose Metric Parameters (*Continued*)

ESTIMATION OF THE EXPOSURE DISTRIBUTION

The report by Litonjua et al. (2002) indicates only that the median home contamination with cockroach allergen was 0.05 U/g of dust and that 6.2% had amounts greater than 2 U/g. If we assume the contamination levels are lognormally distributed, then the geometric mean (GM) is approximately the median, ~0.05 U/g, and if 2 U/g is approximately two geometric standard deviations (GSD) above the GM, then ln(GM) + 2ln(GSD) = ~2 U/d, and the GSD is ~6.3. Given that distribution, the 75th percentile for cockroach allergen will be ~0.2 U/g, which is midway between the median and the 100th percentile. In Figure 14.1 we have plotted the logistic relationships for three possible mean contamination levels, 0.1, 0.2, and 0.5 U/g, that might have been associated with the 30% prevalence that are roughly consistent with the exposure distribution and a baseline prevalence of 15%. Clearly, these three dose metric curves are not precise enough to be applicable for risk assessment of this population. However, they do provide some indication of how one might use the available data to assess what the parameters might be.

ESTIMATION OF LOGISTIC PARAMETERS

There are very limited data with which to estimate the logistic curve that we would use to calculate a dose metric for each exposed group. Given the baseline annual incidence of 15%, or 0.15 attacks per person per year in the group with no detectable exposures, then we can use the logistic relationship in Equation 14.1 to solve for the value of $\tilde{\alpha}$ when exposure is zero, $\tilde{\alpha}$ = In(0.15/1–0.15), and $\tilde{\alpha}$ is –1.73. We can use Equation 14.1 again with the value of $\tilde{\alpha}$ to estimate the value of $\tilde{\beta}$, as shown following. The estimated value of Bla g 1 and g 2 allergen quantities above the median can be used to estimate $\tilde{\beta}$. The approximate level of Bla allergens was ~0.1 μg/cm², and the annual risk was 2.2 times the baseline 0.15, or 0.33 per year.

$$\tilde{\beta} = \left(\ln\left(\frac{p[att]}{1 - p[att]} \right) - \tilde{\alpha} \right) \Big/ \left(C_{expo} \Delta t_{ETI} \right)$$

$$\tilde{\beta} = \left(\ln\left(\frac{0.33}{1 - 0.33} \right) + 1.73 \right) \Big/ \left(0.1 \mu g/cm^2 \times 1\, day \right).$$

$$\tilde{\beta} = 10.4\, cm^2/\mu g\, day$$

Those values yield an estimate of 10.4 cm²/ug day for $\tilde{\beta}$. Note that the units of this parameter are unusual because they are the inverse of $C_{expo}\Delta t_{ETI}$ so that the units will cancel out giving a unitless value to match the α value. These estimates for the two parameters, $\tilde{\alpha}$ and $\tilde{\beta}$, will allow us to calculate a metric that is proportional to the expected number of attacks for groups with measured levels of allergen exposure.

Our proposed dose metric for a discrete reversible effect is based on the joint distribution of the exposure and exposure-response curve:

$$p[\text{att}]_{\Delta t} = \sum p[C_{expo}] \times p[\text{att} | C_{expo}]$$

$$DM = N_{exp} \times p[\text{att}]_{\Delta t}$$

Equation 14.3 Dose metric (DM) for a discrete response

Even with the limited data available from the Litonjua study, we can estimate the approximate exposure distribution; it has a GM of 0.05 U/g with a GSD of 6.3. Using the simple logistic model with the estimated $\tilde{\alpha}$ and $\tilde{\beta}$, we can estimate the probability of an attack given a level of home contamination. Given the exposure probability distribution, we can estimate the expected number of attacks in a period of exposure, N_{exp}, for each dose group. The resulting dose metrics can be used in the full epidemiologic analysis, such as a logistic regression, along with all the other relevant factors, such as age and genetic status. In the case of the Litonjua study, because we can only have two dose groups, the result will be the same as the simpler exposure yes/no analysis.

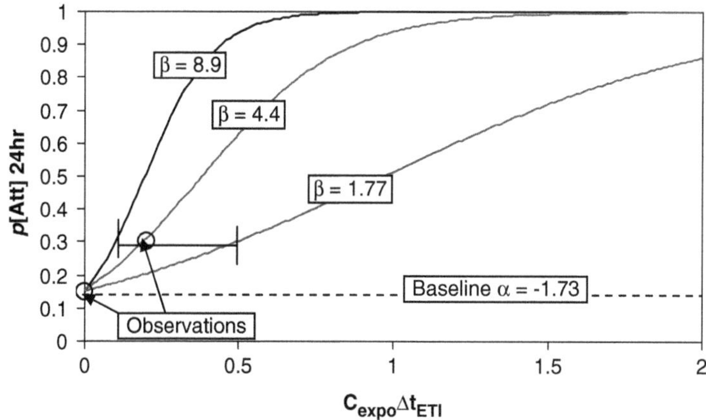

Figure 14.1 Two-point logistic model for dose metric, derived from observations by Litonjua et al. 2002 (see text for derivation).

Limitations of the Litonjua Study

The design of the Litonjua et al. (2002) study has several limits that affect the quality of this dose-response analysis. First, the subjects' personal exposures were implicitly assumed to be proportional to the concentration of cockroach allergen in the dust on the vacuumed surfaces. The investigators did data analyses with both the concentrations of agents in the dust and total amounts of agent and found no evident differences. This implies either that the dust levels and airborne dust generation processes were roughly the same in all subjects' homes or else that the quantitation of exposure was too crude to detect a difference; the latter is more likely. Second, the home contamination values were measured only once. Therefore,to have a meaningful estimate of the dose-response relationship, we must assume that exposures during the observation period had a stable distribution over a long period of time. Third, it was implicitly assumed that the frequency of attacks was approximately stable over time. Any year-to-year variation in either exposure or wheezing was attributed to random variation. If the goal of the study was to obtain evidence that cockroach or other allergen exposures increased risk, it was successful. However, the study provides little information on the dose-response relationship—how much allergen produced how much risk.

Design of a Follow-Up Study

Assuming that we wish to perform a follow-up study with the goal of better describing the dose-response relationship, several refinements are needed to obtain more detailed information on the exposures and responses.

1. *Exposure sampling.* A calibration study is needed to define the relationship between the easy-to-collect surface vacuum samples and the distribution of personal breathing zone exposure levels for the subjects, which are most likely to cause asthma attacks. Surface sampling in the bedroom, living room, and kitchen should be repeated several times within a year—at least in the summer and winter—to define any seasonal variation in dust and allergen components. Within the same season, a subset of homes should be resampled in the same rooms to observe the temporal variation in dust-settled levels. Additionally, room air sampling and some personal sampling should be conducted during various common activities that are likely to produce airborne dust, such as vacuuming and dusting for housekeeping and children running through the rooms and roughhousing on the furniture or beds. The goal is to define the stability of allergen distributions over time and to improve the precision of the exposure estimates so as to define a set of dose groups covering the range of exposures.
2. *Response characterization.* More detail is needed on the attacks of wheezing so they may be better linked with exposure conditions. With a low frequency of only a few days per year, perhaps monthly or quarterly questionnaires could be obtained to better locate the temporal occurrence of wheezing episodes. The questionnaires might also ask about the exposure events occurring at the time of the episode, such as cleaning. Combining these data with the enhanced characterization of exposures will provide a strong basis for assessing the dose-response relationship. Our objective is to sharpen the temporal definition of the exposure-response process and its pattern across the population. We want to use the prior data to guide our choices for data intensification.

The data needed to calculate the dose metric also can be a guide for this further sampling because they show the relative importance of the components and how they relate to each other. The definition of the baseline attack rate without exposure is a critical element, because that and the size of the population determine how precisely we can define the effect of exposure. If the effect of exposure on the attack rate is small relative to the baseline rate, a large population will need to be studied to detect the effects. Clearly, repeated measures over time will sharpen our ability to link exposures and attacks. Usually we are most interested in the responses to low-level exposures, which are the most difficult to define. Similarly, the fraction of the population that is susceptible to attacks is critical. As noted earlier, a population with a long history of exposure will tend to have a low fraction of susceptibles and a low response rate, whereas a group with little prior exposure will often have the most susceptibles. Thus the history of exposure for the population is very important, and similar populations with different histories and more or less time for selection out of the exposed group will have different dose-response relationships.

14.3. DERMATITIS FROM MACHINING FLUID EXPOSURES

Hypersensitivity skin reactions to chemicals, such as isocyanates, nickel, and acid anhydrides, are the result of the chemicals reacting with proteins to form antigens. The skin is a common target organ for reactive chemicals. Contact dermatitis is one of the most frequently reported occupational illnesses (Wassenius, Jarvholm et al. 1998). Under the label of "dermatitis," a variety of responses may occur, including eczema, itching, inflammation, and irritation (Braunwald E, Fauci AS et al. 2002). The onset of the allergic response is often delayed, so the subject may not be aware of exactly when the response began or what exposure produced it. For the allergenic type, once the exposure to the antigen stops, the antigen-antibody complexes are destroyed or removed, and the inflammation and cell recruitment responses are damped down by inhibitory processes until they dissipate. Skin with dermatitis is slow to recover when exposure stops, but with suitable drug treatment recovery may be reasonably rapid. Thus the responses are reversible and decline over hours or days after exposure stops and treatment begins. Severe responses can be extremely unpleasant and even life-threatening and may take a long time to resolve without treatment. Individuals who are not sensitized will not have an immune response, but they can still have nonspecific irritation and inflammatory responses. Thus there are two types of contact dermatitis—irritant and allergic—and the irritant form can lead to the immune response in some individuals. The cellular damage from the irritant or toxic effects of a reactive chemical tends to occur at much higher exposure levels than immune responses. Sensitized individuals will vary widely in their responsiveness to specific chemicals, and some can respond to extremely low concentrations.

The normal long delay in dermatitis development makes it hard to identify the nature of exposure associated with the response in complex or varied exposure settings. It is common that individuals are exposed for months or years before they develop dermatitis. This implies that everyday exposures do not initiate dermatitis in most cases. One hypothesis might be that mechanical trauma and chemical irritation injure the skin and allow allergenic components to penetrate the skin barrier and initiate an immune response in the dermal tissues. If the exposure stops and the skin heals, then simple, low-level exposures without damage to the skin may not produce a recurrence of dermatitis.

Skin Exposures to Metalworking Fluids

Metalworking fluids (MWF), sometimes called cutting fluids or cutting oils, are used to cool and lubricate metal when it is cut. In many metal-cutting operations such as turning, drilling, and grinding, a stream of MWF is directed at the cutting tool and the part being cut. Some of this liquid is thrown into the air as a mist or aerosol during this operation. There is considerable evidence that people breathing MWF aerosols for long periods may suffer from respiratory health problems and increased cancer risk as a result of MWF exposures (Eisen, Tolbert et al. 1992; Wegman 1996; Kriebel, Sama et al. 1997; Rosenman, Reilly et al. 1997; Calvert, Ward et al. 1998; Schneider, Vermeulen et al. 1999; Eisen, Smith et al. 2001; Vermeulen, Stewart et al. 2002). In addition, skin contact has long been recognized as creating a risk of contact dermatitis.

Many workers in metal machining industries are exposed to MWF. Exposure may begin from the first day on the job when workers may have contact with MWF because the metal parts being handled are wet with MWF, or because the machining operation produces a spray of fluid that wets the hands, arms, and clothes. MWF are complex mixtures used as coolants and lubricants during operations in which metal pieces are shaped into desired forms. MWF come in several types: straight oils (predominantly petroleum-based long chain hydrocarbons), soluble oils (mineral oil emulsions in water), synthetics (chemicals in water), and semisynthetics (mixtures of soluble oils and synthetics). The latter three consist mostly of water, with added lubricants, emulsifiers such as detergents, anticorrosion agents such as triethanolamines (alkaline), and biocides, including triazines, which are formaldehyde-releasing agents. Straight oils also have additives for similar purposes, plus high-temperature stabilizers, such as chlorinated hydrocarbons. Many of the added components are irritating to skin, and some may react with skin proteins.

MWFs commonly break down during use because of the heat generated by the cutting processes, and the waste or by-products accumulate in the used fluids. The fluids are expensive, and they must be disposed of as hazardous waste, and so they are almost always recirculated—often for a very long time. Metal particles, dirt, and grease from the parts and caustic hydraulic fluids from machine leaks accumulate in the recirculated MWF. The organic materials in water-based fluids are an attractive growth medium for bacteria and fungi. Counts of microorganisms as high as 10^8 per cm^3 can be found in used MWF (Thorne, DeKoster et al. 1996). Various types of biocides are added to kill the microorganisms, and some of these are skin and respiratory sensitizers. Biocides can be effective, and bacterial counts can be knocked down by many orders of magnitude, but they can also grow back within a few days. It is very difficult to completely sterilize or clean MWF systems. Organisms can form biofilms on surfaces throughout the MWF systems, and these are very hard to remove. They provide a reservoir of organisms for regrowth.

During machining, MWF is applied to the tool and the part being worked. Common operations are drilling, broaching, turning, and grinding. High-speed operations will produce considerable heat, splashing, spray, and direct aerosolization of the MWF. Exposure controls are common and should be universal. Splash guards, enclosures, and ventilation systems are placed to reduce or eliminate operator exposures. Despite these safeguards, control of emissions is often incomplete, and skin and respiratory exposures still result.

Many machining activities can wet the skin with MWF: handling wet parts, splashing, machine sprays, cleaning, and so forth (Sprince, Palmer et al. 1996). These will produce a surface concentration of each of the specific components initially equal to the bulk fluid. Once the skin is wet, additional MWF will not change the concentration, unless the critical agent can rapidly evaporate or is quickly absorbed through the skin. Therefore, the concentration of toxic components in the MWF, the skin area covered and its permeability, and the duration of contact are the critical dimensions of exposure. Longer duration and higher concentrations potentially allow more material to enter the skin and cause more skin damage that can also increase penetration. Depending on how the recycled MWF that feeds each machine is maintained, the composition of the MWF to which the worker is exposed will vary over time.

Skin Responses to MWF

There are a number of possible skin irritants and sensitizers in MWF (Sprince 1996). Irritant dermatitis can be caused by emulsifiers and alkalinity and tiny cuts and abrasions by chips and sharp bits of metal removed during shaping. Allergenic responses have been suspected for fragrances, emulsifiers, corrosion inhibitors in water-based MWF, and chlorinated extreme pressure additives in straight oils. Workers with exposures can have a variety of skin responses. Acne and folliculitis are commonly associated with exposures to straight oils. Reports of irritative and allergic contact dermatitis are associated with soluble oils and synthetics.

The time course of the recovery process is important for defining any disease process model. Thus determining the incidence of attacks requires knowledge of the biology of the recovery process. Clinical data or subject reports can be used to define the time course. Occupational dermatitis cases in metalworking industries will generally recover within a few days or weeks with treatment (Braunwald E, Fauci AS et al. 2002; Suuronen, Aalto-Korte et al. 2007). Recovery dynamics appear to vary considerably among individuals and may also depend on the period of time that agents are retained in the skin. Modeling this recovery process can be handled in an analogous fashion to the recovery from reversible continuous responses by setting a half-time of recovery, but it is more variable and idiosyncratic. In essence, the immune system needs to reset before it can respond again. Given the wide variation in responsiveness across individuals, it may be more useful to define a population half-time of recovery within which half of the subjects will recover after exposure stops.

Allergic dermatitis is initiated when the allergen penetrates through the outer skin layers into the dermis to produce a Type IV reaction. Any skin effect that enhances penetration will increase the risk, such as abrasions or other skin damage. Animal data suggest that delayed Type IV responses may be stimulated by short-duration intense exposures (Burns-Naas, Meade et al. 2001). The delayed immune response requires a period of time to develop, unlike the other three types, which are immediate. The delay also prolongs exposure, increasing the uptake of the agent and the possibility of more skin damage and adds to uncertainty about the nature of the exposure that triggered the response. In that situation, it may not be possible to clearly distinguish the effects of allergic and irritant components.

Epidemiologic Study: Dermatitis among Auto Industry Machinists

Our objective for the examination of a dermatitis case is to explore how we could apply the disease process model to determine a dose-response relationship. As noted earlier, this is difficult because of uncertainty about the nature of the response and its timing. However, we will show how the approach can still be useful.

Evaluation of Dermatitis

A cross-sectional study was conducted in a large auto transmission manufacturing plant by Sprince and coworkers (Sprince, Palmer et al. 1996). The study was designed to determine whether the prevalence of symptoms of contact dermatitis was different among exposed machine operators (n = 158) and unexposed assemblers (n = 51). Symptom questionnaires and dermatologist interviews and examinations were used to identify cases of "skin rash" as the primary outcome variable. Skin exposures were assessed using MWF wetness on the skin and metals deposited by MWF in patches placed on the exposed skin. Workers were exposed to two types of MWF: soluble oil (an emulsion) and semisynthetic fluids that contain oil and chemicals. The researchers did not ask whether the individual had had prior dermatitis or when the current condition began. The investigators assumed that current exposure was the cause of the current rash, an assumption that does not eliminate the possibility that some effects were due to different conditions earlier because dermatitis has a long development period. Cases were identified as "definite" or "probable" by the dermatologist based on the presence of moderate or severe erythema, scaling, or lichenification on the hand, wrist, or forearm, but the researchers did not distinguish potential irritant responses from allergic types of responses. Subject reports also were used to identify cases: a "yes" response to the question "Do you have a skin rash now?" and the report of a rash specifically on the hand, wrist, or forearm. Although the type of response, irritant or allergic, also may vary with intensity of exposures or with variations in relative amounts of mixture components, some of which are more allergenic, these were not separately identified. These differences across individuals may reduce the strength of the observed relationships because irritant and allergenic responses may have different time courses (irritation develops quicker, but allergic responses last longer) and intensities of response per unit of exposure (allergenic responses occur at much lower exposures).

Skin Exposure Measurements

Skin exposures to metals (cobalt, chromium, and nickel) and skin wetting with MWF were the indices of exposure in the Sprince et al. (1996) study. The Environmental Protection Agency's (EPA) protocol for skin exposure to pesticides was adapted to make the measurements. A dermal dosimeter (an absorbent cotton patch) was applied to the back of the forearm for a work shift, approximately 8 hours. The patch was extracted with acid (1% concentrated nitric acid in water) to remove soluble metals, which were quantified by atomic absorption spectroscopy. The amount of metal per unit of patch area per hour of patch exposure ($ng/cm^2/hr$) was calculated. A large fraction of the patches had less than detectable concentrations, so the epidemiologic analysis was based on a dichotomous variable (above or below the limit of detection). Bulk samples of MWF were also analyzed to determine their metals content. A wetness index was calculated from the ratio of the patch level to the bulk MWF level, expressed as $mL/cm^2/hr$. Because of the many samples below the limit of detection (LOD), this was also used as a dichotomous variable, like the concentration. Each subject was asked about the average hours of MWF "wetness" on his skin and clothing that was experienced per day to evaluate the degree of contact. The nature of wetness was not defined, but it could range from a little damp to visible liquid on the skin or clothing.

Dose Metrics

The Sprince et al. (1996) study used several measured dose metrics, including the amounts of cobalt, chromium, and nickel deposited on a patch sampler placed on the mid-forearm. The amount on the patch also was divided by the bulk concentration of metal in the MWF and used to indicate the quantity of fluid deposited per unit of skin area. Questionnaire data on years of work as a machinist, years of MWF exposure, and years since starting as a machinist were used to estimate long-term cumulative exposure. Short-term exposure was indicated by daily hours of skin wetness. All of these were used in the data analysis.

Findings of the Sprince et al. Study

Although the number of subjects was modest—158 machine operators and 51 unexposed assemblers—several interesting findings were obtained. First, machine operators had significantly more definite and possible dermatitis, 27.2%, than the control group of assemblers, 13.7%. There were several personal factors that modified risk, including light skin color and current smoking. Second, machinists who reported more than 1 hour per day of skin or clothing wetness were more likely to have dermatitis. However, none of the long-term cumulative exposure variables, such as years of work as a machinist, years of coolant exposure, or latency since starting work as a machinist, showed a relationship with risk of dermatitis. Given a reversible response, this is to be expected, because the effect is not cumulative. Third, semisynthetic MWF was more strongly associated with self-reported dermatitis than soluble oil, but it was not strongly associated with a dermatologist's evaluation of dermatitis. The latter finding is puzzling because the dermatologist's assessment is considered the gold standard. Finally, use of gloves or barrier cream to protect hands was not useful in reducing the prevalence of dermatitis.

The Sprince et al. study was small and had only limited gradations in exposure to assess dose-response relationships. Hours of daily exposure via wetted skin showed evidence of an increase in dermatitis with longer exposures. No relationship was seen for the quantitative exposure markers, concentration of metals, or amount of MWF on the skin. There was some suggestion that chromium may be more potent than nickel or cobalt, but the investigators did not report the concentrations of the metals. The type of dermatitis was not identified as irritant or allergic, and self-selection for nonallergic workers (a healthy worker effect) may have reduced the ability of the study to detect an allergenic response.

Application of the Disease-process Model to Design a Follow-up Study for MWF and Contact Dermatitis

In this section we present our disease process approach and show how some of the limitations of the Sprince study might be overcome.

Etiologic Time Interval (Δt_{ETI})

An agent of dermatitis must penetrate through the upper layers of skin into the dermis in sufficient concentration to trigger the Langerhans dendritic cells that initiate the immune response. Nickel solutions applied to sensitized human subjects required 4–5 days to stimulate a response with repeated daily exposures (Fischer, Johansen et al. 2007). A study of methylchloroisothiazolinone and methylisothiazolinone (MCI/MI) was conducted with skin tests using two consecutive exposures 4 weeks apart and found that 7 of 25 sensitized individuals responded 16 days after the start of the first exposure and 14 of 25 responded 12 days after the second exposure (Zachariae, Lerbaek et al. 2006). Clearly there is no single Δt_{ETI} that will work for every agent, and extended periods of intense exposure may shorten the period. Garabrant (1985) studied press workers using aziridine containing inks and found that workers preparing inks from concentrate reacted on average 3 months after the initiation of exposure and that pressmen using the diluted inks responded in 6 months. In this study, it is not known what the actual exposures were, but the findings imply either that lower concentrations take longer to produce a sufficient dose or, more likely, that a longer period was necessary because most exposures were too low to be relevant; but there was some low probability of a sufficiently intense concentration to initiate a response. Based on our review of the literature, it would appear that the Δt_{ETI} for allergic contact dermatitis will be in the range of 4–16 days, if exposures are sufficiently high. As a practical matter, we choose 14 days as the Δt_{ETI}.

Assessing Effects

Allergic contact dermatitis from MWF is difficult to study for two reasons: (1) it may be confused with irritant dermatitis that can occur concurrently and (2) the subject must be sensitized to be at risk. Irritant dermatitis may also predispose an individual to developing an allergic response. Sensitized individuals can sometimes be detected by skin testing. Our goal is to observe the temporal development of allergic dermatitis associated with exposures. Therefore, we will skin test a group of volunteers and then repeatedly assess the exposures and skin responses of subjects with positive skin reactions. We can also continue to follow the nonresponders to determine the exposures associated with development of irritant dermatitis. Repeated evaluations of the sensitized subjects at 14-day intervals can be designed with periodic skin exams, including photographs, allergen tests, and brief questionnaires, to have better temporal resolution in both exposures and detection of episodes of dermatitis. This would also allow detection of the development of sensitivity in new cases. Allergen concentrations will be minimized in repeated skin testing to minimize the risk of sensitization by the testing itself. Newly hired workers can be followed to determine whether there is selection bias caused by the departure of sensitized individuals.

Assessing Exposure

Laboratory studies with human subjects have shown that the total amount of agent reaching the dermis is critical to triggering the response, which is defined by the concentration of the agent, the skin area covered, and the duration of contact. Duration has limited impact on the amount of agent reaching the target tissue because absorption is eventually balanced by removal. At that point, increasing duration has no effect. Many factors are important determinants of agent penetration through the skin: the size and polarity of the molecule, the hydration and integrity of the stratum cornium, inflammation of the skin, and the presence of other materials that may act as vehicles to assist transport (Semple and Cherrie 2003). Skin abrasions and inflammation and occasional immersion exposures are characteristic of MWF exposures. Depending on a subject's job activities, the skin contact is usually not continuous but can be episodic, and it can be difficult to characterize the temporal variation in intensity and frequency and the distribution of between-subject variation in sensitization. Repeated sampling over time will be needed to make a precise estimate of the mean exposure and to adequately characterize the infrequent high exposures.

Measurement of biomarkers, such as urinary products or metabolites, can better estimate total exposure and give an indication of the total internal dose. This is a major advantage when the penetration through skin is so variable. However, biomarkers can smooth out some of the exposure variation and obscure brief high concentrations by averaging them with longer periods of low concentration. Data on the degree and location of skin exposures are important for interpretation and for weighting biomarker findings.

Probabilistic Dose Metrics

We need to use a measurement strategy that indicates the temporal and between-subject variation in exposure to estimate the exposure distribution. The logistic dose metric shown in Equation 14.2 estimates the number of responses in a sensitized group expected during a period of exposure, $N_{exp} \times \Delta t_{ETI}$. However, this approach may not work for the *irritant* response, which will follow a reversible proportional dose model discussed in Chapter 10.

14.4. CONCLUSIONS

Reversible discrete diseases have complex temporal behavior in exposed populations because only a brief intense exposure may be needed to provoke a response. We have proposed that a dose metric that represents the full distribution of exposures during the ETI is important to capture the effects of infrequent exposures. If the response follows a logistic dose-response curve, then the nonlinearity of that curve can increase the contribution of the upper tail of the exposure distribution. Temporal data, repeated measures of exposures, and the occurrence of discrete responses are required to resolve the disease process. We have shown two examples of different types of discrete responses, asthma and allergic dermatitis, to show how the approach can be applied.

15 Irreversible Discrete Disease Processes: Silica and Lung Cancer

15.1. OVERVIEW

The introduction to irreversible discrete processes in Chapter 13 noted that there are two important examples of groups of diseases that follow this pattern: immune sensitization and carcinogenesis. Unfortunately, much less is known about the intermediate steps in sensitization, and few examples of biologically based models have been applied to studies of immune diseases in humans, as was noted. In this chapter we discuss carcinogenesis, which has been more extensively studied.

Perhaps the best-known disease process models in epidemiology are discrete irreversible models of cancer. The multistage and two-stage carcinogenesis models first proposed by Armitage and Doll (Armitage and Doll 1954) and Moolgavkar and Knudson (1981), respectively, have been used to study the carcinogenic effects of a number of different exposures, including tobacco (Hazelton, Clements et al. 2005), radon (Luebeck, Heidenreich et al. 1999), arsenic (Hazelton, Luebeck et al. 2001) and coke oven emissions (Moolgavkar, Luebeck et al. 1998). But, although the models have been fairly widely used, there are actually only a few examples of analyses in which such a model has been applied to a dataset containing quantitative exposure data on the individual subjects. More frequently, duration of exposure has been used as the measure of exposure.

The two-stage clonal expansion model (TSCE), also known as the Moolgavkar-Knudson two-stage model, has been shown to describe the observed age-incidence curves for a variety of cancers, to correctly describe the changes in risk after cessation of exposure to a first or second stage carcinogen, and other important temporal dynamics of human carcinogenesis (Moolgavkar and Luebeck 1990; Moolgavkar and Luebeck 2003). This model also mimics certain key behaviors of mammalian cells undergoing precancerous transformations in vitro, and yet it is quite simple, with only four unknown parameters.

Carcinogenesis is often modeled by assuming that it develops through a series of low-probability events occurring over time in a particular order. In the classic multistage and two-stage models, these events were assumed to be mutations, but they need only be rare and irreversible in order for these models to be approximately correct. Such a model assumes that there is a low probability of irreversible DNA damage (a mutation) with each unit of exposure ($C \times \Delta T$). Over a sufficiently long period of exposure, the overall probability of a change is the sum of individual unit probabilities, and the summed probability may be high enough to be observable in a sufficiently large group.

Smoking and Lung Cancer

Cigarette smoke-induced lung cancer is probably the best studied of all environmental cancer risks (Doll and Peto 1976; Auerbach, Hammond et al. 1979; Pathak, Samet et al. 1986; Law, Morris et al. 1997; Doll, Peto et al. 2004). The identity of the specific carcinogenic agent or agents in cigarette smoke is not clear, because tobacco smoke contains thousands of different potentially toxic chemicals, some as gases and vapors and others as particles. The smoke tar particles contain many known human carcinogens and cancer promoters. Evidence implicating some components of tobacco tar comes from studies showing that lung cancer risk per cigarette was reduced by smoking filtered cigarettes, reducing the amount of tar per cigarette reaching the lungs without affecting the gases or vapors (Pathak, Samet et al. 1986).

Cigarette smoking is a series of peak exposures. Each inhaled puff (~ 50 mL inhaled, ~ 1 per min) contains dense smoke of very small particles and high concentrations of gases. Each puff produces a sharp spike of nicotine and carbon monoxide in the smoker's blood and deposits particles of tar in the smoker's airways. The delivered dose depends on the tar produced per cigarette and the puff characteristics (volume, frequency, and duration held in the lungs). The daily dose to the airways is roughly proportional to the number of cigarettes smoked each day, typically in the range of 5 to 40. However, the choice of cigarette (tar) and the puffing behavior of the individual smoker are extremely variable (Scherer 1999), Despite this complexity, the number of cigarettes smoked per day is a reasonably stable value for an individual smoker, which makes it a good semiquantitative dose metric for epidemiological studies. Of course, the number of cigarettes per day is only a surrogate of the true dose, as the actual delivered dose of carcinogens will vary among populations—British physicians compared with U.S. coal miners, for example—because of differences in the type of cigarette smoked and smoking behaviors.

Population Models

The dose response for irreversible discrete effects is a quantal (having only two states for each subject: disease yes or no) probabilistic relationship for a population of exposed individuals. In the simplest case, we would have large groups exposed uniformly at different concentrations, such as different cigarette smoking levels, and we could determine the fraction of subjects with lung cancer cases at each concentration. We would expect to see the characteristic lower end of the sigmoid curve, as shown in Figure 15.1, with data from a very large cohort that

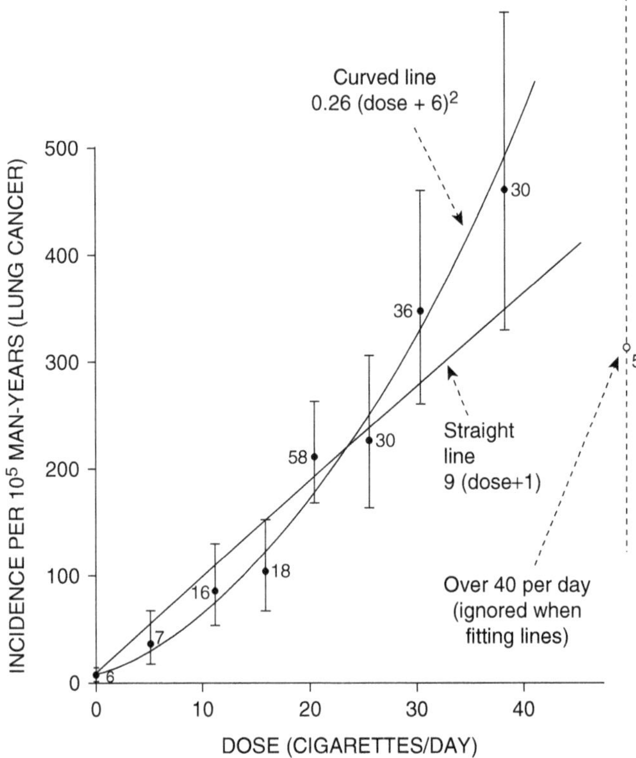

Figure 15.1 Dose-response relationship of lung cancer against daily cigarette consumption, standardized for age, from the British Doctors Study. The numbers of cases in each group are given, and 90% confidence intervals are plotted. From: N.E. Day, Epidemiologic Methods. In: Toxicologic Risk Assessment. Volume II. Clayson, Krewski & Munro.

had smoked steadily at a wide range of smoking rates over a long period of time. The British Physicians Study by Doll and Peto (1978) was such a study. Many other studies have been done using packs smoked per day times years of smoking, "pack years," as the dose metric (assuming 20 cigarettes/pack). This metric implies that smoking half a pack a day for 20 years has the same biological effects as smoking two packs a day for 5 years, which is an overly simplistic view, contradicted by the data. Instead, there is evidence that risk rises more rapidly with the duration of smoking than with the number of cigarettes smoked per day (Doll and Peto 1978).

Dose Metrics

The dose metric for a stochastic process with a yes/no, or quantal, distribution should be proportional to the fraction of the subjects who will respond at a given level of exposure within a given time interval (equivalent to the probability of getting the disease). Cumulative exposure (CE) may be an approximately correct dose metric in this case, although, as we explain, it may be possible to improve on CE with dose metrics that weight duration and intensity of exposure differently.

As we discussed in Chapter 12, CE tends to be a good dose metric for irreversible proportional disease processes—those in which each unit of dose increases risk by a constant amount. For example, it is approximately true that each pack year of smoking causes a fixed amount of loss in pulmonary function. But for some solid tumors, there is evidence suggesting that risk rises more rapidly with the time exposed (e.g., to tobacco smoke, as noted previously) than with the intensity of that exposure, in contradiction to the cumulative exposure-risk proportionality (Doll and Peto 1978). As shown following, this may be related to the reproductive kinetics of initiated cells.

15.2. THE TWO-STAGE CANCER MODEL

Overview of the Two-Stage Model

Experimental evidence now makes it clear that for most types of cancer, there are more than two irreversible steps in the process of complete cell transformation (Weiss 1990; Little 1995); (Boland and Ricciardello 1999). Furthermore, if the number of steps or stages is variable, meaning that there are numerous pathways from an initial carcinogenic exposure to formation of a tumor, then the very concept of a model based on discrete steps may be incorrect. Despite this complexity, it is clear that there are one or more temporal processes of cell modifications in carcinogenesis can be affected by environmental exposures in some cases. Thus there may be public health utility in the distinction between agents that act early in the carcinogenic process and those that act late. More detailed subdivision of the process, even if it were possible to achieve with epidemiologic data, may not have much public health relevance (e.g., determining whether an agent acts at the fifth or sixth of eight stages). For these reasons a two-stage model may be quite useful even when the true process is much more complex. In particular, late-stage carcinogens ("promoters") have been poorly studied in environmental and occupational epidemiology, partly because of the strong tendency to assume that lag periods of a decade or more are the rule in cancer epidemiology. Thus a model that explicitly investigates the early/late distinction could be a useful addition to the epidemiologist's standard repertoire.

The two-stage model can be schematized as shown in Figure 13.6. According to this view, tumors develop after two sequential and irreversible steps. In the first, usually envisioned as a mutation, a normal stem cell is transformed into an intermediate cell. This event is presumed to be rare and is modeled as a Poisson process. The intermediate cell may differentiate or die, removing it from the pool of target cells for the second transformation, or it may divide, increasing this pool. A second rare event, possibly a mutation, leads to the development of a malignant cell, which may begin a clone, leading ultimately to a clinically diagnosed tumor.

A strength of the two-stage model is that if there are more than two exposures acting on cancer risk, their potential interaction can be evaluated in more detail than is possible in standard statistical models. For example, if two different exposures are found to act at two different stages, then the timing of these exposures would be expected to have an important effect on risk. These types of temporal interactions are difficult to implement and investigate with standard models. Empirical time windows have been used, but it is awkward, and the number of possible models rapidly expands, making this approach impractical.

Exposure to a carcinogen could, according to this model, act in one of several ways: through increasing the transition or mutation rate from normal to intermediate cells (increasing α_1); increasing the transition rate from intermediate to malignant cells (increasing α_2); or increasing the proliferation of intermediate cells by altering either the birth (b) or death (d) rate of these cells. By studying the effects of different time courses of exposure and their relationship to risk in an exposed population, it may be possible to determine which of these modes of action appear to be relevant. Also, different carcinogenic exposures that occur in the same environment may act via different pathways, and these pathways may be distinguishable mathematically when they have different time courses. In this model, a tumor initiator is postulated to act by increasing α_1. A promoter should act by increasing the proliferation of intermediate stage cells through altering the balance of (b-d) so as to increase the pool of intermediate cells ready to receive the second "hit" (α_2), whereas increased immune surveillance may increase d and decrease the pool. The two-stage model has been used to investigate cancer risk in cohorts exposed to a variety of environmental carcinogens, including tobacco smoke, radon, cadmium, and coke oven emissions (Table 15.1).

Table 15.1 The two-stage clonal expansion model has been used in a number of epidemiologic studies with quantitative exposure data

Author	Year	Exposure	Cancer Site	Comments
Moolgavkar	1993	Radon	Lung	Interaction with smoking: between additive and multiplicative
Stayner et al.	1995	Cadmium	Lung	Stronger effect on first mutation than on intermediate cell proliferation
Moolgavkar et al.	1998	Coke oven emissions	Lung	Best fit with exponential link function for exposure
Luebeck et al.	1999	Radon	Lung	Effects on proliferation of intermediate cells and on second mutation; little effect on first mutation
Hazelton et al.	2001	Arsenic, radon, tobacco smoke	Lung	Combined effects of all three exposures were studied in a large cohort of Chinese tin miners
Hazelton et al.	2005	Tobacco smoke	Lung	Both initiation and promotion effects, the latter more important than the former
Richardson	2009	Benzene	Leukemia	Benzene acts as a promoter of intermediate cells

A Dosimetric Approach to Fitting the Two-stage Model[1]

The two-stage cancer model assumes that in the first stage of malignant transformation, formation of intermediate cells (Y) occurs through a Poisson process as a result of the first mutation rate α_1 acting on the pool of normal susceptible cells X:

$$E[y](\Delta t) = \alpha_1 * X,$$

where $E[y(\Delta t)]$ is the expected number of new intermediate cells formed during Δt by the first mutation rate α_1, which is used as the parameter for the Poisson process:

$$y(\Delta t) \sim \text{Poisson}\ (\alpha_1 * X),$$

Equation 15.1 Poisson formation of new intermediate cells

We can modify this model to make α_1 a function of exposure, such that exposure has a linear effect on the first stage mutation rate, but other forms can also be used (see Box 15.1). Following this, if the combined rate of initiated cell replication and new initiated cell formation by mutations is larger than the death rate, a pool of intermediate cells, $Y(t)$, may be created. This pool can grow or die as the rates change, as described:

$$Y(t+\Delta t) = Y(t)+y(\Delta t)+Y(t) * \delta$$

Equation 15.2 Temporal difference equation for the number of intermediate cells

where $Y(t + \Delta t)$ is the intermediate cell pool at the end of the time interval, $y(\Delta t)$ are the additional new intermediate cells formed during Δt by the Poisson random mutations, and $Y(\Delta t)*\delta$ represents the cell dynamics of the intermediate cell pool itself as a result of cell proliferation (b in Figure 15.2), cell death (d in Figure 15.2), or losses by additional mutations (α_2 in Figure 15.2), all summarized in one net change parameter, delta (δ) in our model. The probability of additional mutation(s) is assumed to be very small, and so the main contribution to the parameter δ comes from the balance of the proliferation and death rates of the intermediate cells. This overall rate is very important because the risk of forming a transformed malignant cell depends on the number of initiated cells that might make the second mutation. If the initiated cells are growing over time, then the probability of a second mutation will also be growing. Even a modest doubling rate will lead to a nonlinear increase in risk with time.

Box 15.1 Two-Stage Model Dose Metric: Estimation Process for Intermediate and Transformed Cells

The model was implemented in S+, using a standard optimization routine. The objective function was the logarithm of the unconditional logistic likelihood:

$$G = \sum_{i=1}^{n} y_i \log(\hat{p}_i) + (1-y_i)\log(1-\hat{p}_i),$$

where:

y_i = disease status 1/0 for the *i*th subject

and

$$\hat{p}_i = \frac{1}{1+e^{-(\beta_0+\beta_1 Z_i)}}$$

and β_0 and β_1 are unknown parameters to be estimated from the data and Z_i = the dose metric for the *i*th subject. As noted in the text, we did not estimate the second mutation rate, α_2, but rather assumed that it was a constant. This meant that we were assuming that progression could not be independently affected by exposure. We justified this simplifying assumption after investigating a wide range of different models in which α_2 was never found to be affected by silica exposure.

The fitting process proceeded in the following steps:

1. Initial values of the two-stage model parameters: α_1 and α were chosen.
2. For each subject, the equations of the two-stage model were iteratively solved every tenth of a year from birth until risk age, repeatedly calculating Y and Z. The value of Z at risk age minus 3.5 was that person's dose metric.
3. The set of dose metrics for all subjects were combined with their disease status information, and fit with the logistic regression model, to estimate values of β_0 and β_1, above. The log likelihood, G, was calculated.
4. The optimization routine repeated this process, searching for values of α_1 and δ which maximized G. When a maximum was found, the search process stopped, and the final values of α_1 and δ were recorded.
5. This iterative process was repeated for several different forms of the dosimetric model. These included a baseline form in which subjects' exposure histories were not considered. Next a series of runs were performed, with exposure modifying the first stage parameter, α_1. Different parameterizations of the exposure effect on α_1 were investigated, because we lack biologic knowledge about the correct form of this relation. We investigated a set of five functional forms of the exposure effect:

Linear: $P = P1 * (1 + P2 * C_{exp})$
Power: $P = P1 * (1 + P2 * C_{exp}^{\ P3})$
Log: $P = P1 * (1 + \ln(1 + P2 * C_{exp}))$
Negative Exponential: $P = P1 * (1 + (1 - e^{(-P2 * Cexp)}))$
Exponential: $P = P1 * e^{P2 * Cexp}$

where P is one of the two parameters of the model ($\forall\alpha_1$, or δ), $P1$ is a baseline parameter, and $P2$ and $P3$ are additional parameters representing the unit change in P as a function of exposure. In each link function, absence of exposure, or of an effect of exposure on the parameter, causes $P_1 = P$.

Finally, a parallel series of runs were performed in which exposure was allowed to modify δ, the net proliferation rate of intermediate cells. The same five parameterizations were investigated. In sum, 11 different versions of the two-stage model were constructed: a baseline model and two sets of five different ways that exposure could affect the dose metric. The log likelihoods of these different runs were compared, and the better fitting models were judged to be more consistent with the observed cancer data.

The two-stage model proposed by Moolgavkar considers two main steps in the malignant pathway, an early, or initiation, "hit" as a stochastic process with a Poisson probability (first mutation rate) and a second step, promotion, or an effect through increased clonal expansion of the initiated cells, which creates a pool of susceptible cells, at risk of acquiring a second mutation leading to full transformation. The final step—the second mutation rate ("progression" in Figure 13.6)—is assumed to act on one of the cells in this pool of intermediate cells. This second mutation is probably in actuality many additional steps, rather than just one. Through a binomial probability function, one malignant cell is formed out of the pool of intermediate cells, $Z(t)$:

$$Z(t) = Y(t) * \alpha_2,$$

Equation 15.3 Formation of a Malignant Cell from Intermediate cells

where $Z(t)$ is the mean or expected number of fully transformed cells and α_2 represents the probability of transformation leading to a malignant cell. The dose metric estimated for each cohort member is the expected value of $Z(t)$ based on his or her exposure history. There should be a lag or delay between the time that a fully transformed cell occurs and a tumor is diagnosed to account for the time needed for tumor growth. Theoretically, this lag could be estimated from an epidemiologic dataset, but to keep the problem manageable, one can also assume some fixed number of years from the time of full cell transformation until the appearance of the tumor. In his study of lung cancer risk among Chinese tin miners, Hazelton empirically estimated this lag and found the most likely value to be 3.3 years (Hazelton, Luebeck et al. 2001). In the following example on silica and lung cancer, we assumed a fixed lag of 3.5 years.

In this example, we also did not attempt to estimate the variances of the model coefficients. For this investigation of the method, we were more interested in the stage information and the overall improvement in model fit than in the precision of the individual parameters.

Fitting the Two-stage Cancer Model to Epidemiologic Data

We present an example of the application of the 2-stage cancer model to a previously-published study of silica exposure and lung cancer mortality (Checkoway, Heyer et al. 1997; Seixas, Heyer et al. 1997). For this example, we hypothesized that exposure could affect two of the three model parameters: α_1, the first mutation rate, and δ, the net proliferation rate of intermediate cells (Zeka 2003). A previous study using different modeling methods failed to find any evidence of exposure effects on the second mutation rate, α_2, and so we did not investigate this parameter further. Exposure was incorporated into the model as a time-varying variable changing each year. An iterative process was carried out to estimate the expected number of premalignant cells (Y) and malignant cells (Z) over each individual's lifetime as a function of the model parameters. This parameter estimation method is summarized in Box 15.1.

Table 15.2 Interpretation of log-likelihood values

Weight of Evidence	Reduction in –2 LL	Associated p-value
Weak	3.8–2.1	$0.05 < p < 0.15$
Modest	6.6–3.8	$0.01 < p < 0.05$
Strong	> 6.6	$P < 0.01$

Each of the models in which an exposure effect was hypothesized could be compared with the baseline model using the difference in the –2 log likelihood. The baseline model is "nested" within any of the models with an exposure effect, and so the net reduction in –2LL has a chi-squared distribution, with degrees of freedom equal to the difference in the number of parameters. For example, the linear parameterization shown in Box 15.1 adds one additional parameter, indicated as P2, and so there would be one degree of freedom for the chi-squared distribution for the amount of reduction in –2 log likelihood.

We did not perform formal hypothesis testing, as it was not appropriate for this methods development research. However, p-values do provide a way to judge what a "large" improvement in goodness of fit is for an exposure model compared with a baseline model. We used the criteria in Table 15.2 for the weight of evidence indicated by a reduction in –2 log likelihood (the table assumes one degree of freedom difference between the exposure and baseline models):

15.3. APPLYING THE TWO-STAGE CANCER MODEL TO EPIDEMIOLOGIC DATA ON SILICA AND LUNG CANCER

To illustrate the use of the two-stage cancer model in epidemiology, we used a dataset in which the association between silica exposure and lung cancer has been previously reported (Checkoway, Heyer et al. 1997; Seixas, Heyer et al. 1997). Checkoway and colleagues studied a cohort of workers exposed to crystalline silica during the mining and processing of diatomaceous earth (DE). The DE cohort included 2,342 white males (23% Hispanic) who were employed for at least 12 months, including at least 1 day between January 1, 1942, and December 31, 1987, in the DE mining and processing industry (Checkoway, Heyer et al. 1997; Seixas, Heyer et al. 1997). Follow-up covered the period from 1942 to 1994; vital status was determined for 91% of the cohort, and cause of death was ascertained for 716 of 749 (96%) of identified deaths. There have been several published reports of the exposure-response associations between lung cancer and silica exposure based on 77 lung cancer deaths (65 excluding pleural cancers); (Rice, Park et al. 2001; Steenland, Mannetje et al. 2001). Cumulative exposures to respirable dust and respirable crystalline silica have been computed based on historical reconstruction of exposures for all subjects (Seixas, Heyer et al. 1997). The crystalline silica exposure index incorporated data on percentages of crystalline silica in the various product mixtures and secular changes in DE production at the plant. These data provide quantitative estimates of mg/m³-yrs of respirable dust and respirable crystalline silica exposure that can be incorporated into exposure-response models.

A nested case-control analysis of these data found a strong linear association between cumulative exposure (with a 10-year lag) and risk. In a logistic regression model, the odds ratio for cumulative exposure was 1.05 (95% CI: 1.00 to 1.10) per mg/m³-yr. A spline curve revealed a fairly strong linear association between cumulative exposure and log odds, although it appeared to plateau at higher exposures (Figure 15.2). Little weight should be placed on the shape of the curve above about 15 mg/m³-yrs because the data became very sparse in this high-exposure region (the "rug" along the bottom of the graph indicates the locations of the data points, and one can see that there are only a handful of data points above 15 mg/m³-yrs).

Results of Fitting the Dosimetric Model to the Silica and Lung Cancer Data

We fit the two-stage cancer model as described previously to a subset of these data consisting of all lung cancer deaths and an equal number of controls, chosen to be

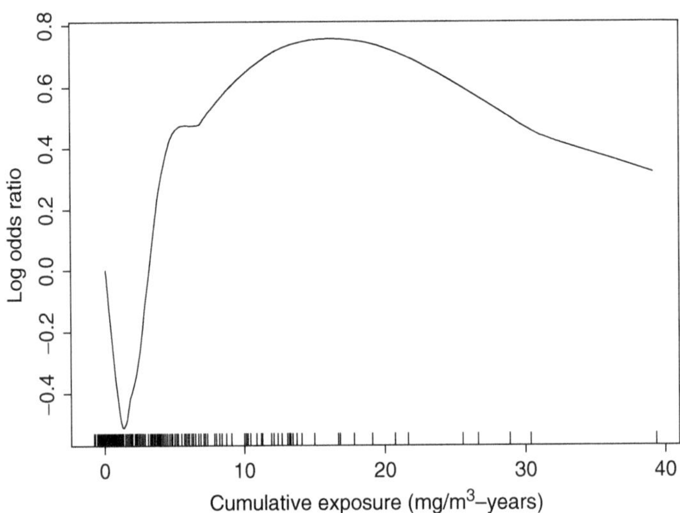

Figure 15.2 Spline curve representation of association between log odds of lung cancer mortality and cumulative silica exposure (mg/m³ –years) (Checkoway, Heyer et al. 1997).

alive and at risk at the age of death (risk age) of each case. There was clear evidence of an exposure effect on the first stage. The only models with exposure terms that provided convergent solutions (iteration values did not expand out of control) were those in which exposure acted on the first mutation rate (Table 15.3). The model with an exponential exposure effect on α_1 provided a good deal better fit to the data than did the baseline model: the improvement in –2LL was 6.6 (416.3 – 409.7), which we would consider modest to strong evidence for an exposure effect. In contrast, models in which exposure was allowed to modify the proliferation of intermediate cells (promotion) did not fit the data well. The best fitting model included silica exposure modifying the first mutation rate, α_1, in an exponential fashion:

$$\alpha_1 = 0.02 * e^{2.0 * Cexp}$$

Equation 15.4 Coefficients of the best-fitting dosimetric model

In other words, the probability of creating an intermediate cell was positively associated with a subject's annual exposure (Figure 15.3). Because the model is iteratively fit each tenth of a year, from birth until the 3.5 years prior to death from cancer, the curve in Figure 15.3 is not a conventional exposure-response curve but instead represents the amount that each annual unit of exposure contributes to the first stage of cancer development. At least two more steps are hypothesized before a cancer cell will result from exposure, according to this model (Figure 13.6).

We can try to illustrate how different this model is from a standard epidemiologic approach by comparing the dose metric to lifetime cumulative exposure. We generated the final dose metric for each subject, with the maximum likelihood values of α_1 (shown previously) and δ (0.06). We then fit a spline curve to the relation between this dose metric (Z) and the log odds of lung cancer (Figure 15.4). Unlike in the relationship between cumulative exposure and log odds (Figure 15.2), the relationship was consistently linear across the entire range of the dose metric. We interpreted this to mean that this metric, using the two-stage model as the weights on each subject's exposure history, more accurately captured the relation between exposure and risk than did cumulative exposure.

Table 15.3 Results of fitting the dosimetric two-stage model to silica and lung cancer data

Model	α_1 baseline	$\alpha_1$exposure	δ	–2LnL
Baseline	0.02	–	0.26	416.3
Linear	0.02	3.4	0.26	416.3
Exponential	0.02	2.0	0.06	409.7

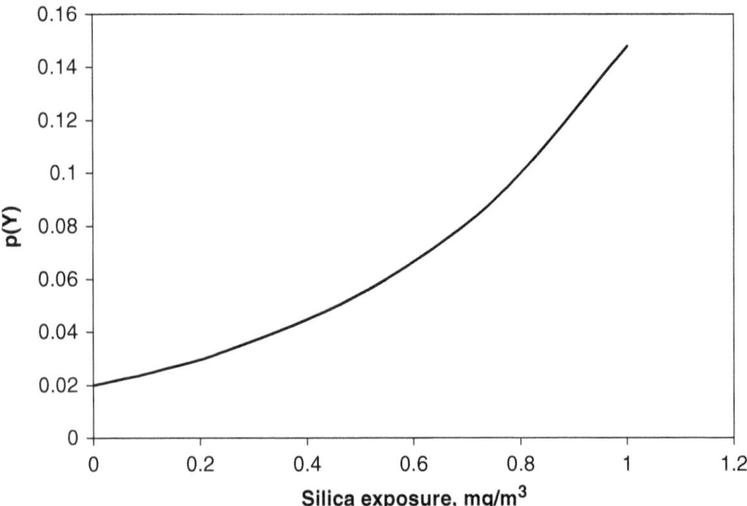

Figure 15.3 Estimated effect of silica exposure on probability of cancer initiation. The vertical axis represents the probability of an intermediate cell occurring for a given annual exposure intenstity (mg/m³).

The correlation between cumulative exposure at risk age (with a 10-year lag) and Z, the two-stage model dose metric (Figure 15.4), was 0.5—not very high. We wanted to try to understand why Z differed from CE, and taking a graphical approach, we constructed Figure 15.5 to permit a visual inspection of the discrepancies between the two metrics. Four subjects were chosen who represent different combinations of CE and Z: (1) high Z and low CE; (2) high Z and high CE; (3) low Z and high CE; and (4) low Z and low CE. For each subject, the individual exposure profile of intensity over time is shown in Figure 15.5.

Subject 1, in the upper left corner of Figure 15.5, has one of the highest values of Z, but only a moderate level of CE. Notice that the exposure profile is characterized by high-intensity exposures in the subject's younger years (16–24 years of age), followed by much lower exposures for the rest of his career. The early high exposures can induce a large population of initiated cells, which are maintained by the lower continuing exposures; during this long period there are both a relatively large population of initiated cells and time for the transformation process to make them into malignant cells. Contrast this profile with that of subject 2, who has nearly the same value of the dose metric but considerably higher CE. One can see that the more sustained period of high exposure for subject 2 would produce more initiated cells, but it is shifted toward later ages, so there would be less time for the transformation process—a larger initiated cell population for a shorter time could be offsetting and give nearly the same dose metric. If the duration of the high-intensity exposure of subject 2 had been the same as that of subject 1, it would have come too late in life to be effective for the second stage of

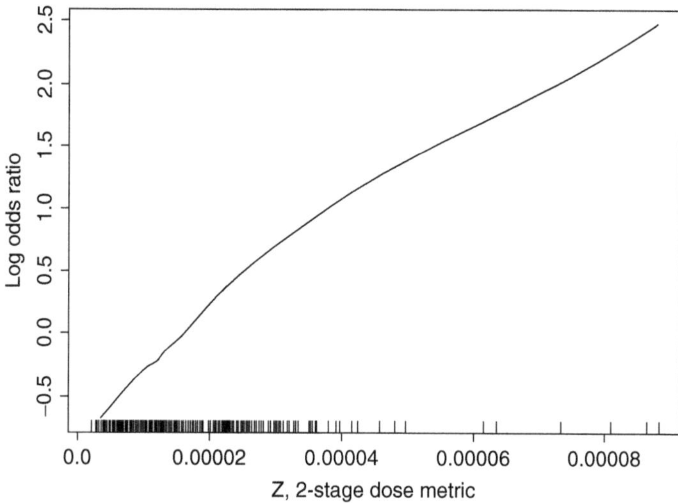

Figure 15.4 Spline curve representation of association between log odds of lung cancer mortality and silica 2-stage dose metric.

carcinogenesis. That is, the area under the intensity-time curve of subject 2 does not contribute as much to risk, according to the two-stage model. Thus the higher CE for subject 2 did not result in a higher value of Z than for subject 1.

Now compare subjects 2 and 3, who have nearly identical CE, as well as similar risk ages and durations of exposure. Despite these similarities, subject 2 has a considerably higher value of Z than subject 3. Why? One can speculate that much of the area under the intensity-time curve for subject 3 comes too late to allow time for the second stage of carcinogenesis. Finally, compare subjects 1 and 4, who have very different values of the dose metric but similar cumulative exposures. Notice that subject 4's exposure profile is shifted to later ages, compared with subject 1, and so it would seem that the exposure was less effective at increasing second-stage risk.

Silica as a Lung Cancer Initiator

Standard published epidemiologic models using time windows show strong evidence that silica increases lung cancer risk after a latency of 10 to 15 years (Rice, Park et al. 2001; Steenland, Mannetje et al. 2001). We interpret our findings with the two-stage model to be consistent with this observation of an early effect of exposure that requires time to develop into risk. The dosimetric modeling approach yielded a clear result for an early stage effect and no evidence for a later stage effect. The dosimetric approach also allowed us to investigate the correlation between cumulative exposure and the two-stage dose metric. This comparison

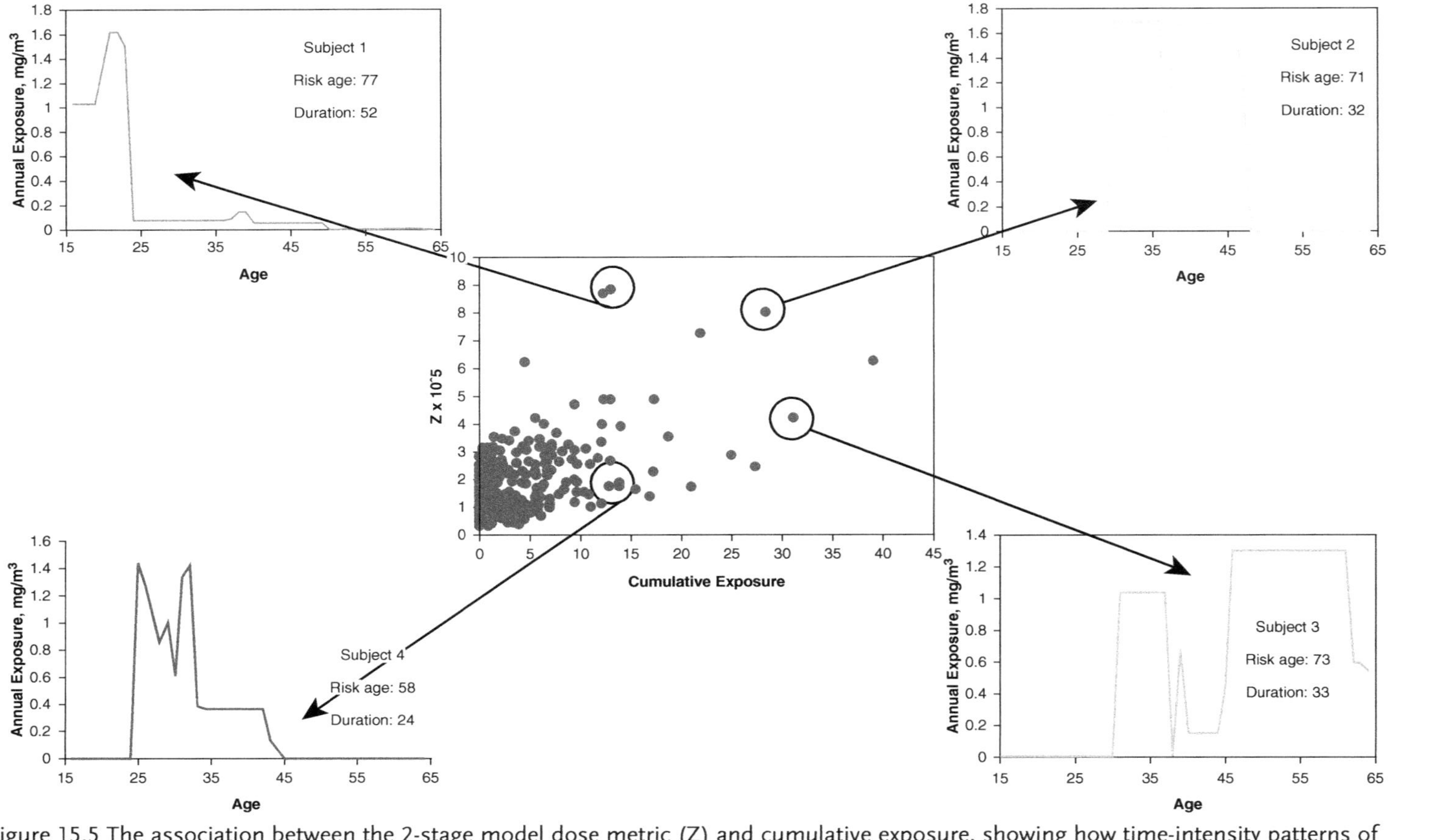

Figure 15.5 The association between the 2-stage model dose metric (Z) and cumulative exposure, showing how time-intensity patterns of exposure vary for four selected subjects. Central graph shows the association between cumulative exposure and the 2-stage model dose metric (Z); the four peripheral graphs represent exposure histories for four actual study subjects from age 15 to 65.

revealed interesting differences between the information captured by cumulative exposure and the dose metric, Z.

15.4. SUMMARY AND CONCLUSIONS

Some epidemiologists are skeptical that epidemiologic data can provide detailed biologic information, such as the stage of action of a carcinogen, because of the inevitable errors and approximations in the data. In one of the few formal investigations of the performance of epidemiologic data in a biologic model, Thomas (Thomas 1987) conducted simulations which showed that the stage of action of a carcinogen could be correctly identified even when substantial exposure misclassification was present. It is probably inappropriate to propose that complex models such as the two-stage cancer model should be used as the only epidemiologic analysis. Greenland has argued that the ideal approach probably involves several different models, each representing a different compromise between validity and precision (Robins and Greenland 1986; Greenland 2000).

Another concern about the utility of the two-stage model, noted earlier, is the likelihood that there are, in fact, more than two stages in carcinogenesis. Hazelton and colleagues were able to provide some useful insights on this through their extensive modeling of the Chinese tin miners' lung cancer data (Hazelton et al. 2001). They found that these data were consistent with only one, or at most two, sequential mutations in the initiation process. In contrast, the final step, transformation, was likely due to multiple sequential mutations. As long as these later mutations are independent of the exposure under study, then there should be no loss of utility from summarizing these multiple steps as one. Indeed, in the model described earlier for silica and lung cancer, we have made this simplifying assumption by allowing exposure to modify only the first mutation rate and the proliferation of intermediate cells.

The potential utility of biologically motivated models such as the two-stage cancer model has been well summarized by Prentice and Thomas (1993):

> With the rapid growth in our understanding of the fundamental biology of cancer, further development of methods to...incorporate [this understanding] into the analysis of epidemiologic data would be worthwhile. Most of the models that have been considered seriously are sufficiently general that some parameter values can be found to provide an adequate fit to epidemiologic data sets. Thus, these models are not easily falsified as a class, and it is unlikely that one could choose among them on purely statistical grounds. Instead, their utility lies in the types of comparisons that can be made within the context of a particular model—whether a carcinogen acts at an early or a late stage...for example. Their real value, therefore, lies in their ability to organize a complex set of hypotheses into a unified framework and to suggest empirical tests, in populations of humans or animals, of mechanistic ideas suggested by observations at the cellular level.

A practical limitation of the modeling methods described here is that many of the alternative parameterizations failed to converge. This means that the models frequently did not have optimum solutions, or else the optima were far from the starting values that we used. We investigated the sensitivity of results to alternative starting values, and for the models that fit well, there was fairly consistent convergence on the same optima. Models that fit less well sometimes showed sensitivity to initial values and would find different "optima" depending on where the routines were started. Thus the approach to the two-stage model described here needs further technical improvements. Moolgavkar and colleagues have many years' experience with an alternative method for fitting the two-stage model, and their approach tended to run more quickly and converge more consistently (Heidenreich, Luebeck et al. 1997; Hazelton, Clements et al. 2005). The approach we have presented here fits more easily into the dosimetric model paradigm that we advocate, but Moolgavkar's approach has much to recommend it, and we have found it to produce similar results. Richardson (2008, 2009) has more recently developed a convenient approach for fitting the 2-stage model in SAS.

NOTE

1. We are grateful to Drs. Ariana Zeka and Rebecca Gore for assistance with the development of this approach to fitting the 2-stage model.

16 Where Do We Go From Here?

In the past chapters we have laid out a systematic approach for constructing disease process models and their related dose metrics, which we believe can advance the integration of exposure assessment and epidemiology. At this time, there is no systematic way to integrate exposure data, toxicologic mechanisms, and population investigations that takes advantage of the strengths of each approach. More important, our structured approach naturally leads to the formulation of testable hypotheses as new information is integrated. In this final chapter we identify the many possible future directions for developing this approach for specific problems and as a broad theoretical framework.

16.1. DISEASE MODEL STRUCTURE

We noted earlier that success in applying disease models depends on formulation of appropriate model structures. One of the major limitations of empirical modeling is that large amounts of data are required to determine whether one model is statistically "better" than another.

Disease process models are not empirical. They are constructed to describe the observed time course of adverse human effects. When using disease process modeling, we require that the model structure be based on what is known about the underlying basic biological processes, such as cell kinetics, and the time course of the adverse effect or disease. As we have repeatedly noted, this requirement is very constraining, even when there are only limited data on tissue responses and their temporal variations, because the model must simultaneously fit all of it. There are existing biomathematical models of many common biological processes, such as cell and enzyme kinetics, chaotic processes of heart and respiratory dynamics, predator-prey relationships, and infectious disease behavior in populations. Many of these models can be adapted to newly identified cellular processes, as we saw in Chapter 14 for reversible proportional FEV1 changes caused by ozone damage, and in Chapter 15 for irreversible carcinogenesis. We recommend that investigators study the implications of model selection by noting where there are reasonable alternative choices and examining their temporal behavior. Sometimes a reasonable form cannot fit the data, so we must discard it. Other times, we find that a model fits, and its behavior suggests that there are additional things we can look for to test the model. For example, a model developed for acute effects can be extrapolated to longer time scales or visa versa, conditions that may not have been studied. Most of the models we have presented have been relatively simple. We can represent more complex behavior by linking models together, such as an acute reversible response model and a cumulative damage model. We have also shown that widely used empirical dose metrics, such as cumulative exposure, are consistent with some disease process models, but not with others.

16.2. DETERMINING MODEL UNCERTAINTIES

Fitting and estimating process model parameters can require sophisticated statistical procedures and may be computationally intensive when dose metrics must be calculated for each subject's long history of exposure. Bayesian Monte Carlo fitting procedures have advantages because they can estimate individual subject parameters while simultaneously estimating population parameters and their variation. This type of fitting concurrently estimates all of the parameters, so it takes account of interactions among the variables. Uncertainties that derive from the choice of statistical research methods and mathematical models are especially in need of investigation because so little work has been done in this area.

As noted in Chapter 9, dosimetric model parameters are often assumed to represent population averages, and only the model input—the exposure vector (Equation 9.1)—is allowed to vary. But it is important to be aware that the assumption that the model parameters do not vary among individuals is usually made because we have no choice, lacking any data with which to investigate individual variability in such characteristics as a clearance half-time for an aerosol in the lungs or a heavy metal in the blood. In reality, it is likely that all of the physiologic processes described by the dosimetric model *do vary individually* and that our assumption of constancy is a necessary expedient. A consequence of assuming population averages for model parameters is probably to introduce nondifferential misclassification into the dose metrics. A future goal should be to develop methods and tools that will allow us to develop personalized time courses and model parameters for each subject in our epidemiologic studies.

16.3. STUDY DESIGN

One of the major contributions of the disease process model approach is the guidance it provides for study design. In addition to the usual aspects of epidemiologic study design, such as type of study, choice of exposure and outcome variables, and power, the process model approach requires us to explicitly define the underlying biology and its implications for how we are going to test our hypotheses. Because the models are dynamic, we will often want to use repeated-measures approaches so we can see how the human biological system responds to changes in exposure. Repeated measurements on each subject allow us to personalize the dose models; same form is used but the parameters are adjusted.

An important limitation of the greater detail needed on each subject is that retrospective studies may not be possible, and this tends to limit the number of subjects who can be studied. However, in some cases, the much greater dose resolution may allow us to have fewer subjects, especially if we have high-resolution information on the nature of the health effects.

16.4. DISEASE PROCESS MODELS PROVIDE THE LINK WITH RISK ASSESSMENT

The physiologically based pharmacokinetic (PBPK) models have found growing use in risk assessment because they potentially provide a means of extrapolating toxicologic findings from animals to humans. This provides a means of predicting human risks from environmental exposures based on tests of laboratory animals, which is much less expensive and time-consuming than large epidemiologic studies. However, the promise of this approach has not been realized, because there are important differences between animals and humans, especially metabolism and cellular mechanisms, that have confounded the extrapolations. These are perhaps best illustrated by the history of concern about inorganic arsenic. Laboratory scientists were unable to produce cancers in rodents, despite years of trying, whereas epidemiologic studies very clearly demonstrated that inorganic arsenic is a lung and skin carcinogen with airborne dust and water exposures (NRC 1983).

For each toxic response, there is a dose metric that best describes the quantitative dose-response relationship. Unfortunately, there is no single dose metric that is satisfactory for all types of responses. For example, it is common that receptor-based drug effects are well described by the logarithm of the AUC. On the other hand, anesthetic effects of gases on the central nervous system are a function of the immediate tissue concentration, and anesthesia is achieved immediately once an adequate CNS concentration is achieved. Secondary toxic effects and systemic responses may be identified as the toxic effects of an agent, which will expand and modify the initial form of the dose-response relationship. The key observation is that the "best" form of the dose metric depends on the mechanism of the effects, and the mechanism is rarely known, even for laboratory studies. This is a very active area of current research in toxicology.

16.5. LINKAGE OF EPIDEMIOLOGIC AND TOXICOLOGIC STUDIES

Better linkage of epidemiologic and toxicologic research is needed. A parallel research strategy needs to be developed in which laboratory mechanistic studies of early effects are used to generate hypotheses for epidemiologic studies with good dosimetry. Mechanistic research is advancing quickly to describe the processes by which toxic materials cause human effects. As these are elucidated, then field studies of exposed populations can be conducted to verify their predictive value.

How can the dosimetric model estimates of tissue concentration be validated? There are currently a number of biomarkers, such as hemoglobin and DNA adducts, that can be used to provide estimates of the concentrations of activated metabolites in the blood and tissues, respectively. In some cases, parent and activated materials can be measured directly in blood samples. Although it may be analytically difficult to measure some types of materials, there has been considerable progress in developing these methods. Proteomic methods also provide a means of identifying reaction products in which active agents, such as epoxides, metabolites of environmental chemicals, react with proteins. These products also may be useful as biomarkers of exposure and dose.

Tissue dose of the active agent provides the key link enabling extrapolation of animal study results to humans. Laboratory experiments using appropriate target tissue doses can be used to study mechanisms and effects in vitro. Conversely, animal findings can be better extrapolated to human tissue doses, and then doses can be extrapolated to exposure situations likely to produce them.

The risk management paradigm for occupational and environmental regulatory intervention is designed to minimize or eliminate risk by controlling exposure, which is usually achieved by controlling emissions from sources. However, regulations have been set by establishing exposure limits which are like speed limits—one is presumed to be safe below the limit and at risk above it. This is based on an overly simplistic dose metric which assumes that everyone who receives a period of exposure at the same concentration has about the same risk of an adverse effect. Additional periods of exposure increase the risk in a linear fashion. This is clearly not true, but generally we have not had the dose-response data needed to set a more appropriate set of limits that deal with the variation in risk by age, gender, and susceptibility factors. As a result, our exposure limits may not be protective for all individuals.

16.6. RISK ASSESSMENT: EXTRAPOLATING ANIMAL DOSE-RESPONSE RELATIONSHIPS TO HUMANS

Toxicity and direct mechanistic data on adverse effects of exposure to toxic materials are generally only available from animal and tissue testing. These data can be very valuable in showing that an agent can produce effects, which can in turn be used to aid formulation of the descriptive disease process model. But there are several limitations for the extrapolation of animal toxicity and dose data to human environmental exposures. First, the administered dose ranges are often widely different, frequently by several orders of magnitude, ranging from mg/kg for animals down to ng/kg or lower for human exposures. Second, the fractions of subjects affected by those exposures are substantially different, ranging from low to high percentages for animals compared with one per thousand or less for humans. Third, there are often large species differences when whole animals are tested. Studies to determine the causes of interspecies variability have regularly found large differences in tissue levels produced by the same administered doses, differences in metabolic products, and substantial differences in tissue responses.

1,3-Butadiene (BD) is an excellent example of these kinds of differences among test species, which were extensively reviewed by Himmelstein and coworkers (Himmelstein, 1996). Rats are three orders of magnitude less sensitive to carcinogenic effects of inhaled butadiene than mice, and tumors formed at different sites in rats and mice. Female rats and mice were more sensitive to BD than males of both species. High exposures (625 ppm for 6 hr/d) for short duration (26 wk) were more effective in producing tumors in mice than lower (312 ppm for 6 hr/d), longer exposures (52 wk), which represent the same administered dose ($C \times T$). This may have occurred because mice are susceptible to depletion of glutathione when exposed to high levels of BD. The mouse-rat cancer risk differences were attributed to the fact that mice produce much larger concentrations of the highly mutagenic diepoxy butane than rats, but this does not explain the male-female differences.

Animal models are a favored methodology for studying mechanisms of effects (Klaassen 2001). Unfortunately, it is often difficult to determine whether a species of test animal and humans have the same mechanisms. For example, studies of drugs to control hyperlipidemia have used a variety of animals, including mice, rats, hamsters, and monkeys (Krause and Princen 1998). However, none of these animal systems have given responses comparable to human responses for lipid-lowering drugs. New studies are being done with transgenetic mice that have human genes previously identified as part of the causal pathway. These mice have shown responses comparable to those of humans, and it was concluded that the new genetic mouse models may be better, more focused tools to detect new drugs that might be useful for humans. The same may be true for toxic effects.

16.7. FUTURE CHALLENGES

Two major advantages of the disease process model are that they can use mechanistic data directly and that they can investigate the population level implications of molecular and cellular level findings from the lab. We have explored several examples in which cell kinetics were important. Continued rapid development of our understanding of genetic and related processes will improve our mechanistic models.

Gene-environment Interactions

Toxicogenomics

It has long been recognized that DNA was a target of reactive chemical species. We now recognize that another important effect is changes in gene expression, specifically the up-and-down regulation of genes. Considerable effort is now going into developing methods and assessing gene responses to toxic effects, both to define toxic responses and to identify susceptibility. Primarily, this has been investigated in animal studies.

Epigenetic Effects

Factors affecting micro-RNAs and DNA methylation are now recognized as sources of adverse effects because of alteration of gene activity. At this stage, we are just learning how effects by these mechanisms are operating. The environmental sources of these effects are as yet poorly understood. We can expect a rapid development of understanding in the next few years.

Biomechanical Injury Processes

Biomechanical injury processes are important for ergonomic and musculoskeletal diseases (MSD), and our understanding of them has developed largely independently from toxicologic paradigms of mechanism. Investigators have begun to use disease model concepts applied to physical forces as exposures and the agents of adverse effects. Clearly, the dose of physical energy to a susceptible tissue can cause damage, as indicated when someone undergoes an intense period of physical exercise at the limit of his or her capacity. Within a day or two of the event, the individual may experience muscle soreness associated with microtears, damage to the muscles. This may pass as the damage is repaired, and repeated episodes may increase the strength of the muscles. Too much damage may take longer to repair and will not strengthen the muscles. What biological models are appropriate for this effect? Time to failure? How to quantify effects? This is an important area for further development of disease process models.

Behavioral and Psychosocial Factors

We know that psychological factors can modify damage responses and alter susceptibility. Work-related stress associated with repetitious jobs in which the individual has little control over the pace of the work can increase risk of heart disease [Karasek 1990]. What is the appropriate measure of the "dose" of stress? Again, we believe that the disease process modeling approach can be profitably applied to this area of research.

16.8. FINAL THOUGHTS

Exposure biology is a fairly recent term for the study of the interactions between humans and hazards in the environment, especially the chemical and physical hazards associated with human activities. It is quite consistent with the approach advocated in this book. Research in exposure biology covers the integrative study of the physical, chemical, and behavioral dimensions that define external exposure. It also includes the biological interactions that define the uptake and dose to the internal target tissues, which are the proximal cause of adverse effects. The ultimate goal is to define personal determinants of exposure, agent uptake, internal pharmacokinetics, and target tissue dosimetry so that we can define who is at risk, and the dose-response relationship, and the points for effective intervention.

Exposure biology provides an integrative framework of concepts, relationships, and methods that bring together portions of many fields, including exposure assessment, industrial hygiene, radiation health, toxicology, environmental epidemiology, and environmental health. We hope that disease process models can become a central theme of exposure biology and that each of these frameworks enhances the other.

We hope to encourage more researchers to investigate the disease process modeling approach and to seek new data that are suitable for modeling. Continued use of simple exposure-response epidemiologic models which take no account of the temporal dynamics of physiologic processes may lead to a self-fulfilling prophecy: the health risks of toxins whose mechanisms can be adequately described by static models will be accurately estimated and appropriate standards will be adopted; those with more complex mechanisms may be missed entirely, further reinforcing the idea that the static models are adequate. Furthermore, toxicologic processes that are not adequately described by static epidemiologic models may not be adequately regulated by the standard sorts of legal exposure limits—the average ozone level in urban air, for example. If a person's *history* of exposure is an important determinant of the impact of a given unit of exposure, then an exposed individual may achieve a certain level of target tissue dose by a wide range of patterns of exposure intensity and duration. Thus short "peaks" of exposure may, in some disease processes, carry extra risk or may modify the impact of successive exposures. Adequate protection may require limits specifying not only average exposures but also *patterns* of exposure.

This approach to epidemiology provides a framework within which *all* available data—including experimental and laboratory animal studies—exploring the mechanisms causing risk by a toxic substance can be utilized. Epidemiologists have often been reluctant to use animal studies for anything more than a source of hypotheses. But some of these examples have shown that animal studies may be quite useful in suggesting and evaluating the structures of dosimetric models (because these structures may reflect fundamental physiologic processes that are likely to be qualitatively, if not quantitatively, similar in similar species) and sometimes also for quantification of these models when the parameters are judged to be similar or scalable from one species to another.

In summary, then, we have tried to show that standard epidemiologic models cannot capture important aspects of the temporal dynamic behavior of a toxin as it enters the body, is transported, interacts with target tissues, triggers repair mechanisms, and so on. To model the exposure-response relationship more accurately, one must make use of what is known about the underlying physiologic and pathologic processes. Especially important to capture in the modeling are key features of the *temporal dynamics* (e.g., reversibility/irreversibility, accumulation, repair, attenuation, and sensitization) of the physiologic processes under study.

ANNOTATED REFERENCE BOOKS

American Conference of Governmental Industrial Hygienists (2001). *Air Sampling Instruments*, 9th edition, American Conference of Governmental Industrial Hygienists, Akron, OH. This excellent, highly detailed professional reference covers the full range of environmental and occupational instruments and has excellent background chapters on practical technical issues, applications, validation, calibration, and sampling strategy.

Boleij, J. S. M., E. Buringh, D. Heederik, and H. Kromhout (1995). *Occupational Hygiene of Chemical and Biological Agents*, Elsevier Science B.V., Amsterdam. This text provides a focused introduction to occupational hygiene and specifically biological agents, exposure modeling, and the methods used in occupational epidemiology.

Checkoway, H., N. Pearce, and D. Kriebel (2004). *Research Methods in Occupational Epidemiology*. 2nd edition, Oxford University Press, New York. This is a very useful reference for the epidemiologist and occupational hygienist interested in the methods used in occupational epidemiology, including exposure assessment and dose estimation.

Hinds, W. (1999). *Aerosol Technology: Properties, Behavior, and Measurement of Airborne Particles*, 2nd edition, Wiley-Interscience, Philedelphia. This is a highly technical reference book on all of the aspects of aerosol formation, behavior, and measurement. It will be most useful for someone with a strong mechanical engineering or physical science background.

Klaassen, C. D., Ed. (2001). *Casarett and Doull's Toxicology: The basic science of poisons*. McGraw-Hill Medical, New York. This broad detailed introduction to toxicology is a good reference for the clear presentation of the concepts, methods, and common toxic materials and contains many nice examples.

Lippmann, M., B. S. Cohen, and R. B. Shlesinger (2003). *Environmental Health Science*, Oxford University Press, New York. A broad introduction to environmental health problems, concepts, and methods.

Manahan, S. E. (2004). *Environmental Chemistry*, Lewis Publishing, Boston. This is an in-depth technical presentation on the chemical aspects of environmental hazards, with good coverage of the background science underlying the problems of air, water, and soil contamination.

Nieuwenhuijsen, M. Ed. (2003). *Exposure Assessment in Occupational and Environmental Epidemiology*, Oxford University Press. This is a collection of essays by leading authorities that provides a good survey of the state-of-the-art practical methodology in this area.

Rothman, K., S. Greenland and J. Lasch. (2008). *Modern Epidemiology*, 3rd edition, Lippincott, Williams, and Wilkins, Philadelphia. This is the third edition of the most influential epidemiology textbook of our day. It is the standard reference to which all others are compared. It is not an easy read—beginners should start with a more basic text (we recommend A. Aschengrau and G.R. Seage III. *Essential Epidemiology in Public Health*. Jones and Bartlett, 2nd edition, 2008).

REFERENCES

Agalliu, I., E. A. Eisen, D. Kriebel, M. M. Quinn, and D. H. Wegman. (2005). A biological approach to characterizing exposure to metalworking fluids and risk of prostate cancer (United States). *Cancer Causes and Control* 16 (4): 323–31.

Ainsworth, B. E., D. R. Jacobs, A. S. Leon, M. T. Richardson, and H. J. Montoye. (1993). Assessment of the accuracy of physical-activity questionnaire occupational data. *Journal of Occupational and Environmental Medicine* 35 (10): 1017–27.

Akaike, H. (1973). Information theory and an extension of the maximum likelihood principle. In *Second international symposium on information theory*, ed. B. N. Petrov and F. Csaki, 267–281. Budapest, Hungary: Akademiai Kiado.

Akpinar-Elci, M., W. D. Travis, D. A. Lynch, and K. Kreiss. (2004). Bronchiolitis obliterans syndrome in popcorn production plant workers. *European Respiratory Journal* 24 (2): 298–302.

American Conference of Governmental Industrial Hygienists. (1961). *Threshold limit values*. Akron, OH: American Conference of Governmental Industrial Hygienists.

Andersen, M. E. (2003). Toxicokinetic modeling and its applications in chemical risk assessment. *Toxicology Letters* 138 (1-2): 9–27.

Andersen, M. E., J. E. Dennison, R. S. Thomas, and R. B. Conolly. (2005). New directions in incidence-dose modeling. *Trends in Biotechnology* 23 (3): 122–7.

Armitage, P., and R. Doll. (1954). The age distribution of cancer and a multistage theory of carcinogenesis. *British Journal of Cancer* 8: 1–12.

Armstrong, B. G. (1992). Confidence intervals for arithmetic means of lognormally distributed exposures. *American Industrial Hygiene Association Journal* 53 (8): 481–5.

_____. (1995). Study design for exposure assessment in epidemiological studies. *Science of the Total Environment* 168: 187–194.

_____. (1996). Optimizing power in allocating resources to exposure assessment in an epidemiologic study. *American Journal of Epidemiology* 144 (2): 192–7.

_____. (1998). "Effect of measurement error on epidemiological studies of environmental and occupational exposures. *Occupational and Environmental Medicine* 55 (10): 651–6.

Armstrong, B. K., E. White, and R. Saracci. (1992). *Principles of exposure measurement in epidemiology*. New York: Oxford University Press.

Astrand, I. (1983). Effect of physical exercise on uptake, distribution, and elimination of vapors in man. In: *Modeling of Inhalation Exposure to Vapors: Uptake, Distribution, and Elimination. V. Fiserova-Bergerova*. Boca Raton, FL, CRC Press. **2:** 108–130.

Auerbach, O., E. C. Hammond, and L. Garfinkel. (1979). Changes in bronchial epithelium in relation to cigarette smoking, 1955–1960 vs. 1970–1977. *New England Journal of Medicine* 300 (8): 381–5.

Axelson, O., and K. Steenland. (1988). Indirect methods of assessing the effects of tobacco use in occupational studies. *American Journal of Industrial Medicine* 13 (1): 105–18.

Bailar, J. I., and A. J. Bailer. (1999). Risk assessment: The mother of all uncertainties: Disciplinary perspectives on uncertainty in risk assessment. *Annals of the New York Academy of Sciences* 895: 273–285.

Ballew, M. A., D. Kriebel, and T. J. Smith. (1995). Epidemiologic application of a dosimetric model of dust overload. *American Journal of Epidemiology* 141 (7): 690–696.

Becking, G. C. (1995). Use of mechanistic information in risk assessment for toxic chemicals. *Toxicology Letters* **77**(1-3): 15–24.

Becklake, M. R. (1995). Relationship of acute obstructive airway change to chronic (fixed) obstruction. *Thorax* 50, Suppl. no. 1: S16–21.

Bennet, G. (1962). Ozone contamination of high altitude aircraft cabins. *Aerospace Medicine* 33: 969–73.

Bennett, J. S., C. E. Feigley, D. W. Underhill, W. Drane, T. A. Payne, P. A. Stewart, R. F. Herrick, D. F. Utterback, and R. B. Hayes. (1996). Estimating the contribution of individual work tasks to room concentration: Method applied to embalming. *American Industrial Hygiene Association Journal* 57 (7): 599–609.

Berhane, K., W. J. Gauderman, D. O. Stram, and D. C. Thomas. (2004). Statistical issues in studies of the long-term effects of air pollution: The Southern California Children's Health Study. *Statistical Science* 19 (3): 414–34.

Bjornberg, K. A., M. Vahter, B. Berglund, B. Niklasson, M. Blennow, and G. Sandborgh-Englund. (2005). Transport of methylmercury and inorganic mercury to the fetus and breast-fed infant. *Environmental Health Perspectives* 113 (10): 1381–5.

Bjornberg, K. A., M. Vahter, K. Petersson-Grawe, A. Glynn, S. Cnattingius, P. O. Darnerud, S. Atuma, M. Aune, W. Becker, and M. Berglund. (2003). Methyl mercury and inorganic mercury in Swedish pregnant women and in cord blood: Influence of fish consumption. *Environmental Health Perspectives* 111 (4): 637–41.

Black, K. G., and R. A. Fenske. (1996). Dislodgeability of chlorpyrifos and fluorescent tracer residues on turf: Comparison of wipe and foliar wash sampling techniques. *Archives of Environmental Contamination and Toxicology* 31 (4): 563–70.

Blair, A., and P. Stewart. (1992). Do quantitative exposure assessments improve risk estimates in occupational studies of cancer? *American Journal of Industrial Medicine* 21: 53–63.

Boland, C., and L. Ricciardello. (1999). How many mutations does it take to make a tumor? *Proceedings of the National Academy of Sciences* 96 (26): 14675–14677.

Boleij, J. S. M., E. Buringh, D. Heederik, and H. Kromhout. (1995). *Occupational hygiene of chemical and biological agents*. Amsterdam: Elsevier.

Bradford Hill, A. (1965). The environment and disease: Association or causation. *Proceedings of the Royal Society of Medicine* 58: 295–300.

Brain, J. D., B. D. Beck, A. J. Warren, and R. A. Shaikh, eds. (1988). *Variations in susceptibility to inhaled pollutants: Identification, mechanisms and policy implications.* Baltimore: Johns Hopkins University Press.

Brenner, H., and D. Loomis. (1994). Varied forms of bias due to nondifferential error in measuring exposure. *Epidemiology* 5 (5): 510–7.

Buckley, T. J., J. D. Prah, D. Ashley, R. A. Zweidinger, and L. A. Wallace. (1997). Body burden measurements and models to assess inhalation exposure to methyl tertiary butyl ether (MTBE). *Journal of the Air and Waste Management Association* 47 (7): 739–52.

Burns-Naas, L. A., B. Meade, and A. E. Munson. (2001). Toxic responses of the immune system. In *Casarett and Doull's toxicology: The basic science of poisons*, ed. C. D. Klaassen, New York: McGraw-Hill, 123–126.

Burstyn, I., H. Kromhout, T. Kauppinen, P. Heikkila, and P. Boffetta. (2000). Statistical modelling of the determinants of historical exposure to bitumen and polycyclic aromatic hydrocarbons among paving workers. *Annals of Occupational Hygiene* 44 (1): 43–56.

Busse, W. W., and H. Mitchell. (2007). Addressing issues of asthma in inner-city children. *Journal of Allergy and Clinical Immunology* 119 (1): 43–9.

Cairns, J., and E. Smith. (1996). Uncertainties associated with extrapolating from toxicologic responses in laboratory systems to the responses of natural systems. In *Scientific uncertainty and environmental problem solving*, ed. J. Lemons. Cambridge, MA: Blackwell Science.

Calvert, G. M., E. Ward, T. M. Schnorr, and L. J. Fine. (1998). Cancer risks among workers exposed to metalworking fluids: A systematic review. *American Journal of Industrial Medicine* 33 (3): 282–92.

Carlin, B., and T. Louis. (1996). *Bayes and empirical Bayes methods for data analysis.* New York: Chapman and Hall.

Carson, R. (1962). *Silent spring.* Boston: Houghton Mifflin.

Carvalho, F. M., F. Lima, and D. Kriebel. (2004). Re: On John Snow's unquestioned long division. *American Journal of Epidemiology* 159 (4): 422.

Centers for Disease Control. (2002). Fixed obstructive lung disease in workers at a microwave popcorn factory: Missouri, 2000–2002. *Morbidity and Mortality Weekly Report* 51: 345–7.

Chan, E. Y. Y., S. M. Griffiths, and C. W. Chan. (2008). Public-health risks of melamine in milk products. *Lancet* 372 (9648): 1444–5.

Chang, F. K., M. L. Chen, S. F. Cheng, T. S. Shih, and I. F. Mao. (2007). Dermal absorption of solvents as a major source of exposure among shipyard spray painters. *Journal of Occupational and Environmental Medicine* 49 (4): 430–6.

Chatterjee, N., and S. Wacholder. (2002). Validation studies: Bias, efficiency, and exposure assessment [editorial]. *Epidemiology* 13: 5036.

Checkoway, H., N. J. Heyer, N. S. Seixas, E. A. Welp, P. A. Demers, J. M. Hughes, and H. Weill. (1997). Dose-response associations of silica with nonmalignant respiratory disease and lung cancer mortality in the diatomaceous earth industry. *American Journal of Epidemiology* 145 (8): 680–8.

Checkoway, H., N. Pearce, and D. Kriebel. (2004). *Research methods in occupational epidemiology, 2nd ed.* Oxford, UK: Oxford University Press.

Chen, J. C., W. R. Chang, T. S. Shih, C. J. Chen, W. P. Chang, J. T. Dennerlein, L. M. Ryan, and D. C. Christiani. (2004). Using exposure prediction rules for exposure assessment: An example on whole-body vibration in taxi drivers. *Epidemiology* 15 (3): 293–9.

Chen, N. H., and D. F. Othmer. (1962). New generalized equation for gas diffusion coefficient. *Journal of Chemical and Engineering Data* 7 (1): 37–41.

Churg, A., and F. H. Y. Green, eds. (1998). *Pathology of occupational lung disease.* London: Williams & Wilkins.

Clark, N. N., J. M. Kern, C. M. Atkinson, and R. D. Nine. (2002). Factors affecting heavy-duty diesel vehicle emissions. *Journal of the Air and Waste Management Association* 52 (1): 84–94.

Clewell, H. J., and M. E. Andersen. (2004). Applying mode-of-action and pharmacokinetic considerations in contemporary cancer risk assessments: An example with trichloroethylene. *Critical Reviews in Toxicology* 34 (5): 385–445.

Clewell, H. J., J. M. Gearhart, P. R. Gentry, T. R. Covington, C. B. VanLandingham, K. S. Crump, and A. M. Shipp. (1999). Evaluation of the uncertainty in an oral reference dose for methylmercury due to interindividual variability in pharmacokinetics. *Risk Analysis* 19 (4): 547–58.

Clewell, H. J., J. Teeguarden, T. McDonald, R. Sarangapani, G. Lawrence, T. Covington, R. Gentry, and A. Shipp. (2002). Review and evaluation of the potential impact of age- and gender-specific pharmacokinetic differences on tissue dosimetry. *Critical Reviews in Toxicology* 32 (5): 329–89.

Coughlin, J., S. K. Hammond, and P. H. Gann. (1989). Development of epidemiologic tools for measuring environmental tobacco-smoke exposure. *American Journal of Epidemiology* 130 (4): 696–704.

Corley, R. A., S. M. Gordon, and L. A. Wallace. (2000). Physiologically based pharmacokinetic modeling of the temperature-dependent dermal absorption of chloroform by humans following bath water exposures. *Toxicological Sciences* 53 (1): 13–23.

Crapo, J. D., J. Glassroth, J. B. Karlinsky, and T. E. King, Jr. (2004). *Baum's textbook of pulmonary diseases.* Philadelphia: Lippincott Williams & Wilkins.

Davies, M. J., S. L. deLacey, and R. J. Norman. (2005). Towards less confusing terminology in reproductive medicine. *Human Reproduction* 20 (10): 2669–71.

Davis, M. E., T. J. Smith, F. Laden, J. E. Hart, L. Ryan, and E. Garshick. (2006). Modeling particle exposure in U.S. trucking terminals. *Environmental Science and Technology* 40 (13): 4226–32.

Day, G. A., N. A. Esmen, and T. A. Hall. (1999). Sample size-based indication of normality in lognormally distributed populations. *Applied Occupational and Environmental Hygiene* 14 (6): 376–83.

de Cock, J., D. Heederik, H. Kromhout, J. S. Boleij, F. Hoek, H. Wegh, and N. E. Tjoe. (1998). Exposure to captan in fruit growing. *American Industrial Hygiene Association Journal* 59 (3): 158–65.

Debes, F., E. Budtz-Jorgensen, P. Weihe, R. F. White, and P. Grandjean. (2006). Impact of prenatal methylmercury exposure on neurobehavioral function at age 14 years. *Neurotoxicology and Teratology* 28 (3): 363–75.

Department of Health and Human Services. (1990). The health benefits of smoking sessation. Bethesda, MD: U.S. Government Printing Office.

Devlin, R. B., W. F. McDonnell, S. Becker, M. C. Madden, M. P. McGee, R. Perez, G. Hatch, D. E. House, and H. S. Koren. (1996). Time-dependent changes of inflammatory mediators in the lungs of humans exposed to 0.4 ppm ozone for 2 hr: A comparison of mediators found in bronchoalveolar lavage fluid 1 and 18 hr after exposure. *Toxicology and Applied Pharmacology* 138 (1): 176–85.

Diggle, P., P. Heagerty, K.-Y. Liang, and S. Zeger. (2002). *Analysis of longitudinal data.* Oxford, UK: Oxford University Press.

Doll, R., and R. Peto. (1976). Mortality in relation to smoking: 20 years' observations on male British doctors. *British Medical Journal* 2 (6051): 1525–36.

_____. (1978). Cigarette smoking and bronchial carcinoma: Dose and time relationships among regular smokers and lifelong non-smokers. *Journal of Epidemiology and Community Health* 32 (4): 303–13.

Doll, R., R. Peto, J. Boreham, and I. Sutherland. (2004). Mortality in relation to smoking: 50 years' observations on male British doctors. *British Medical Journal* 328 (7455): 1519.

Dosemeci, M., S. Wacholder, and J. H. Lubin. (1990). Does nondifferential misclassification of exposure always bias a true effect toward the null value? *American Journal of Epidemiology* 132 (4): 746–8.

Droz, P. O. (1992). Quantification of biological variability. *Annals of Occupational Hygiene* 6 (3): 295–306.

_____. (1993). Pharmacokinetic modeling as a tool for biological monitoring. *International Archives of Occupational and Environmental Health* 65, Suppl. no. 1: S53–9.

Durant, J. L., J. Chen, H. F. Hemond, and W. G. Thilly. (1995). Elevated incidence of childhood leukemia in Woburn, Massachusetts: NIEHS Superfund Basic Research Program searches for causes. *Environmental Health Perspectives* 103: 93–8.

Ehrenberg, L., S. Osterman-Golkar, and M. Tornqvist. (1986). Macromolecule adducts, target dose and risk assessment. *Progress in Clinical and Biological Research* 209B:253–60.

Eisen, E. A., P. E. Tolbert, et al. (1992). "Mortality studies of machining fluid exposure in the automobile industry I: A standardized mortality ratio analysis." *American Journal of Industrial Medicine* 22 (6): 809–24.

Eisen, E. A., T. J. Smith, et al. (2001). Respiratory health of automobile workers and exposures to metal-working fluid aerosols: lung spirometry. *American Journal of Industrial Medicine* 39 (5): 443–53.

eMedicine. (2007). WebMD, http//www.eMedinine.com.

Environmental Protection Agency. (1996). *Air quality criteria for ozone and related photochemical oxidants.* Washington, DC: U.S. Environmental Protection Agency, National Center for Environmental Assessment.

____. (1994). *Methods for Derivation of Inhalation Reference Concentrations and Application of Inhalation Dosimetry.* Washington, DC, Office of Health and Environmental Assessment, Environmental Protection Agency.

Fauci, A. S., E., Braunwald, D. L. Kasper, S. L. Hauser, D. L. Longo, J. L. Jameson, J. Loscalzo, Eds. (2008). *Harrison's principles of internal medicine. 17th edition.* New York: McGraw-Hill.

Fenske, R. A., and S. G. Birnbaum. (1997). Second generation video imaging technique for assessing dermal exposure (VITAE system). *American Industrial Hygiene Association Journal* 58 (9): 636–45.

Ferguson, W. S., W. C. Koch, L. B. Webster, and J. R. Gould. (1977). Human physiological response and adaption to ammonia. *Journal of Occupational Medicine* 19: 319–26.

Feychting, M., A. Ahlbom, and D. Savitz. (1998). Electromagnetic fields and childhood leukemia. *Epidemiology* 9(3): 225–6.

Fingerhut, M., W. Halperin, D. A. Marlow, L. A. Piacitelli, P. A. Honchar, M. H. Sweeney, A. L. Greife, P. A. Dill, K. Steenland, A. J. Suruda. (1991). Cancer mortality in workers exposed to 2,3,7,8-tetrachlorodibenzo-p-dioxin. *New England Journal of Medicine* 324: 212–8.

Fischer, L. A., J. D. Johansen, and T. Menne. (2007). Nickel allergy: Relationship between patch test and repeated open application test thresholds. *British Journal of Dermatology* 157 (4): 723–9.

Fiserova-Bergerova, V. (1995). Extrapolation of physiological parameters for physiologically based simulation models. *Toxicology Letters* 79:77–86.

Forster, M., and E. Sober. (1994). How to tell when simpler, more unified, or less ad hoc theories will provide more accurate predictions. *British Journal for the Philosophy of Science* 45: 1–35.

Freijer, J. I., J. C. H. van Eijkeren, and L. van Bree. (2002). A model for the effect on health of repeated exposure to ozone. *Environmental Modelling and Software* 17 (6): 553–62.

Fruin, S., D. Westerdahl, T. Sax, C. Sioutas, and P. M. Fine. (2008). Measurements and predictors of on-road ultrafine particle concentrations and associated pollutants in Los Angeles. *Atmospheric Environment* 42 (2): 207–19.

Funtowicz, S., and J. Ravetz. (1990). *Uncertainty and quality in science for policy.* Dordrecht, Netherlands: Kluwer Academic.

Ganong, W. F., ed. (2005). *Review of medical physiology.* New York: McGraw-Hill.

Garabrant, D. H. (1985). Dermatitis from aziridine hardener in printing ink. *Contact Dermatitis* 12 (4): 209–12.

Gargas, M. L., M. A. Medinsky, et al. (1995). Pharmacokinetic modeling approaches for describing the uptake, systematic distribution, and disposition of inhaled chemicals. *Critical Reviews in Toxicology* **25** (3): 237–254.

Gardiner, K., and J. M. Harrington. (2005). *Occupational hygiene.* Malden, MA: Blackwell.

Gerhardsson, L., L. Dahlin, R. Knebel, and A. Schutz. (2002). Blood lead concentration after a shotgun accident. *Environmental Health Perspectives* 110: 115–17.

Ginevan, M. E., and D. E. Splitstone. (2003). *Statistical tools for environmental quality measurement.* Boca Raton, FL: CRC Press.

Gold, D., H. Burge, V. Carey, D. K. Milton, T. Platts-Mills, and S. T. Weiss. (1999). Predictors of repeated wheeze in the first year of life: The relative roles of cockroach, birth weight, acute lower respiratory illness, and maternal smoking. *American Journal of Respiratory and Critical Care Medicine* 160: 227–36.

Gold, D. R., and R. Wright. (2005). Population disparities in asthma. *Annual Review of Public Health* 26: 89–113.

Grandjean, P., E. Budtz-Jorgensen, R. F. White, P. J. Jorgensen, P. Weihe, F. Debes, and N. Keiding. (1999). Methylmercury exposure biomarkers as indicators of neurotoxicity in children aged 7 years. *American Journal of Epidemiology* 150 (3): 301–05.

Grandjean, P., and R. F. White. (1999). Effects of methylmercury exposure on neurodevelopment. *Journal of the American Medical Association* 281 (10): 896; author reply 897.

Greenland, S. (1996). Basic methods for sensitivity analysis of biases. *International Journal of Epidemiology* 25: 1107–16.

_____. (1999). Relation of probability of causation to relative risk and doubling dose: A methodologic error that has become a social problem. *American Journal of Public Health* 89 (8): 1166–9.

_____. (2000). Principles of multilevel modeling. *International Journal of Epidemiology* 29 (1): 158–167.

_____. (2001). Sensitivity analysis, Monte Carlo risk analysis, and Bayesian uncertainty assessment. *Risk Analysis* 21: 579–83.

Greenland, S., and H. Morgenstern. (2001). Confounding in health research. *Annual Review of Public Health* 22: 189–212.

Greenland, S., and J. M. Robins. (1988). Conceptual problems in the definition and interpretation of attributable fractions. *American Journal of Epidemiology* 128 (6): 1185–97.

Gregus, Z. and C. D. Klaassen. (2001). Mechanisms of toxicity. In *Casarett and Doull's Toxicology: The Basic Science of Poisons.* Ed. C. D. Klaassen. New York: McGraw-Hill.

Gruchalla, R. S., J. Pongracic, M. Plaut, R. Evans III, C. M. Visness, M. Walter, E. F. Crain, et al. (2005). Inner City Asthma Study: Relationships among sensitivity, allergen exposure, and asthma morbidity. *Journal of Allergy and Clinical Immunology* 115 (3): 478–85.

Hammond, S. K., J. Coughlin, P. H. Gann, P. L. Skipper, and S. R. Tannenbaum. (1993). Relationship between environmental tobacco-smoke exposure and carcinogen hemoglobin adduct levels in nonsmokers: Response. *Journal of the National Cancer Institute* 85 (20): 1695–96.

Harada, M. (1995). Minamata Disease: Methylmercury poisoning in Japan caused by environmental pollution. *Critical Reviews in Toxicology* 25 (1): 1–24.

Hazelton, W. D., E. G. Luebeck, W. F. Heidenreich, and S. H. Moolgavkar. (2001). Analysis of a historical cohort of Chinese tin miners with arsenic, radon, cigarette smoke, and pipe smoke exposures using the biologically based two-stage clonal expansion model. *Radiation Research* 156 (1): 78–94.

Hazelton, W. D., M. S. Clements, and S. H. Moolgavkar. (2005). Multistage carcinogenesis and lung cancer mortality in three cohorts. *Cancer Epidemiology, Biomarkers, and Prevention* 14 (5): 1171–81.

Health Effects Institute. (1995). A Critical Analysis of Emissions, Exposure, and Health Effects. A Special Report of the Institute's Diesel Working Group. Cambridge, MA, Health Effects Institute.

Hedberg, E., A. Kristensson, M. Ohlsson, C. Johansson, P.-A. Johansson, E. Swietlicki, V. Veselya, U. Wideqvist, and R. Westerholm. (2002). Chemical and physical characterization of emissions from birch wood combustion in a wood stove. *Atmospheric Environment* 36 (30): 4823–37.

Heederik, D., H. Pouwels, H. Kromhout, and D. Kromhout. (1989). Chronic non-specific lung disease and occupational exposures estimated by means of a job exposure matrix: The Zutphen Study. *International Journal of Epidemiology* 18(2): 382–9.

Heidenreich, W. F., E. G. Luebeck, and S. H. Moolgavkar. (1997). Some properties of the hazard function of the two-mutation clonal expansion model. *Risk Analysis* 17 (3): 391–9.

Heij, C. (1989). *Deterministic identification of dynamical systems*. Vol. 127, Lecture Notes in Control and Information Sciences, ed. A. W. M. Thoma. Berlin: Springer-Verlag.

Held, T., Q. Ying, M. J. Kleeman, J. J. Schauer, and M. P. Fraser. (2005). A comparison of the UCD/CIT air quality model and the CMB source-receptor model for primary airborne particulate matter. *Atmospheric Environment* 39 (12): 2281–97.

Helsel, D. R. (2005). More than obvious: Better methods for interpreting nondetect data. *Environmental Science and Technology* 39 (20): 419A–423A.

Hinds, W. C. (1999). *Aerosol technology: Properties, behavior, and measurement of airborne particles*. New York: Wiley Interscience.

Hodgson, J., and R. Jones. (1990). Mortality of a cohort of tin miners 1941–86. *British Journal of Industrial Medicine* 47: 665–76.

Holford, T., and C. Stack. (1995). Study design for epidemiologic studies with measurement error. *Statistical Methods in Medical Research* 4: 339–58.

Hornung, R. W., R. F. Herrick, P. A. Stewart, D. F. Utterback, C. E. Feigley, D. K. Wall, D. E. Douthit, and R. B. Hayes. (1996). An experimental design approach to retrospective exposure assessment. *American Industrial Hygiene Association Journal* 57 (3): 251–6.

Hornung, R. and L. D. Reed. (1990). Estimation of average concentration in the presence of non-detectable values. *Applied Occupational and Environmental Hygiene* 5: 46–51.

Horvath, E. P., G. A. doPico, R. A. Barbee, and H. A. Dickie. (1978). Nitrogen dioxide-induced pulmonary disease: Five new cases and a review of the literature. *Journal of Occupational Medicine* 20 (2): 103–10.

Hu, H., M. Rabinowitz, and D. Smith. (1998). Bone lead as a biological marker in epidemiologic studies of chronic toxicity: Conceptual paradigms. *Environmental Health Perspectives* 106 (1): 1–8.

Hu, H., R. Shih, S. Rothenberg, and B. S. Schwartz. (2007). The epidemiology of lead toxicity in adults: Measuring dose and consideration of other methodologic issues. *Environmental Health Perspectives* 115 (3): 455–62.

Hubbs, A., A. Battelli, W. T. Goldsmith, D. W. Porter, D. Frazer, S. Friend, D. Schwegler-Berry, et al. (2002). Necrosis of nasal and airway epithelium in rats inhaling vapors of artificial butter flavoring. *Toxicology and Applied Pharmacology* 185: 128–135.

Hunter, D. (1969). *The diseases of occupations*. London: English Universities Press.

International Center for Environmental Technology Transfer. (1998). (ICETT). *Approaches to water pollution control (case study-2)*. Minimata City, Kumamoto Prefecture Yokkaichi.

International Commission on Radiological Protection. (1975). *Report of the Task Group on Reference Man*. Oxford, UK: Pergamon Press.

International Life Sciences Institute. (1994). *Physiological parameter values for PBPK models*. Washington, DC: International Life Sciences Institute, Risk Sciences Institute.

Johanson, G., A. Nihlen, and A. Lof. (1995). Toxicokinetics and acute effects of MTBE and ETBE in male volunteers. *Toxicology Letters* 82–83: 713–18.

Jurek, A. M., S. Greenland, G. Maldonado, and T. R. Church. (2005). Proper interpretation of non-differential misclassification effects: Expectations vs observations. *International Journal of Epidemiology* 34(3): 680–7.

Jurek, A. M., G. Maldonado, S. Greenland, and T. R. Church. (2006). Exposure-measurement error is frequently ignored when interpreting epidemiologic study results. *European Journal of Epidemiology* 21 (12): 871–6.

Kales, S. N., K. L. Huyck, and R. H. Goldman. (2006). Elevated urine arsenic: Un-speciated results lead to unnecessary concern and further evaluations. *Journal of Analytical Toxicology* 30 (2): 80–85.

Karasek, R. and T. Theorell. (1990). *Healthy work*. New York: Basic Books.

Kerr, M. S., J. W. Frank, H. S. Shannon, R. W. Norman, R. P. Wells, W. P. Neumann, and C. Bombardier. (2001). Biomechanical and psychosocial risk factors for low back pain at work. *American Journal of Public Health* 91 (7): 1069–1075.

Kezic, S., W. J. A. Meuling, and I. Jakasa. (2004). Free and total urinary 2-butoxyacetic acid following dermal and inhalation exposure to 2-butoxyethanol in human volunteers. *International Archives of Occupational and Environmental Health* 77 (8): 580–86.

Kile, M. L., E. A. Houseman, E. Roderigues, T. J. Smith, Q. Quamruzzaman, M. Rahman, G. Mahiuddin, L. Su, and D. C. Christiani. (2005). Toenail arsenic concentrations, GSTT1 gene polymorphisms, and arsenic exposure from drinking water. *Cancer Epidemiology, Biomarkers, and Prevention* 14 (10): 2419–26.

Kim, D., M. E. Andersen, and L. A. Nylander-French. (2006). A dermatotoxicokinetic model of human exposures to jet fuel. *Toxicological Sciences* 93 (1): 22–33.

King, T. E., Jr. (1998). Update in pulmonary medicine. *Annals of Internal Medicine* 129 (10): 806–12.

Klaassen, C. D., ed. (2001). *Casarett and Doull's toxicology: The basic science of poisons*. New York: McGraw-Hill.

Kleinbaum, D. G., and M. Klein. (2002). *Logistic regression: A self-learning text*. New York: Springer-Verlag.

Krajcarski, S., and R. Wells. (2008). The time variation pattern of mechanical exposure and the reporting of low back pain. *Theoretical Issues in Ergonomics Science* 9 (1): 45–71.

Krause, B. R. and H. M. G. Princen. (1998). Lack of predictability of classical animal models for hypolipidemic activity - a good sign for mice. *Atherosclerosis* **140** (1): 15–24.

Kraus, T., A. Pfahlberg, P. Zöbelein, O. Gefeller, and H. J. Raithel. (2004). Lung function among workers in the soft tissue paper-producing industry. *Chest* 125: 731–6.

Kreiss, K., A. Gomaa, G. Kullman, K. Fedan, E. Simoes, and P. Enright. (2002). Clinical bronchiolitis obliterans in workers at a microwave popcorn plant. *New England Journal of Medicine* 347: 330–8.

Kriebel, D. (1994). The dosimetric model in occupational and environmental epidemiology. *Occupational Hygiene* 1 (1): 55–68.

Kriebel, D., H. Checkoway, and N. Pearce. (2007). Exposure and dose modelling in occupational epidemiology. *Occupational and Environmental Medicine* 64 (7): 492–8.

Kriebel, D., D. Myers, M. Cheng, S. Woskie, and B. Cocanour. (2001). Short-term effects of formaldehyde on peak expiratory flow and irritant symptoms. *Archives of Environmental Health* 56 (1): 11–18.

Kriebel, D., S. R. Sama, et al. (1997). A field investigation of the acute respiratory effects of metal working fluids. I. Effects of aerosol exposures. *American Journal of Industrial Medicine* 31(6): 756–66.

Kriebel, D., and T. J. Smith. (1990). A nonlinear pharmacologic model of the acute effects of ozone on the human lungs. *Environmental Research* 51 (2): 120–46.

Kriebel, D., A. Zeka, E. A. Eisen, and D. H. Wegman. (2004). Quantitative evaluation of the effects of uncontrolled confounding by alcohol and tobacco in occupational cancer studies. *International Journal of Epidemiology* 33 (5): 1040–5.

Kromhout, H., D. P. Loomis, R. C. Kleckner, and D. A. Savitz. (1997). Sensitivity of the relation between cumulative magnetic field exposure and brain cancer mortality to choice of monitoring data grouping scheme. *Epidemiology* 8 (4): 442–5.

Kromhout, H., D. P. Loomis, G. J. Mihlan, L. A. Peipins, R. C. Kleckner, R. Iriye, and D. A. Savitz. (1995). Assessment and grouping of occupational magnetic field exposure in five electric utility companies. *Scandinavian Journal of Work, Environment and Health* 21 (1): 43–50.

Kromhout, H., E. Symanski, and S. M. Rappaport. (1993). A comprehensive evaluation of within- and between-worker components of occupational exposure to chemical agents. *Annals of Occupational Hygiene* 37 (3): 253–70.

Kullman, G., R. Boylstein, W. Jones, C. Piacitelli, S. Pendergrass, and K. Kreiss. (2005). Characterization of respiratory exposures at a microwave popcorn plant with cases of bronchiolitis obliterans. *Journal of Occupational and Environmental Hygiene* 2: 169–78.

Lagakos, S. W. (1988). Effects of mismodelling and mismeasuring explanatory variables on tests of their association with a response variable. *Statistics in Medicine* 7 (1–2): 257–274.

Lagakos, S. W., B. J. Wessen, and M. Zelen. (1986). An analysis of contaminated well water and health effects in Woburn, Massachusetts. *Journal of the American Statistical Association* 81 (395): 583–96.

Langholz, B., and D. C. Thomas. (1990). Nested case-control and case-cohort methods of sampling from a cohort: A critical comparison. *American Journal of Epidemiology* 131 (1): 169–176.

Law, M. R., J. K. Morris, H. C. Watt, and N. J. Wald. (1997). The dose-response relationship between cigarette consumption, biochemical markers and risk of lung cancer. *British Journal of Cancer* 75 (11): 1690–3.

Lesley, J. E. (2005). *Immunology for life sciences*. New York: John Wiley & Sons, Ltd.

Levins, R. (1995). Preparing for uncertainty. *Ecosystem Health* 1: 47–57.

Lewontin, R. (1972). The analysis of variance and the analysis of causes. *American Journal of Human Genetics* 26: 400–11.

Lide, D. R., ed. (2007). *CRC handbook of chemistry and physics*. Boca Raton, FL: CRC Press.

Links, J., B. Schwartz, and D. Simon. (2001). Characterization of toxicokinetics with linear systems theory: Application to lead-associated cognitive decline. *Environmental Health Perspectives* 109: 361–368.

Lippmann, M., B. S. Cohen, and R. B. Schlesinger. (2003). *Environmental health science: Recognition, evaluation, and control of chemical and physical health hazards*. New York: Oxford University Press.

Litonjua, A. A., D. K. Milton, J. E. Celedon, L. Ryan, S. T. Weiss, and D. R. Gold. (2002). A longitudinal analysis of wheezing in young children: The independent effects of early life exposure to house dust endotoxin, allergens, and pets. *Journal of Allergy and Clinical Immunology* 110: 736–42.

Little, M. (1995). Are two mutations sufficient to cause cancer? Some generalizations of the two-mutation model of carcinogenesis of Moolgavkar, Venzon and Knudson, and of the multistage model of Armitage and Doll. *Biometrics* 51: 1278–91.

Lloyd, J. W. (1971). Long-term mortality study of steelworkers: V. Respiratory cancer in coke plant workers. *Journal of Occupational and Environmental Medicine* 13 (2): 53–68.

Loomis, D., A. Salvan, H. Kromhout, and D. Kriebel. (1999). Selecting indices of occupational exposure for epidemiologic studies. *Occupational Hygiene* 5: 55–68.

Loomis, D. P., L. A. Peipins, S. R. Browning, R. L. Howard, H. Kromhout, and D. A. Savitz. (1994). Organization and classification of work history data in industry-wide studies: an application to the electric power industry. *American Journal of Industrial Medicine* 26 (3): 413–25.

Luebeck, E. G., W. F. Heidenreich, W. D. Hazelton, H. G. Paretzke, and S. H. Moolgavkar. (1999). Biologically based analysis of the data for the Colorado uranium miners cohort: Age, dose and dose-rate effects. *Radiation Research* 152 (4): 339–51.

Maclure, M. (1991). The case-crossover design: A method for studying transient effects on the risk of acute events. *American Journal of Epidemiology* 133 (2): 144–53.

MacMahon, B., and T. Pugh. (1970). *Epidemiology: Principles and methods*. Boston: Little, Brown.

Maldonado, G. and S. Greenland. (1993). Interpreting model coefficients when the true model form is unknown. *Epidemiology* 4: 310–18.

_____. (1996). Impact of model-form selection on the accuracy of rate estimation. *Epidemiology* 7 (1): 46–54.

Manahan, S. E. (2004). *Environmental chemistry*. Boca Raton, FL: CRC Press.

Marsh, G. M., A. O. Youk, R. A. Stone, J. M. Buchanich, M. J. Gula, T. J. Smith, and M. M. Quinn. (2001). Historical cohort study of U.S. man-made vitreous fiber production workers: I. (1992) fiberglass cohort follow-up: Initial findings. *Journal of Occupational and Environmental Medicine* 43 (9): 741–756.

McConnell, R., K. Berhane, F. Gilliland, S. J. London, T. Islam, W. J. Gauderman, E. Avol, H. G. Margolis, and J. M. Peters. (2002). Asthma in exercising children exposed to ozone: A cohort study. *Lancet* 359: 386–91.

McDonnell, W. F. (1993). Utility of controlled human exposure studies for assessing the health effects of complex mixtures and indoor air pollutants. *Environmental Health Perspectives* 101, Suppl. no. 4: 199–203.

McDonnell, W. F., and M. V. Smith. (1994). Description of acute oxone response as a function of exposure rate and total inhaled dose. *Journal of Applied Physiology* 76 (6): 2776–84.

McDonnell, W. F., P. W. Stewart, and M. V. Smith. (2007). The temporal dynamics of ozone-induced FEV1 changes in humans: An exposure-response model. *Inhalation Toxicology* 19 (6–7): 483–94.

McNamee, R. (2005). Regression modelling and other methods to control confounding. *Occupational and Environmental Medicine* 62 (7): 500–6.

Mehendale, H. M. (2005). Tissue repair: An important determinant of final outcome of toxicant-induced injury. *Toxicologic Pathology* 33 (1): 41–51.

Menzel, D. B. (1995). "Extrapolating the future: research trends in modeling." *Toxicol Lett* 79(1-3): 299–303.

Middendorf, P. J. (2004). Surveillance of occupational noise exposures using OSHA's Integrated Management Information System. *American Journal of Industrial Medicine* 46 (5): 492–504.

Miettinen, O. (1985). *Theoretical epidemiology*. New York: Wiley.

Miller, F. J., P. M. Schlosser, and D. B. Janszen. (2000). Haber's rule: A special case in a family of curves relating concentration and duration of exposure to a fixed level of response for a given endpoint. *Toxicology* 149 (1): 21–34.

Moolgavkar, S. H. (1978). The multistage theory of carcinogenesis and the age distribution of cancer in man. *Journal of the National Cancer Institute* 61 (1): 49–52.

_____. (1991). Carcinogenesis models: An overview. *Basic Life Sciences* 58: 387–96; discussion 396–9.

Moolgavkar, S. H., and A. G. Knudson, Jr. (1981). Mutation and cancer: A model for human carcinogenesis. *Journal of the National Cancer Institute* 66 (6): 1037–52.

Moolgavkar, S. H., and E. G. Luebeck. (1990). Two-event model for carcinogenesis: Biological, mathematical, and statistical considerations. *Risk Analysis* 10 (2): 323–41.

_____. (2003). Multistage carcinogenesis and the incidence of human cancer. *Genes, Chromosomes, and Cancer* 38 (4): 302–6.

Moolgavkar, S. H., E. G. Luebeck, et al. (1993). Radon, cigarette smoke, and lung cancer: a re-analysis of the Colorado Plateau uranium miners' data. [see comments]. *Epidemiology* 4(3): 204–17.

Moolgavkar, S. H., E. G. Luebeck, and E. L. Anderson. (1998). Estimation of unit risk for coke oven emissions. *Risk Analysis* 18 (6): 813–25.

Morabia, A. (2004). *History of epidemiologic methods and concepts*. Basel: Birkhauser Verlag.

Morgan, M., and M. Henrion. (1992). *Uncertainty: A guide to dealing with uncertainty in quantitative risk and policy analysis*. Cambridge, UK: Cambridge University Press.

Murata, K., P. Grandjean, and M. Dakeishi. (2007). Neurophysiological evidence of methylmercury neurotoxicity. *American Journal of Industrial Medicine* 50 (10): 765–71.

Myers, G. J., P. W. Davidson, C. Cox, C. F. Shamlaye, M. A. Tanner, D. O. Marsh, E. Cernichiari, L. W. Lapham, M. Berlin, and T. W. Clarkson. (1995). Summary of the Seychelles child development study on the relationship of fetal methylmercury exposure to neurodevelopment. *Neurotoxicology* 16 (4): 711–16.

National Institute for Occupational Safety and Health. (2002). *NIOSH evaluates worker exposures at a popcorn plant in Missouri. NIOSH Report No.2002-128*. Washington, DC: National Occupational Safety and Health Institute.

National Institute for Occupational Safety and Health and World Health Organization. (1998). *International Chemical Safety Cards, U.S. national version*. Washington, DC: National Occupational Safety and Health Institute.

National Research Council (1983) (U.S NRC). Committee on the Institutional Means for Assessment of Risks to Public Health. *Risk assessment in the federal government: managing the process*. Washington, D.C: National Academy Press.

______. (2000). National Research Council (U.S.). Committee on Toxicological Effects of Methylmercury. *Toxicological Effects of Methylmercury*. Washington, D.C., National Academy Press.

Navidi, W., and F. Lurmann. (1995). Measurement error in air-pollution exposure assessment. *Journal of Exposure Analysis and Environmental Epidemiology* 5 (2): 111–24.

Newton, E., and R. Rudel. (2007). Estimating correlation with multiply censored data arising from the adjustment of singly censored data. *Environmental Science and Technology* 41 (1): 221–8.

Neyman, J., and E. Pearson. (1928). On the use and interpretation of certain test criteria for purposes of statistical inference. *Biometrika* 28: 175–240.

Nieuwenhuijsen, M. (2003a). Introduction to exposure assessment. In *Exposure assessment in occupational and environmental epidemiology*, ed. M. Nieuwenhuijsen. New York: Oxford University Press.

_____. (2003b). Questionnaires. In *Exposure assessment in occupational and environmental epidemiology*, ed. M. Nieuwenhuijsen. New York: Oxford University Press.

_____, ed. (2003c). *Exposure assessment in occupational and environmental epidemiology*. New York: Oxford University Press.

Nihlen, A., A. Lof, and G. Johanson. (1998). Controlled ethyl tert-butyl ether (ETBE) exposure of male volunteers. I. Toxicokinetics. *Toxicological Sciences* 46 (1): 1–10.

Norman, R., R. Wells, P. Neumann, J. Frank, H. Shannon, and M. Kerr. (1998). A comparison of peak vs. cumulative physical work exposure risk factors for the reporting of low back pain in the automotive industry. *Clinical Biomechanics (Bristol, Avon)* 13 (8): 561–73.

Ostro, B. D., W. Y. Feng, R. Broadwin, B. J. Malig, R. S. Green, and M. J. Lipsett. (2008). The impact of components of fine particulate matter on cardiovascular mortality in susceptible subpopulations. *Occupational and Environmental Medicine* 65 (11): 750–6.

Park, R., F. Rice, L. Stayner, R. Smith, S. Gilbert, and H. Checkoway. (2002). Exposure to crystalline silica, silicosis, and lung disease other than cancer in diatomaceous earth industry workers: a quantitative risk assessment. *Occupational and Environmental Medicine* 59 (1): 36–43.

Parkes, W. R. (1983). *Occupational lung disorders*. London: Butterworths.

Pathak, D. R., J. M. Samet, C. G. Humble, and B. J. Skipper. (1986). Determinants of lung cancer risk in cigarette smokers in New Mexico. *Journal of the National Cancer Institute* 76 (4): 597–604.

Payne, M. P., and L. C. Kenny. (2002). Comparison of models for the estimation of biological partition coefficients. *Journal of Toxicology and Environmental Health: Part A* 65 (13): 897–931.

Pearce, N., H. Checkoway, and D. Kriebel. (2006). Bias in occupational epidemiology studies. *Occupational and Environmental Medicine* 64: 562–8.

Peters, J. L., J. I. Levy, C. A. Rogers, H. A. Burge, and J. D. Spengler. (2007). Determinants of allergen concentrations in apartments of asthmatic children living in public housing. *Journal of Urban Health* 84 (2): 185–97.

Peters, J. M., E. Avol, W. J. Gauderman, W. S. Linn, W. Navidi, S. J. London, H. Margolis, et al. (1999). A study of twelve southern California communities with differing levels and types of air pollution: II. Effects on pulmonary function. *American Journal of Respiratory and Critical Care Medicine* 159 (3): 768–75.

Peters, J. M., E. Avol, W. Navidi, S. J. London, W. J. Gauderman, F. Lurmann, W. S. Linn, et al. (1999). A study of twelve southern California communities with differing levels and types of air pollution: I. Prevalence of respiratory morbidity. *American Journal of Respiratory and Critical Care Medicine* 159 (3): 760–7.

Peto, R., S. Darby, H. Deo, P. Silcocks, E. Whitley, and R. Doll. (2000). Smoking, smoking cessation, and lung cancer in the UK since 1950: Combination of national statistics with two case-control studies. *British Medical Journal* 321 (7257): 323–9.

Pinto, S. S., V. Henderson, and P. E. Enterline. (1978). Mortality experience of arsenic-exposed workers. *Archives of Environmental Health* 33 (6): 325–31.

Platt, J. R. (1964). Strong inference. *Science* 146: 347–53.

Prah, J., D. Ashley, B. Blount, M. Case, T. Leavens, J. Pleil, and F. Cardinali. (2004). Dermal, oral, and inhalation pharmacokinetics of methyl tertiary butyl ether (MTBE) in human volunteers. *Toxicological Sciences* 77 (2): 195–205.

Pratt, W. B., and P. Taylor. (1990). *Principles of drug action*. Philadelphia: Churchill Livingstone.

Prentice, R.L. and D. Thomas. (1993). Methodologic research needs in environmental epidemiology: Data analysis. *Environmental Health Perspectives* 101, Suppl, no. 4: 39–48.

Price, P. S., R. B. Conolly, C. F. Chaisson, E. A. Gross, J. S. Young, E. T. Mathis, and D. R. Tedder. (2003). Modeling interindividual variation in physiological factors used in PBPK models of humans. *Critical Reviews in Toxicology* 33 (5): 469–503.

Quinn, M. M., D. Kriebel, K. Geiser, and R. Moure-Eraso. (1998). Sustainable production: A proposed strategy for the work environment. *American Journal of Industrial Medicine* 34 (4): 297–304.

Quinn, M. M., T. J. Smith, E. A. Eisen, D. H. Wegman, and M. J. Ellenbecker. (2000). Implications of different fiber measures for epidemiologic studies of man-made vitreous fibers. *American Journal of Industrial Medicine* 38 (2): 132–139.

Quinn, M. M., T. J. Smith, A. O. Youk, G. M. Marsh, R. A. Stone, J. M. Buchanich, and M. J. Gula. (2001). Historical cohort study of US man-made vitreous fiber production workers: VIII. Exposure-specific job analysis. *Journal of Occupational and Environmental Medicine* 43 (9): 824–34.

Quinn, M. M., T. J. Smith, T. Schneider, E. A. Eisen, and D. H. Wegman. (2005). Determinants of airborne fiber size in the glass fiber production industry. *Journal of Occupational and Environmental Hygiene* 2 (1): 19–28.

Rappaport, S. (1985). Smoothing of exposure variability at the receptor: Implications for health standards. *Annals of Occupational Hygiene* 29: 201–14.

Rappaport, S. M., H. Kromhout, and E. Symanski. (1993). Variation of exposure between workers in homogeneous exposure groups. *American Industrial Hygiene Association Journal* 54 (11): 654–62.

Rappaport, S. M., L. L. Kupper, and Y. S. Lin. (2005). On the importance of exposure variability to the doses of volatile organic compounds. *Toxicological Sciences* 83 (2): 224–36.

Rappaport, S. M., and S. Selvin. (1987). A method for evaluating the mean exposure from a lognormal distribution. *American Industrial Hygiene Association Journal* 48 (4): 374–79.

Rappaport, S. M., and R. C. Spear. (1988). Physiological damping of exposure variability during brief periods. *Annals of Occupational Hygiene* 32 (1): 21–33.

Rappaport, S. M., S. Waidyanatha, K. Yeowell-O'Connell, N. Rothman, M. T. Smith, L. Zhang, Q. Qu, R. Shore, G. Li, and S. Yin. (2005). Protein adducts as biomarkers of human benzene metabolism. *Chemico-Biological Interactions* 153-154: 103–9.

Reilly, M. J., K. D. Rosenman, J. H. Abrams, Z. Zhu, C. Tseng, V. Hertzberg, and C. Rice. (1995). Ocular effects of exposure to triethylamine in the sand core cold box of a foundry. *Occupational and Environmental Medicine* 52 (5): 337–43.

Reis, J. P., K. D. Dubose, B. E. Ainsworth, C. A. Macera, and M. M. Yore. (2005). Reliability and validity of the occupational physical activity questionnaire. *Medicine and Science in Sports and Exercise* 37 (12): 2075–83.

Rice, F. L., R. Park, L. Stayner, R. Smith, S. Gilbert, and H. Checkoway. (2001). Crystalline silica exposure and lung cancer mortality in diatomaceous earth industry workers: A quantitative risk assessment. *Occupational and Environmental Medicine* 58 (1): 38–45.

Richardson, D. B. (2008). "A simple approach for fitting linear relative rate models in SAS." *American Journal of Epidemiology* 168 (11): 1333–8.

Richardson, D. B. (2009). "Multistage modeling of leukemia in benzene workers: a simple approach to fitting the 2-stage clonal expansion model. *American Journal of Epidemiology* 169 (1): 78–85.

Rimm, E. B., E. L. Giovannucci, M. J. Stampfer, G. A. Colditz, L. B. Litin, and W. C. Willett. (1992). Reproducibility and validity of an expanded self-administered semiquantitative food frequency questionnaire among male health-professionals." *American Journal of Epidemiology* 135 (10): 1114–26.

Roach, S. A. (1966). A more rational basis for air sampling programs. *American Industrial Hygiene Association Journal* 27: 1–12.

_____. (1977). A most rational basis for air sampling programmes. *Annals of Occupational Hygiene* 20 (1): 65–84.

Robins, J. M., and S. Greenland. (1986). The role of model selection in causal inference from nonexperimental data. *American Journal of Epidemiology* 123 (3): 392–402.

_____. (1989). Estimability and estimation of excess and etiologic fractions. *Statistics in Medicine* 8 (7): 845–59.

Rose, G. (2001). Sick individuals and sick populations. *International Journal of Epidemiology* 30 (3): 427–32; discussion 433–4.

Rosen, G. (1993). *A history of public health.* Baltimore: Johns Hopkins University Press.

Rosenman, K. D., M. J. Reilly, and D. Kalinowski. (1997). Work-related asthma and respiratory symptoms among workers exposed to metal-working fluids. *American Journal of Industrial Medicine* 32 (4): 325–31.

Rosner, B. (2006). *Fundamentals of biostatistics*, 6th ed. Pacific Grove, CA: Brooks/Cole Cengage Learning.

Rosner, B., D. Spiegelman, and W. C. Willett. (1990). Correction of logistic regression relative risk estimates and confidence intervals for measurement error: The case of multiple covariates measured with error. *American Journal of Epidemiology* 132: 734–45.

Rothman, K. J., S. Greenland and Lash. (2008). *Modern epidemiology*, 3rd *ed.* Philadelphia: Lippincott-Raven.

_____. (2005). Causation and causal inference in epidemiology. *American Journal of Public Health* 95, Suppl. no. 1: S144–50.

Rowland, M., and T. N. Tozer. (1995). *Clinical pharmacokinetics.* Baltimore: Williams and Wilkins.

Rozman, K. K. (2000). The role of time in toxicology or Haber's C x *T* product. *Toxicology* 149 (1): 35–42.

Salvan, A., L. Stayner, K. Steenland, and R. Smith. (1995). *Selecting an exposure lag period. Epidemiology* 6 (4): 387–90.

Salvan, A., K. Thomaseth, P. Bortot, and N. Sartori. (2001). Use of a toxicokinetic model in the analysis of cancer mortality in relation to the estimated absorbed dose of dioxin (2,3,7,8-tetrachlorodibenzo-p-dioxin, TCDD). *Science of the Total Environment* 274: 21-35.

Samet, J. M., F. Dominici, F. C. Curriero, I. Coursac, and S. L. Zeger. (2000). Fine particulate air pollution and mortality in 20 US cities, 1987–1994. *New England Journal of Medicine* 343 (24): 1742–49.

Sander, R. (1999). A compilation of Henry's Law constants for inorganic and organic species of potential importance in environmental chemistry., version 3, (2007), http://www.henrys-law.org.

Sarangapani, R., P. R. Gentry, T. R. Covington, J. G. Teeguarden, and H. J. Clewell III. (2003). Evaluation of the potential impact of age- and gender-specific lung morphology and ventilation rate on the dosimetry of vapors. *Inhalation Toxicology* 15 (10): 987–1016.

Savitz, D. A. (2001). Occupational exposure to magnetic fields and brain cancer. *Occupational and Environmental Medicine* 58 (10): 617–8.

_____. (2002). Magnetic fields and miscarriage. *Epidemiology* 13 (1): 1–4.

_____. (2003). Health effects of electric and magnetic fields: Are we done yet? *Epidemiology* 14 (1): 15–7.

Scherer, G. (1999). Smoking behaviour and compensation: A review of the literature. *Psychopharmacology (Berl)* 145 (1): 1–20.

Schneider, T., R. Vermeulen, D. H. Brouwer, J. W. Cherrie, H. Kromhout, and C. L. Fogh. (1999). Conceptual model for assessment of dermal exposure. *Occupational and Environmental Medicine* 56 (11): 765–73.

Seixas, N. S., and H. Checkoway. (1995). Exposure assessment in industry specific retrospective occupational epidemiology studies. *Occupational and Environmental Medicine* 52 (10): 625–33.

Seixas, N. S., N. J. Heyer, E. A. Welp, and H. Checkoway. (1997). Quantification of historical dust exposures in the diatomaceous earth industry. *Annals of Occupational Hygiene* 41 (5): 591–604.

Seixas, N. S., T. G. Robins, and L. H. Moulton. (1988). The use of geometric and arithmetic mean exposures in occupational epidemiology. *American Journal of Industrial Medicine* 14 (4): 465–77.

Semple, S. (2004). Dermal exposure to chemicals in the workplace: Just how important is skin absorption? *Occupational and Environmental Medicine* 61 (4): 376–82.

Semple, S., and J. W. Cherrie. (2003). Dermal exposure assessment. In *Exposure assessment in occupational and environmental epidemiology*, ed. M. Nieuwenhuijsen. New York: Oxford University Press.

Sexton, K., M. A. Callahan, et al. (1995). "Estimating exposure and dose to characterize health risks: the role of human tissue monitoring in exposure assessment." *Environ Health Perspect 103 Suppl 3*: 13–29.

Shamlaye, C. F., D. O. Marsh, G. J. Myers, C. Cox, P. W. Davidson, O. Choisy, E. Cernichiari, A. Choi, M. A. Tanner, and T. W. Clarkson. (1995). The Seychelles child development study on neurodevelopmental outcomes in children following in utero exposure to methylmercury from a maternal fish diet: background and demographics. *Neurotoxicology* 16 (4): 597–612.

Shusterman, D., E. Matovinovic, and A. Salmon. (2006). Does Haber's Law apply to human sensory irritation? *Inhalation Toxicology* 18 (7): 457–71.

Siemiatycki, J., S. Wacholder, R. Dewar, L. Wald, D. Bégin, L. Richardson, K. Rosenman, and M. Gérin. (1988). Smoking and degree of occupational exposure: Are internal analyses in cohort studies likely to be confounded by smoking status? *American Journal of Industrial Medicine* 13 (1): 59–69.

Simons, E., J. Curtin-Brosnan, T. Buckley, P. Breysse, and P. A. Eggleston. (2007). Indoor environmental differences between inner-city and suburban homes of children with asthma. *Journal of Urban Health* 84 (4): 577–90.

Sjogren, M., H. Li, C. Banner, J. Rafter, R. Westerholm, and U. Rannug. (1996). Influence of physical and chemical characteristics of diesel fuels and exhaust emissions on biological effects of particle extracts: A multivariate statistical analysis of ten diesel fuels. *Chemical Research in Toxicology* 9 (1): 197–207.

Smith, T. J. (1985). Development and application of a model for estimating alveolar and interstitial dust levels. *Annals of Occupational Hygiene* 29: 495–516.

_____. (1992). Occupational exposure and dose over time: Limitations of cumulative exposure. *American Journal of Industrial Medicine* 21(1): 35–51.

Smith, T. J., F. Y. Bois, Y.-S. Lin, C. Brochot, S. Micallef, D. Kim, and K. T. Kelsey. (2008). Quantifying heterogeneity in exposure–risk relationships using exhaled breath biomarkers for 1,3-butadiene exposures. *Journal of Breath Research* 2: 037018.

Smith, T., S. Hammond, F. Laidlaw, and S. Fine. (1984). Respiratory exposures associated with silicon carbide production: Estimation of cumulative exposures for an epidemiological study. *British Journal of Industrial Medicine* 41: 100–8.

Smith, T. J., S. K. Hammond, and O. Wong. (1993). Health effects of gasoline exposure: I. Exposure assessment for U.S. distribution workers. *Environmental Health Perspectives* 101, Suppl. no. 6: 13–21.

Smith, T. J., M. M. Quinn, G. M. Marsh, A. O. Youk, R. A. Stone, J. M. Buchanich and M. J. Gula. (2001). Historical cohort study of U.S. man-made vitreous fiber production workers: VII. Overview of the exposure assessment. *Journal of Occupational and Environmental Medicine* 43 (9): 809–23.

Smith, T. J., P. A. Stewart, and R. F. Herrick. (1995). Retrospective exposure assessment. In *Occupational hygiene*, ed. J. M. Harrington and K. Gardiner, 308–. London: Blackwell Scientific.

Snow, J. (1936). *Snow on cholera: being a reprint of two papers* / by John Snow; together with a biographical memoir by B.W. Richardson and an introduction by Wade Hampton Frost. New York: The Commonwealth Fund; London: H. Milford, Oxford University Press.

Spiegelman, D., A. McDermott, and B. Rosner. (1997). Regression calibration method for correcting measurement-error bias in nutritional epidemiology. *American Journal of Clinical Nutrition* 65, Suppl. no. 4: 1179S-1186S.

Spiegelman, D., S. Schneeweiss, and A. McDermott. (1997). Measurement error correction for logistic regression models with an "alloyed gold standard". *American Journal of Epidemiology* 145 (2): 184–96.

Spiegelman, D., and B. Valanis. (1998). Correcting for bias in relative risk estimates due to exposure measurement error: A case study of occupational exposure to antineoplastics in pharmacists. *American Journal of Public Health* 88: 406–12.

Sprince, N. L., J. A. Palmer, W. Popendorf, P. S. Thorne, M. I. Selim, C. Zwerling, and E. R. Miller. (1996). Dermatitis among automobile production machine operators exposed to metal-working fluids. *American Journal of Industrial Medicine* 30 (4): 421–29.

Stayner, L., A. Bailer, R. Smith, S. Gilbert, F. Rice, and E. Kuempel. (1999). Sources of uncertainty in dose-response modeling of epidemiological data for cancer risk assessment. *Annals of the New York Academy of Sciences* 895: 212–22.

Stayner, L., R. Smith, A. J. Bailer, E. G. Luebeck, S. H. Moolgavkar. (1995). Modeling epidemiologic studies of occupational cohorts for the quantitative assessment of carcinogenic hazards. *American Journal of Industrial Medicine* 27 (2): 155–70.

Steele, R., and K. Mummery. (2003). Occupational physical activity across occupational categories. *Journal of Science and Medicine in Sport* 6 (4): 398–407.

Steenland, K., and J. A. Deddens. (2004). A practical guide to dose-response analyses and risk assessment in occupational epidemiology. *Epidemiology* 15 (1): 63–70.

Steenland, K., A. Mannetje, P. Boffetta, L. Stayner, M. Attfield, J. Chen, M. Dosemeci, et al. (2001). Pooled exposure-response analyses and risk assessment for lung cancer in 10 cohorts of silica-exposed workers: An IARC multicentre study. *Cancer Causes and Control* 12 (9): 773–84.

Su, H. H. (2003). Particulate matter. In *Exposure assessment in occupational and environmental epidemiology*, ed. M. Nieuwenhuijsen. New York: Oxford University Press.

Sussman, R., B. Cohen, and M. Lippmann. (1991). Asbestos fiber deposition in a human tracheobronchial cast: II. Empirical model. *Inhalation Toxicology* 3 (2): 161–179.

Suuronen, K., K. Aalto-Korte, R. Piipari, T. Tuomi, and R. Jolanki. (2007). Occupational dermatitis and allergic respiratory diseases in Finnish metalworking machinists. *Occupational Medicine-Oxford* 57 (4): 277–83.

Symanski, E., L. L. Kupper, H. Kromhout, and S. M. Rappaport. (1996). An investigation of systematic changes in occupational exposure. *American Industrial Hygiene Association Journal* 57 (8): 724–35.

Tan, Y. M., K. H. Liao, and H. J. Clewell III. (2007). Reverse dosimetry: Interpreting trihalomethanes biomonitoring data using physiologically based pharmacokinetic modeling. *Journal of Exposure Science and Environmental Epidemiology* 17 (7): 591–603.

Tan, Y. M., K. H. Liao, R. B. Conolly, B. C. Blount, A. M. Mason, and H. J. Clewell. (2006). Use of a physiologically based pharmacokinetic model to identify exposures consistent with human biomonitoring data for chloroform. *Journal of Toxicology and Environmental Health, Part A* 69 (18): 1727–56.

Tardif, R., P. O. Droz, G. Charest-Tardif, G. Pierrehumbert, and G. Truchon. (2002). Impact of human variability on the biological monitoring of exposure to toluene: I. Physiologically based toxicokinetic modelling. *Toxicology Letters* 134 (1–3): 155–63.

Thier, R., T. Bruning, P. H. Roos, H. P. Rihs, K. Golka, Y. Ko, and H. M. Bolt. (2003). Markers of genetic susceptibility in human environmental hygiene and toxicology: The role of selected CYP, NAT and GST genes. *International Journal of Hygiene and Environmental Health* 206 (3): 149–171.

Thomas, D. C. (1983). Statistical methods for analyzing effects of temporal patterns of exposure on cancer risks. *Scandinavian Journal of Work, Environment, and Health* 9: 353–66.

_____. (1987). Pitfalls in the analysis of exposure-time-response relationships. *Journal of Chronic Diseases* 40, Suppl no. 2: 71S-78S.

Thomaseth, K., and A. Salvan. (1998). Estimation of occupational exposure to 2,3,7,8-tetrachlorodibenzo-p-dioxin using a minimal physiologic toxicokinetic model. *Environmental Health Perspectives* 106, Suppl. no. 2: 743–53.

Thompson, D., D. Kriebel, M. M. Quinn, D. H. Wegman, and E. A. Eisen. (2005). Occupational exposure to metalworking fluids and risk of breast cancer among female autoworkers. *American Journal of Industrial Medicine* 47(2): 153–60.

Thompson, M. L., J. E. Myers, and D. Kriebel. (1998). Prevalence odds ratio or prevalence ratio in the analysis of cross sectional data: What is to be done? *Occupational and Environmental Medicine* 55 (4): 272–7.

Thorne, P. S., J. A. DeKoster, and P. Subramanian. (1996). Environmental assessment of aerosols, bioaerosols, and airborne endotoxins in a machining plant. *American Industrial Hygiene Association Journal* 57 (12): 1163–7.

Thurston, G., M. Lippmann, M. Scott, and J. Fine. (1997). Summertime haze air pollution and children with asthma. *American Journal of Respiratory and Critical Care Medicine* 155: 654–60.

Tingen, C., J. B. Stanford, and D. B. Dunson. (2004). Methodologic and statistical approaches to studying human fertility and environmental exposure. *Environmental Health Perspectives* 112: 87–9.

U. S. Department of Health and Human Services. (1990). *The health benefits of smoking cessation*. Bethesda, MD: U.S. Department of Health and Human Services.

U.S. Environmental Protection Agency. (1994). *Methods for derivation of inhalation reference concentrations and application of inhalation dosimetry*. Washington, DC: Environmental Protection Agency Office of Health and Environmental Assessment.

Vandenbroucke, J. P. (1987). Should we abandon statistical modeling altogether? *American Journal of Epidemiology* 126 (1): 10–13.

Vermeulen, R., P. Stewart, and H. Kromhout. (2002). Dermal exposure assessment in occupational epidemiologic research. *Scandinavian Journal of Work, Environment, and Health* 28 (6): 371–85.

Verschueren, K., ed. (2000). *Handbook of environmental data on organic chemicals*. New York: Wiley.

Vineis, P., and D. Kriebel. (2006). Causal models in epidemiology: Past inheritance and genetic future. *Environmental Health* 5: 21.

Walker, W., P. Harremoës, J. Rotmans, J. P. van der Sluijs, M. B. A. van Asselt, P. Janssen, M. P. Krayer von Krauss. (2003). Defining uncertainty: A conceptual basis for uncertainty management in model-based decision support. *Integrated Assessment* 4: 5–17.

Wassenius, O., B. Jarvholm, T. Engstrom, L. Lillienberg, and B. Meding. (1998). Variability in the skin exposure of machine operators exposed to cutting fluids. *Scandinavian Journal of Work, Environment, and Health* 24 (2): 125–9.

Wegman, D. H. (1996). Machining operations and associated machining fluid exposures: Issues for health and safety intervention in manufacturing. *American Journal of Industrial Medicine* 29 (4): 397–401.

Weihe, P., P. Grandjean, F. Debes, and R. White. (1996). Health implications for Faroe islanders of heavy metals and PCBs from pilot whales. *Science of the Total Environment* 186 (1-2): 141–8.

Weiss, K. M. (1990). Cancer models and cancer genetics. *Epidemiology* 1 (6): 486–90.

Wells, R., R. Norman, P. Neumann, D. Andrews, J. Frank, H. Shannon, and M. Kerr. (1997). Assessment of physical work load in epidemiologic studies: Common measurement metrics for exposure assessment. *Ergonomics* 40 (1): 51–61.

Wenzl, T. B., D. Kriebel, E. A. Eisen, and M. J. Ellenbecker. (1995). Comparisons between magnetic field exposure indices in an automobile transmission plant. *American Industrial Hygiene Association Journal* 56 (4): 341–8.

Wessels, D., D. B. Barr, and P. Mendola. (2003). Use of biomarkers to indicate exposure of children to organophosphate pesticides: Implications for a longitudinal study of children's environmental health. *Environmental Health Perspectives* 111 (16): 1939–46.

Wigle, D., T. Arbuckle, M. Walker, M. Wade, S. Liu, and D. Krewski. (2007). Environmental hazards: Evidence for effects on child health. *Journal of Toxicology and Environmental Health, Part B* 10 (1 & 2): 3–39.

Witheridge, W. N., and C. P. Yaglou. (1939). *Transactions of the American Society for Heating and Ventilation Engineering* 45: 509.

Woskie, S. R., E. E. Eisen, D. H. Wegman, X. Hu, and D. Kriebel. (1998). Worker sensitivity and reactivity: indicators of worker susceptibility to nasal irritation. *American Journal of Industrial Medicine* 34 (6): 614–622.

Yu, O., L. Sheppard, T. Lumley, J. Q. Koenig, and G. G. Shapiro. (2000). Effects of ambient air pollution on symptoms of asthma in Seattle-area children enrolled in the CAMP Study. *Environmental Health Perspectives* 108: 1209–14.

Yu, R. C., and S. M. Rappaport. (1997). A lung retention model based on Michaelis-Menten-like kinetics. *Environmental Health Perspectives* 105 (5): 496–503.

Zachariae, C., A. Lerbaek, P. M. McNamee, J. E. Gray, M. Wooder, and T. Menne. (2006). An evaluation of dose/unit area and time as key factors influencing the elicitation capacity of methylchloroisothiazolinone/methylisothiazolinone (MCI/MI) in MCI/MI-allergic patients. *Contact Dermatitis* 55 (3): 160–6.

Zeger, S. L., D. Thomas, F. Dominici, J. M. Samet, J. Schwartz, D. Dockery, and A. Cohen. (2000). Exposure measurement error in time-series studies of air pollution: Concepts and consequences. *Environmental Health Perspectives* 108 (5): 419–26.

Zeka, A. (2003). Aerodigestive Cancer Risk and Metal Working Fluids. Department of Work Environment. University of Massachusetts Lowell, Lowell MA. Doctoral Dissertation.

Zhang, K. M., A. S. Wexler, Y. F. Zhu, W. C. Hinds, and C. Sioutas. (2004). Evolution of particle number distribution near roadways Part II. The "road-to-ambient process". *Atmospheric Environment* 38 (38): 6655–65.

Zhu, Y. F., W. C. Hinds, S. Kim, S. Shen, and C. Sioutas. (2002). Study of ultrafine particles near a major highway with heavy-duty diesel traffic. *Atmospheric Environment* 36 (27): 4323–35.

INDEX

Note: Page numbers followed by "*f*" and "*t*" denote figures and tables, respectively.

Absorption, 4, 5, 6, 29
  entry routes, 30, 107, 142
  gastrointestinal, 4
  GI, 37
  hair, 309, 311
  PBTK, 139*f*
  rates of absorption, 105
  skin, 36, 37–38, 47, 54*t*, 60, 104, 111–12, 117, 131
  toxicity, 29
  toxin uptake, 5, 6
  UV light, 57
Acclimatization, 167, 254, 256, 292
  adaptation, 287
  human, 302
  sensitivity, 256, 287
Acquired sensitivities, 336
Acute exposures, 71, 256, 291, 294
Acute reversible response model, 384
Adaptation. *See* Acclimatization
Aerodynamic diameter, 108–09
Aerosol, 29, 38
  "aerosol aging," 3
  artificial butter flavoring, 174
  metalworking fluid (MWF), 358
  in sampling, 56
  silicon carbide, 243
  silicosis, 167
AIC. *See* Akaike's information criterion (AIC)
Air contamination, 38
  asbestos exposure, 41
  contaminants, 110*t*
  emission contact, 40
  environmental transport, 40
  exposure controls, 42
  exposure measurement, 58
  multiple emission, 41
  sources, 39
Akaike's information criterion (AIC), 240
Allergens, 169, 188
  asthma, 322–23
  biomarkers, 350
  in home, 348–49, 352
"All or nothing" phenomenon, 209, 320, 321
Ammonia and eye irritation, 253–56
Ammonia's diffusion rate, 281,
Analysis of variance (ANOVA), 80
  subject variation, 82
ANOVA. *See* Analysis of variance (ANOVA)
Antibodies
  immunological system, 322, 339
  sensitization, 338
  specific protein, 337
Antigen
  asthma, 12
  hypersensitivity, 357
  immunological system, 322
  sensitization, 337, 338–40

Apoptosis, 4, 7, 322
ARE. *See* Asymptotic relative efficiency (ARE)
Asthma, 8, 10, 322, 336–38, 348
  air pollution, 384–49
  biological model, 12
  biomarkers, 350
  cockroach allergen, 352
  etiologic time interval (ETI), 324, 337, 350
  exposure marker, 350–51
  mechanisms, 323–24, 337–38
  "peak" exposures, 335
  probabilistic dose metric, 351–52
  probability calculation, 328, 330
  responses, 349–50
  sensitivity, 352
  sensitization, 323
  urban air pollution, 348–49
Asymptotic relative efficiency (ARE), 241–42
  lead neurobehavioral study, 245t
Attenuation. *See* Acclimatization
Averaging time interval, 56–57

Baseline attack rate, 349–50, 357
BD. *See* 1, 3-Butadiene (BD)
BEF index. *See* Biologically Effective Fibers (BEF) index
Biases, 8, 212
  categories of, 212
  information bias, 213
  selection bias, 213
    index cases, 213
Bioaccumulation, 45
Bioavailability, 45
Biologically Effective Fibers (BEF) index, 234*f*
  adverse effects, 145–46
  biomarker time scale, 142
Biologic model, 6, 11
  asthma, for, 12
  exposure-response relationships, 11
  ozone exposure, 12
Biologic relevance, 35
  exposure, 35–36
Biomarker, 311–12, 137, 138, 386
  causal agent relationship, 142
  characteristics, 141
  cross-sectional measurements, 155
  disease, 140
  empirical modeling, 147, 150*f*, 151f
  entry route, 142–43
  epidemiologic applications, 140–41
  exposure, 138, 147
  functions and use, 137–38, 156
  metabolic products as, 144
  MTBE inhalation exposure, 149*f*
  physiologically-based toxicokinetic (PBTK) model, 154
  population characteristics, 157
  prospective calibration, 156
  reaction product, 144–45
  response, 139–40
  sample type/media differences, 143
  susceptibility, 140
  temporal effects, 152, 153*f*
  temporal relationships, 151–52
  time course, 150*f*, 151*f*
  toxicokinetic models, 147
    limitations of, 154
  two-compartment empirical model, 150*f*
Biomechanical injury processes, 388–89
Biotransformation, 6, 114–15
  activation, 114
  detoxification, 114
  excretion, 115
  processes, 114
  storage, 114
Black box, 4–10, 11–12, 225
BMI. *See* Body mass index (BMI)
BO. *See* Bronchiolitis obliterans (BO)
Body mass index (BMI), 238
Bronchiolitis obliterans (BO), 174, 312
  buttered-flavored popcorn, 313
  clinical picture, 314
  cross-sectional study, 317
  disease process (DP) model application, 315
  epidemiologic study, 315
  lung biopsy, 317
  popcorn workers' lung, 313
  prevention, 318
  toxic flavoring ingredients, 313
  vapor analysis, 313–14, 317
1,3-Butadiene (BD), 387

Campfire, 27, 28–29, 31, 33, 34
  air contaminants, 39
  exposure intensity, 48
  exposure variation, 33
  fuel characteristics, 31
  reversible proportional disease process, 253
Cancer
  carcinogenesis, 344, 367
  cell kinetics, 340, 345
  cigarette smoke-induced, 367–69
  dose metrics, 369
  dose response, 368*f*
  disease process (DP) model, 345
  lung cancer and silica, 376*f*, 379*f*
  silica as initiator, 379
  silica exposure effect, 378*f*

two-stage model, 369
clonal expansion model, 371
dose metric and cumulative exposure (CE), 380*f*
dosimetric approach, 371–74
epidemiologic data, 374–75
estimation, 372–73
silica and lung cancer, 375–81, 377*t*
Carcinogenesis, 8, 344, 367
Causal agent, 4, 8, 56
causation, web of, 188
and exposure biomarkers, 142
CDC. *See* Centers for Disease Control (CDC)
CE. *See* Cumulative exposure (CE)
Centers for Disease Control (CDC), 174
Chemical exposure, 29
composition, 29
dimensions, 29
environmental concentration, 30
physical form, 29–30
Cholera, 3, 4
Clonal expansion model, 371
Cockroach allergen, 352
Coefficient of determination. *See* Asymptotic relative efficiency (ARE)
Coefficient of variation (CV), 72
Compartmental model, 117
characteristics, 122
ingestion, 111
inhalation exposure, 106
inhalation route of entry, 109
kinetic behavior, 125
molecular diffusion and perfusion, 123–24
partition coefficient, 123
skin exposure, 112
Confounding, 189–92, 208
confounder, 189, 208
control, 208
example, 190*t*
for continuous variables, 209
exposure-response relation, 210*f*, 211*f*
gender, 209–10, 211*f*
incomplete control of, 212
stratification, 191
ways to control, 191–92
Contact dermatitis, 357, 363. *See also* Dermatitis
Contaminants
Ammonia, 253–56
cholera, 4
complex mixture of, 139
diacetyl, 171–174, 314–317
lead, 243–46
manmade viterous fibers, 230
mercury, 304, 305–12
metalworking fluid, 358, 362
ozone, 286, 347
silica, 167, 216, 229, 267–271, 374–379
2,3,7,8-Tetrachlorodibenzo-*p*-dioxin (TCDD), 238
Contamination, 4
environmental, 138–39
skin, 33, 46, 47, 61
sources, 43
Counterfactual evidence, 181
Cox proportional hazards model, 193, 215–16
Cross-sectional study, 317
Cumulative exposure (CE), 13, 194, 224, 256, 369
calculation of, 13, 224
exposure profiles and, 14*f*
"time windows of exposure," 15, 16*f*
Cutting fluids. *See* Metalworking fluids (MWF)
Cutting oils. *See* Metalworking fluids (MWF)
CV. *See* Coefficient of variation (CV)

Damage
irreversible, 256–57
rate, 257, 259
rate constant, 263*f*
repair, 252, 301
repair model, 159–60, 262. *See also* Disease process model
repair process, 160–61
Difference equation, 17
DDT. *See* Dichloro-diphenyltrichloroethane (DDT)
DE. *See* Diatomaceous earth (DE)
Dementia, 304
Dermatitis, 8, 357
contact, 357, 363
effects assessment, 363–64
etiologic time interval (ETI), 363
exposure assessment, 364
probabilistic dose metrics, 364–65
dose metrics, 362
epidemiologic study, 360–65
evaluation of, 360
hypersensitivity, 357
MWF, 358–60
responses, 357
Deviance. *See*-2 log likelihood statistics (-2LL)
Diacetyl, 174–175, 314–317
Diatomaceous earth (DE), 375
Dichloro-diphenyltrichloroethane (DDT), 45
Difference equation, 253, 258, 269, 271
model, 276*f*, 277
Diffusion rate, 282
Dimethyl sulfoxide (DMSO), 112
Dioxin. *See* 2, 3, 7, 8-tetrachlorodibenzo-*p*-dioxin (TCDD)

Discrete disease processes, 320, 331
  Asthma, 321–24, 328, 330, 335, 336, 337, 347–52
  Dermatitis, 357–62
    contact, 357, 363
  discrete irreversible responses, 338, 345–46
    sensitization process, 338–41, 339*f*, 342
  discrete responses, 320, 321
  discrete reversible responses, 332, 335, 348
    DM calculation, 334, 353–55
    DM characteristics, 334, 336
    DM two-point logistic model, 355–56
    dose-response relationship, 356–57
    exposure distribution, 333
    immune reactions types, 335
    probabilistic dose metric, 333
    simple dose metrics, 332–33
  immune responses, 322
  irreversible, 336, 366
    acquired sensitivities, 336
    Asthma, 336–38
    DM, 341, 343
    dose response, 368
    lung cancer, 367–69
  probabilistic nature, 320–21
  response dependence, 321
  reversible, 321, 347
    approach, 348
    Asthma, 322–24, 347
    exposure response curve, 326
    exposure-response relationship, 324
    group response, 325–26
    logistic model of discrete, 326
    probability of response, 325*f*
    simulation, 327
    "switch response," 324
Discrete responses, 320, 321
Disease model structure, 383–84
  challenges
    behavioral and psychosocial factors, 389
    biomechanical injury processes, 388–89
    epigenetic effects, 388
    toxicogenomics, 388
  determining uncertainties, 384
  epidemiologic and toxicologic link, 386
    biomarkers, 386
    tissue dose, 386
  in risk assessment, 385, 386, 387-
    animal toxicity extrapolation, 387
  in study design, 385
Disease process (DP), 287
  assumptions, 260–61
  damage, 256
  dose metrics, 261–65
  epidemiologic applications, 265–66
  equation, 260
  irreversible damage, 256–57
  linear difference equation, 257, 258
  pesticides toxicity, 266
  repair, 256, 257
  repair half-time, 257
  steady-state condition, 256
  toxic agent effects, 256
Disease process (DP) models, 6–7, 7*f*, 158, 159. *See also* Epidemiology, dosimetry in
  adverse effects, 159
    basic damage-repair model, 159–60
    time course of effects, 162
  critical elements, 8, 20
  damage-repair process, 160–61
  discrete effects, 10
  diseases prevention, 174–75
  environmental, 165
  epidemiologic model, 228
  epidemiologic study designs, 194
  irreversible discrete, 171
    DP-4 in practice, 172
    key assumptions, 171
    typical temporal patterns, 171
  irreversible proportional, 167
    DP-2 in practice, 168
    key assumptions, 168
    typical temporal patterns, 168
  local independence principle, 9–10
  lung cancer, 20–22
  observed response, 9
  proportional versus discrete response, 21
  rate limiting steps, 161
  repair processes, 161
  response temporal pattern, 18*f*, 19*f*
  response type, based on, 8*t*
  reversible discrete, 169
    DP-3 in practice, 170
    key assumptions, 169
    typical temporal patterns, 169
  reversible proportional, 165
    DP-1 in practice, 166
    key assumptions, 165
    typical temporal patterns, 166
  reversible *vs*. irreversible time course, 20
  time course, 8*t*, 166*f*
DM. *See* Dose metric (DM)
DMSO. *See* Dimethyl sulfoxide (DMSO)
DNA, 7, 362
Dose, 172
  components of, 172
  -effect relationship, 5, 303, 356–57
  response curve, 187*f*
  -response relationship, 5, 303, 356–57

Dose heterogeneity, 135
Dose metric (DM), 7–8, 96, 124, 228, 261–65, 273–75, 299, 312, 319, 332–33, 362, 369
  alternative, 237*f*
  calculation, 334, 353–55
  characteristics, 124
  characteristics, 334, 336
  exposure data, 236*f*
  lead neurobehavioral effects, 243–46
  model parameters, 384
  performance comparison, 245*t*
  probabilistic discrete reversible processes, 173
  proportional effect, 173
  for pulmonary effects, 299–300
  for toxic response, 385
  two-point logistic model, 355–56
  2-stage model versus CE, 380*f*
Dosimetric approach, 371–74
Dosimetric models, 227, 228, 229*t*
  alternative dose metrics comparison, 241
    ARE, 241–42
  association measures comparison, 240–41
  goodness of fit, 239
  interindividual variability, 236
    serum dioxin and lung cancer, 238
  low back pain, 234
    alternative dose metrics, 237*f*
    dose metric, 236*f*
    exposure data, 236*f*
    regression models, 235
  MMVF lung dose, 230–34
    deposition in target tissue, 231
    fiber durability, 232
    tumor formation, 233
  parameters determination, 238–39
  p–values comparison, 241
  qualitative considerations, 242
    dose metrics, 243
DP models. *See* Disease process (DP) models
Dynamic processes, 5
  biological process models, 17. *See also* Disease process (DP) models
  disease process model, 6, 7*f*
  dose-effect relationship, 5
  exposure-dose relationship, 5
  expression in difference equation, 17
  toxicokinetic model, 6*f*

EC. *See* Elemental carbon (EC)
Effect estimator. *See* Measure of association
Effect modification, 192–93. *See also* Confounding
Electromagnetic field (EMF), 96, 200
Elemental carbon (EC), 92
EMF. *See* Electromagnetic field (EMF)
Emission
  "aerosol aging," 39
  contact with, 40
  control devices, 42
  environmental transport, 40
  multiple, 41
  sources, 39
Empirical modeling, 147, 150*f*, 151*f*
Environmental Protection Agency (EPA), 58, 61, 115
Environmental tobacco smoke (ETS), 94
EPA. *See* Environmental Protection Agency (EPA)
Epidemiology, 190–91, 307–08
  approaches, 225–27
  biases, 212
  choices made in, 215*t*
  counterfactual evidence, 181
  data, 22, 23, 374–75
  disease in populations, 181
    characteristics, 182
    risk estimation, 183
    threshold, population perspective on, 185*f*
    toxic response threshold, 185*f*
  disease process model, 194–95
    design options, 196*t*
  disease views, 179–81
  exposure biomarkers, 94
  limitation, 181
  measure of association, 186
    dose-response curve, 187*f*
    John Snow's data, 186*t*
    rate ratio, 186
    relative risk, 186
  studies, 81, 85*t*, 181, 194, 286–87, 315, 360–65
  toxicokinetic modeling, 124
  toxicologic link, 386
  treatment of time, 193
    Cox proportional hazards model, 193
    cumulative exposure, 194
  uncertainty sources, 199*f*
  web of causation, 188*f*
    confounding, 189
    effect modification, 192
Epidemiology, dosimetry in, 223, 228. *See also* Dosimetric models
  interindividual variability, 236
  low back pain, 234
  MMVF lung dose, 230
  model parameters determination, 238
  two-step process, 227
    dose metric, 228
    independent variables, 228
    mathematical forms, 228

Epidemiology, environmental, 3
  black box, 4
  dynamic processes, 5
  exposure-response relationship, 5*f*
  temporal processes, 4
  toxicokinetic processes, 6f
Epidemiology, exposure assessment, 92
  measurement error, 97
  qualitative exposure assignments, 93–95
  quantitative exposure estimates, 96
  semiquantitative measures, 95–96
Epigenetic effects, 388
Ergonomics, 27, 234–236, 388
ETI. *See* Etiologic time interval (ETI)
Etiologic time interval (ETI), 84, 162, 195, 260, 299, 311. *See also* Disease process (DP) models
  exposure-disease processes, 164*f*
ETS. *See* Environmental tobacco smoke (ETS)
Exposure
  acute, 71, 256, 291, 294
  airborne, 58
  asbestos, 41
  assessment, 92–96
  assessor, 97
  association with, 208
  biological relevance, 35
  cumulative (CE), 13, 194, 224
  disease processes, 164*f*
  ingestion, 37
  inhalation, 37, 149*f*
  low back pain data, 236f
  methods comparison, 206
  misclassification, differential, 200
  misclassification, nondifferential, 201
  MTBE inhalation exposure, 149*f*
  profiles and CE, 14*f*
  response pattern, 252
  response relationship, 5*f*, 79, 210*f*, 211*f*, 324
  skin, 37–38, 61–62, 361
  summary measures. *See* Dose metrics
  "time windows of," 15, 16*f*. *See also* Dynamic processes
  tissue concentration relationships, 101
  whole body vibration (WBV), 206
Exposure, biomarkers, 137, 138. *See also* Biomarkers
  complex contaminants, 139
  effects, 139
  environmental contaminant, 138–39
  relationship equation, 156
Exposure-dose models. *See* Toxicokinetic model
  asbestos exposure, 41
  biological relevance, 35–36, 389
  biomarkers, 94, 138
  campfire study, 28
  cornea surface area, 281
  dimensions, 29
  estimation errors, 200
  ingestion exposures, 37
  inhalation exposures, 37
  marker, 51–52, 350–51. *See also* Biomarkers
  measurement, 28, 52, 53–54*t*, 58, 59
  origin, 31
  ozone, 12
  peak exposures, 81, 83
  personal, 79
  potential exposure, 28, 47
  primary exposure routes, 37
  response curve, 326
  response relationship, 5*f*, 11
  silica, 378*f*
  simulation, 291*f*, 298*f*
  skin, 34, 37–38
  stability, 83–84
  temporal sequence, 319
  time-weighted average (TWA), 78
  variability, 262–65
    strategies, 265
  variation, 30, 34–35
  vector, 224
Exposure, environmental data, 88
  extrapolating exposure data, 89
  source-receptor models, 89–90
  statistical exposure model, 89
Exposure estimation, errors in
  calibration regression, 207
  evaluation, 206
  logistic model, 207
  methods comparison, 206
  regression models, 206
  whole body vibration (WBV) exposure, 206
Exposure, ingestion, 107
  compartmental model, 111*f*
Exposure, inhalation, 107
  aerodynamic diameter, 108–09
  air contaminants, 110*t*
  compartmental model, 109*f*, 131*f*
  health effects, 110*t*
  MTBE, 149*f*
  particle deposition, 108
  particle interactions, 108
Exposure measurement, 49
  agent selection, 50–51
  airborne exposures, 58
  approaches, 57
  exposure marker, 51–52
  food contamination, 60–61
  methods, 52, 53–54*t*
    accuracy, 56
    averaging time interval, 56–57

interferences, 55
LOD, 52, 55
precision, 56
response time, 57
selectivity, 55
sensitivity, 55
skin exposure, 61–62
water contamination, 59–60
Exposure misclassification, 200
differential, 200
nondifferential, 201
with categorical data, 202–04
with continuous data, 204
hypothetical data, 203*f*
simulated data, 205*f*
Exposure quality control-quality assurance (QC-QA) program, 62–64
blank samples, 63
blind analyses, 62
interlab sample exchanges, 63
logs, 63
replicate samples, 63
spiked samples, 63
Exposure-response model, 8. *See also* Disease process model
static, 10
"time windows of exposure," 15, 16*f*
Exposures estimation, unmeasured, 86
ambient data, 88
exposures extrapolation, 91
personal data, 87
Exposure, skin, 111
absorption rate, 111–12
compartmental model, 112*f*
particle interactions, 112
Exposure, source-receptor model, 31, 32
basic model, 32
contact with receptor, 33
exposure variation, 34–35
points of intervention, with, 32
source characteristics, 31
variation in contact, 33
Exposures, personal, 38
air contamination, 38
bioaccumulation, 45
bioavailability, 45
food contaminants, 45
skin contamination, 46
water contaminants, 43
Exposures skin, 37–38
Exposure summary measures, 223
duration of exposure, 224
exposure vector, 224
Exposure variability, 71
epidemiologic studies, 81
lognormal distribution, 71, 72
random variation, 71
sources of variability, 97
Exposure variation, 30, 71
characteristics, 66
composition and physical form, 67–68
contaminated water, 34
epidemiologic importance, 30
graphical methods, 73
quantile plot, 73, 75*f*, 76*f*
frequency plots, 74*f*
individuals and groups, 31
intensity, 68
time profile, 70*f*
parametric methods, 77
ANOVA, 80, 82
arithmetic mean, 77
inhaled dose relationship, 77
peak exposures, 81, 83
TWA exposures, 78
skin exposure, 34
temporal variation, 70
time scales, 69*t*
Eye fluid-air partition coefficient, 119
Eye irritation simulation, 255*f*

FDA. *See* Food and Drug Administration (FDA)
Fibrogenesis, 276f
Fibrosis, 268
Food and Drug Administration (FDA), 58
Food contaminants, 32–33, 45
exposure measurement, 60–61
sources, 45–46
Forced vital capacity (FVC), 168, 347
Frequency plots, 74*f*
FVC. *See* Forced vital capacity (FVC)

Gas chromatography (GC), 55
Gastritis, 8
Gastrointestinal (GI), 4, 30
GC. *See* Gas chromatography (GC)
Generally recognized as safe (GRAS), 313
Geometric mean (GM), 72, 77
Geometric standard deviation (GSD), 72, 77
GI. *See* Gastrointestinal (GI)
Glutathione-S-transferase (GST), 154
GM. *See* Geometric mean (GM)
GOF. *See* Goodness of fit (GOF)
Goldilocks principle, 124, 129
Goodness of fit (GOF), 239
AIC, 240
nesting, 240
-2LL, 240
GRAS. *See* "generally recognized as safe" (GRAS)
GSD. *See* Geometric standard deviation (GSD)
GST. *See* Glutathione-S-transferase (GST)

Half-time ($t_{1/2}$), 257, 264, 301
Hydration of skin, 112
  effects of, 112
Hyperresponsiveness 321
Hypersensitivity, 357
Hypothesis, 3, 7, 9

ICD. *See* International Classification of Disease (ICD)
ICRP. *See* International Commission on Radiological Protection (ICRP)
Immune reactions types, 335
Information bias, 213. *See also* Recall bias
Ingestion, 110
  exposures, 37
Inhalation, 4, 5, 6, 107
  exposures, 37
Integrated Risk Information System (IRIS), 118
Interaction. *See* Effect modification
International Classification of Disease (ICD), 195
International Commission on Radiological Protection (ICRP), 115
IRIS. *See* Integrated Risk Information System (IRIS)
Irreversible proportional disease processes, 267, 276, 318
  bronchiolitis obliterans (BO), 312
  characteristics, 306*t*
  dementia, 304
  difference equation, 269, 271
  dose metric, 319
  fibrosis, 268
  in fetus, 319
  mechanisms, 268–69
  mercury effects, 304, 305–12
  model, 269*f*, 270–77
    assumptions, 272–73
    difference equation model, 276*f*, 277
    dose metrics, 273–75
    total damage, 270
    total dose, 273
  silicosis, 267
  temporal sequence exposure, 319
  toxin, artificial butter flavor, 304

LC. *See* Liquid chromatography (LC)
Lead poisoning, 102
  kinetics, 103, 104
  solubilization process, 104–05
  time course of blood lead, 104*f*
LEV. *See* Local exhaust ventilation (LEV)
Limit of detection (LOD), 52, 55, 361
Liquid chromatography (LC), 55
Local exhaust ventilation (LEV), 42
LOD. *See* Limit of detection (LOD)
Lognormal distribution, 71, 72, 73
Low back pain, 234
Lower respiratory symptoms, 160
Lung
  biopsy, 317
  cancer, 20–22, 367–69, 375–81, 373–74, 377*t*
  cancer, and dioxin exposure, 238

Manmade vitreous fibers (MMVF), 230
  conditions for biologic activity, 232*f*
  deposition in target tissue, 231
  fiber durability, 232
  lung dose, 230–34
  tumor formation, 233
Mass spectrometry (MS), 55
Measure of association (rate ratio), 186
Mercury neurobehavioral effects, 304, 305–12
  biomarker, 311–12
  dose metrics, 312
  epidemiology, 307–08
  ETI, 311
  mechanism, 308–11
Metalworking fluids (MWF), 358, 362
  dose metrics, 362
  follow-up study design, 363
  skin exposure measurements, 361
  skin responses to, 359–60
Methyl t-butyl ether (MTBE), 148
  inhalation exposure, 149*f*
Michaelis-Menten enzyme kinetics, 126
MMVF. *See* Manmade vitreous fibers (MMVF)
Model
  alternative tissue response DP, 290–94, 291*f*, 292*f*
  ammonia, 283–84
  biological, 6, 11
  compartmental, 106*f*, 111*f*, 112*f*, 119*f*, 116
  Cox proportional hazards, 215–16
  damage-repair, 290, 293
  empirical, 287–88
  Freijer for ozone, 295–99
  nonlinear, 132–34
  one-compartment, 289
  PBTK, 132, 253*f*, 280*f*-281
  physiologic, 115, 127
Models, dynamic versus static, 10
  biologic models, 11
  dynamic biological process models, 17
  time lags in response, 13
  "time windows of exposure," 15, 16*f*
Molecular diffusion and perfusion, 123–24
Moolgavkar-Knudson two-stage model. *See* Two-stage clonal expansion model (TSCE)
MS. *See* Mass spectrometry (MS)
MSD. *See* Musculoskeletal diseases (MSD)

MTBE. *See* Methyl t-butyl ether (MTBE)
Musculoskeletal diseases (MSD), 27, 388
  low back pain, 234–236
Mutations and cancer, 367
MWF. *See* Metalworking fluids (MWF)

National Institute for Occupational Safety and Health (NIOSH), 58, 174, 187, 216
National Institute of Standards and Technology (NIST), 63
Natural killer cells (NK), 322
Nested statistical models, 240
Neurobehavioral effects, 243–46
Neurological response rate, 282
NIOSH. *See* National Institute for Occupational Safety and Health (NIOSH)
NIST. *See* National Institute of Standards and Technology (NIST)
NK. *See* Natural killer cells (NK)

Occupational Safety and Health Administration (OSHA), 58, 90
Occupational exposures
  ammonia, 253
  crystalline silica
    diatomaceous earth, 375
  electromagnetic radiation, 96
  low back pain, 234
  man-made vitreous fibers (MMVF), 230
  metal working fluids (MWF), 357
  silicon carbide, 243
  2, 3, 7, 8-tetrachlorodibenzo-*p*-dioxin (TCDD), 238
Organophosphate (OP) pesticides, 139
Ozone, 286, 347
  exposure, 12

PAHs. *See* Polycyclic aromatic hydrocarbons (PAHs)
Parallelism of temporal disease patterns, 9
Partition coefficient (PC), 123, 155
  calculation, 119
  for gases, 120–21*t*
Partitioning, Henry's Law coefficient, 281
Pathophysiologic processes, 159
PBF. *See* Percent body fat (PBF)
PBTK model. *See* Physiologically based toxicokinetic (PBTK) model
PC. *See* Partition coefficient (PC)
PCBs. *See* Polychlorinated biphenyls (PCBs)
Peak expiratory flow (PEF), 167
Peak exposures, 81, 83, 335
PEF. *See* Peak expiratory flow (PEF)
Percent body fat (PBF), 238
Pesticides toxicity, 266
Pharmacokinetics, 102. *See also* Toxicokinetic model
Physiologically based toxicokinetic (PBTK) model, 101, 128, 131, 132, 139*f*, 143, 385
  biomarkers, 154
  modeling of, 128
  processes, 133
  spreadsheet parameters, 129*t*
Pneumoconioses, 8, 168
Pollutants
  air-purifying cartridges, 43
  airway inflammation, 21
  arsenic in water, as contaminant, 32, 43
  cockroach allergen, 352
  diacetyl, 313
  exposure, 79, 170
  food contamination, 32
  mercury, 304
  OP pesticides, 139
  ozone, 286, 347
  oxidant, 286
  PAHs, 336
  PCBs, 94
  respiratory sensitization, 336
  reversible proportional response mechanisms, 166
  smoke emissions, 28
Polychlorinated biphenyls (PCBs), 94
Polycyclic aromatic hydrocarbons (PAHs), 55
Polymorphonuclear (PMN) leucocytes, 322
Poorly perfused (PP) compartment, 117, 127
Popcorn workers' lung, 313
Population, 3, 4, 10
PP compartment. *See* Poorly perfused (PP) compartment
Probabilistic responses
  discrete disease processes, 169, 173
  dose metric, 172, 333, 351–52
  model, 331. *See also* Discrete disease processes
  nature, 320–21
Process models of disease, 6, 7*f*, 17
Proliferation of cells, 7
Proliferation signal, 7. *See also* Repair process
Proportional effect, 173
Proportional versus discrete response, 21

QC. *See* Quality control (QC)
Quality control (QC), 62
Quantile plot, 73
  combined distribution of data, 76*f*
  log-transformed concentration data, 75*f*
  normally distributed, 75*f*
  types, 73

Rate
  difference (RD), 186
  limiting steps, 161
  ratio (RR), 186
RBC. *See* Red blood cell (RBC)
RD. *See* Rate—difference (RD); Risk—difference (RD)
Recall bias, 201
Recovery, 252, 256
  half-time, 264
Recruitment signal for cells, 7. *See also* Damage—repair process
Red blood cell (RBC), 139
Regression model, 192
Relative risk (RR), 186
Repair, 4, 6, 7, 8, 9, 10, 257, 258
  half-time, 257
  processes, 161
  rate constant, 263*f*
  rate of repair, 258–59
Response
  modeling, 287
  recovery, 252, 254–56
  time, 57
Reversible proportional disease processes, 252, 300–30
  ammonia and eye irritation, 253–56,
    acclimatization, 254, 256
    ammonia model behavior, 283–84
    ammonia's diffusion rate, 281,
    data collection, 282–83
    diffusion rate, 282
    effects, 279
    exposed cornea surface area, 281
    eye irritation simulation, 255*f*
    field study to followup, 284–86
    Henry's Law coefficient, 281
    neurological response rate, 282
    PBTK model, 253*f*, 280*f*–281
    response-recovery processes, 254–56
    sensory responses, 254
    steady-state concentration, 281, 282
    temporal behavior, 284*f*
    threshold, 254
  biology of effect, 301–02
  characteristics, 279
  damage rate, 259
  damage-repair, 252, 301
  dose-response relationship, 303
  half-time of repair, 301
  ozone, 286, 347
    alternative tissue response DP models, 290–94, 291*f*, 292*f*
    damage-repair model, 290, 293
    dose metric for pulmonary effects, 299–300
    empirical models, 287–88
    epidemiologic studies, 286–87
    etiologic time interval, 299
    exposure simulation, 291*f*, 298*f*
    modified Freijer model, 295–99
    one-compartment model, 289
    responses modeling, 287-
    steady-state fractions, 300
    time as dose variable, 288–89
    tissue ozone, 289–90
  response-recovery, 252
Reversible versus irreversible time course, 20
Risk
  difference (RD), 186
  estimation, 183
RR. *See* Rate ratio (RR); Relative risk (RR)

Sampling strategies, 84
  environmental measurement, 84
  for epidemiologic study, 85*t*
SCBA. *See* Self-contained breathing apparatus (SCBA)
SD. *See* Standard deviation (SD)
Selection bias, 213
Self-contained breathing apparatus (SCBA), 43
Sensitization process, 338–41, 339*f*, 342
Shortness of breath (SOB), 315
Silica, 167, 216, 229, 267–271, 374–379
Silicosis, 9, 267, 269*f*
Skin contamination, 33, 46
  contact with contaminants, 47
  exposure measurement, 61
  sources, 46
Snow, John, 3–4
SOB. *See* Shortness of breath (SOB)
SOD. *See* Superoxide dismutase (SOD)
Standard deviation (SD), 56, 77, 224
Statistical significance, 219–20
Steady-state concentration, 281, 282
Steady-state condition, 256, 300
  time required, 257
Stratification of study population, 191
Superoxide dismutase (SOD), 289
"Switch response," 324

TBA. *See* t-butyl alcohol (TBA)
t-butyl alcohol (TBA), 148
TCDD. *See* 2, 3, 7, 8-tetrachlorodibenzo-*p*-dioxin (TCDD)
Temporal behavior, 284*f*
Temporal patterns, 18*f*, 19*f*, 166, 168, 169, 171
Temporal processes, 4. *See also* Dynamic processes
  exposure-response relationship, 5*f*

2,3,7,8-Tetrachlorodibenzo-*p*-dioxin (TCDD), 238
Threshold
  population view, 185*f*
  toxic response, 185*f*
Time-weighted average (TWA), 56, 78
  calculation, 79
  formula, 78
  personal exposure, 79
Time windows of exposure, 15, 16*f*. *See also* Dynamic processes
  empirical time windows, 15
  limitations, 15
  statistical interaction, 17
Tissue
  dose, 386
  ozone, 289–90
Total damage, 270
Total dose, 273
Toxic
  agent effects, 256
  flavoring ingredients, 313
Toxicity extrapolation, 387–88
Toxicogenomics, 388
Toxicokinetic model, 6*f*, 102, 117. *See also* Dose metrics
  agent kinetics, 125
    basic compartmental equation, 125
    defining model structure, 127
    Michaelis-Menten relationship, 126
  compartmental model, 106*f*, 117, 119*f*
    spreadsheet parameters, 130*t*
  components of model, 116
  Goldilocks principle, 124, 129
  nonlinear models, 132
    organs and tissues reference data, 134
    protein binding, 133
  PBTK models, 132
  physiologic models, 115, 127
    chemical partitioning, 116
    compartment characteristics, 122
    one-compartment model, 119
    standard man parameters, 120–21*t*
Toxicokinetic processes, 6*f*, 102, 105
  biotransformation, 114
    excretion, 115
    storage, 114
  compartmental model, 106*f*
  in epidemiology, 124
  internal transport processes, 113
    blood circulation, 113
    lymph transportation, 114
  processes affecting entry, 107, 117–18
  routes of entry, 107
    combined routes, 113
    ingestion, 110
    inhalation, 107
    skin absorption, 111
  toxicokinetic processes
Toxin, 6, 8
  artificial butter flavor, 304
TSCE. *See* Two-stage clonal expansion model (TSCE)
Tumor formation, 233
TWA. *See* Time-weighted average (TWA)
-2 log likelihood statistics (-2LL), 240
Two-stage clonal expansion model (TSCE), 366
Two-stage model, 366, 369

Ultraviolet (UV) radiation, 55, 347
Uncertainty, risk 198
  biases, 212
  confounding, 208
  errors in outcome variables, 217
    confidence intervals, 220–21
    hypothesis testing, 218
    outcome variability evaluation, 217
    statistical significance limitations, 219–20
  exposure estimation errors, 200
  management, 221
  model form misspecification, 214
    Cox proportional hazards model, 215–16
    logistic, 215–16
  sources, 199*f*

Vapor analysis, 313–14, 317
Variability of exposure, 9
  sources, 97

Water contaminants, 32, 43
  chlorination, 44
  environmental transport, 43–44
  exposure measurement, 59
  sources, 43
  water treatment, 44
Water sample, 59
WBV. *See* Whole-body vibration (WBV)
Web of causation, 188*f*
Well-perfused (WP) compartment, 117
Whole-body vibration (WBV), 206
WP compartment. *See* Well-perfused (WP) compartment

www.ingramcontent.com/pod-product-compliance
Ingram Content Group UK Ltd.
Pitfield, Milton Keynes, MK11 3LW, UK
UKHW021439280726
14060UKWH00001BA/156